Nordic Wittgenstein Studies

Volume 7

The series publishes high-quality studies of Ludwig Wittgenstein's work and philosophy. It is affiliated with The Nordic Wittgenstein Society, The Wittgenstein Archives at the University of Bergen and The von Wright and Wittgenstein Archives at the University of Helsinki. The series welcomes any first rank study of Wittgenstein's philosophy, biography or work, and contributions in the subject areas of philosophy and other human and social studies (including philology, linguistics, cognitive science and others) that draw upon Wittgenstein's work. It also invites studies that demonstrate the philosophical relevance of Wittgenstein's Nachlass as well as purely philological or literary studies of the Nachlass. Each submission to the series, if found eligible by the series editor, is peer reviewed by the editorial board and independent experts. The series accepts submissions in English of approximately 80 000 – 125 000 words. For further information (about how to submit a proposal, formatting etc.), please contact: niklas.forsberg@filosofi.uu.se.

More information about this series at http://www.springer.com/series/13863

Juliet Floyd • Felix Mühlhölzer

Wittgenstein's Annotations to Hardy's Course of Pure Mathematics

An Investigation of Wittgenstein's Non-Extensionalist Understanding of the Real Numbers

Springer

Juliet Floyd
Department of Philosophy
Boston University
Boston, MA, USA

Felix Mühlhölzer
Philosophisches Seminar
Georg-August-Universität Göttingen
Göttingen, Germany

ISSN 2520-1514 ISSN 2520-1522 (electronic)
Nordic Wittgenstein Studies
ISBN 978-3-030-48480-4 ISBN 978-3-030-48481-1 (eBook)
https://doi.org/10.1007/978-3-030-48481-1

Cover Illustration: The cover makes use of Wittgenstein Nachlass MS 115, page 118. The Master and Fellows of Trinity College Cambridge and the University of Bergen have kindly permitted the use of this picture.

This Springer imprint is published by the registered company Springer Nature Switzerland AG.
The registered company address is: Gewerbestrasse 11, 6330 Cham, Switzerland

For Margaux and Marianne

Preface

Some time in 1942–1943 Wittgenstein wrote down annotations in the margins of his copy of G. H. Hardy's book on analysis, *A Course of Pure Mathematics* (hereafter CPM), the eighth edition (1941).[1] Images of his annotations are published, transcribed, and translated by us in Part IV below.[2] These annotations are not very numerous, they are all brief, and they record reactions and philosophical hints rather than presenting real arguments. Nevertheless, for the following reasons we believe they should be published and commented on.

First, they directly refer to assertions of a real mathematician, one whom Wittgenstein knew as a colleague, and a brilliant one at that. Moreover they occur in a book that was extremely influential in Great Britain for a very long time. In his Foreword to the Centenary Edition of CPM, the mathematician T. W. Körner writes that after its first edition in 1908 Hardy's book defined the first analysis course for students in Britain for the next 70 years.[3]

All too frequently Wittgenstein's reflections on mathematics proceed at a great distance from what real mathematicians say, but this case is an exception and therefore has, we think, a special significance. Recent publications of detailed lecture notes taken down by his students and closest colleagues have begun to place Wittgenstein more sharply against the background of his Cambridge context, showing how his ideas evolved from year to year in the face of discussion, not only with philosophers, but with advanced mathematics students.[4] Our examination of his annotations to CPM will round out our understanding of how his ideas are relevant

---

[1] Below we refer to a specific *n*th edition as "CPM_n", and without an index "CPM" refers to CPM_8, the edition in which Wittgenstein made his annotations. Hardy (2008), a reprint of the 10th edition, is currently in print.

[2] When we are quoting from Wittgenstein's annotations in our transcriptions, translations and preceding commentary we always print the annotated words in Courier font, to distinguish them.

[3] See Hardy (2008), p. vii. Körner is himself the author of a much praised textbook on analysis, Körner (2004).

[4] See, e.g., Klagge and Nordmann (2003), Moore (2016), hereafter MWL, Wittgenstein (2017), hereafter SWL; we discuss these transcriptions below in 1.2.

to what mathematicians think, not only those in Cambridge during his lifetime, but also some in our time. The point, in the end, is to explore philosophical ideas about the real numbers, not only Wittgenstein's, but Hardy's and others' as well.

Second, these annotations form the basis for Part V of *Remarks on the Foundations of Mathematics*, hereafter RFM, an edition of selections from Wittgenstein's manuscripts posthumously published by Wittgenstein's literary executors G.H. von Wright, Rush Rhees, and G.E.M. Anscombe.[5] Part V of RFM dates from 1942–1944, just the period of Wittgenstein's annotations to Hardy's CPM. For this reason the annotations, examined as remarks taken up later on in manuscripts, shed interesting light, not only on this posthumously published material but also on the development and nature of Wittgenstein's overall philosophy of mathematics. And this part of Wittgenstein's philosophy, though it was central to him, remains very contentious to this day, and not well understood.

The ultimate rationale for this book is our foray into the philosophy of mathematics through the eyes of these annotations and Wittgenstein's development of them in manuscripts. In the end, we are sometimes not sure of what to say, and we hold somewhat different attitudes toward the ultimate philosophical position. In order to make this clear, we end the volume with two essays, one by Mühlhölzer, one by Floyd. Mühlhölzer's essay is a sort of supplement to Part II of this book, discussing questions concerning Cantor's diagonal method and its relevance to our understanding of the real numbers, questions that are totally factored out in his annotations to CPM and also the corresponding notes published in Part V of RFM. Floyd's essay explores Wittgenstein's way of coming to see Cantor's diagonal method as a proof *technique* with multiple ranges of applicability, pointing out that while the notion of a technique is something invented, according to Wittgenstein novel proof techniques *reveal* new aspects of real numbers that are there to be seen. Floyd closes with an account of the impact of Turing's famous (1936) on Wittgenstein.

Wittgenstein wrote a great deal on the foundations of mathematics, and though he never brought his ideas to any settled form, it is clear that he always envisioned writing a book whose first part would treat the nature of meaning and the proposition, and whose second part would apply this perspective to logic and the foundations of mathematics. The annotations and their incorporation into Wittgenstein's thinking during this period (1942–1943) show us a distinctive direction of thought which remains, all the same, incomplete and not fully developed.

The annotations should be made available for all scholars to use and interpret in their own way, and we have provided images of them below (Part IV).[6] Yet we also

[5] See Wittgenstein (1978). The original publication date was 1956; a revised edition appeared in 1978. The German editions were titled *Bemerkungen über die Grundlagen der Mathematik* (Wittgenstein 1974a, hereafter BGM).

[6] Since the original copy of Wittgenstein's edition of CPM_8 is currently lost, we must work with these photographic images alone, provided to Juliet Floyd by Michael Nedo at the Wittgenstein Archives, Cambridge. We cannot prove that they comprise all the annotations, but there is no special reason to doubt that they do. Wittgenstein's edition of CPM_8 was seen in person as late as

believe that their release *without* commentary and the illumination we provide by juxtaposing them with Wittgenstein's other numerous manuscripts – some of which were clearly derived from the annotations – would leave the annotations themselves rather useless, or, worse, liable to be used as drive-by quotations to prove this or that point about their author's beliefs. In what follows, therefore, we have tried carefully to attend to Wittgenstein's own specific ways of using the annotations in his other writings as we reconstruct and analyze the direct flow of his actual thinking and writing. This exercise proved valuable to us as we discussed the annotations in relation to the manuscripts and some of Wittgenstein's published writings during our academic year together in Göttingen in 2009–2010. We have relied on the Bergen electronic edition of Wittgenstein's *Nachlass* for this manuscript work (Wittgenstein 2015), and citations below take the usual form of typescript or manuscript number from this edition, originating in von Wright (1969).

Part I offers an overview of some of the main issues about real numbers that are investigated by Wittgenstein, with attention to mathematical literature of the time and later. Part II turns to a detailed analysis of the annotations themselves. Here we transcribe, translate, contextualize, and analyze the annotations, discussing their provenance and significance within philosophy of mathematics. Our arguments and the examples will lead in a host of different directions, for we tie the examples discussed in Wittgenstein's annotations to remarks in his accompanying manuscripts and extract argumentation from these in order to bring the philosophical issues to light. Part IV contains images and transcripts of Wittgenstein's annotations.

Our discussions of the material enabled us to sort out many issues of translation and interpretation that have plagued Wittgenstein's readers. We compared the annotations and their reappearance in Wittgenstein's manuscripts in the original German with the edited and translated selections from RFM. Sometimes the English, wrongly transposed and looked at outside of the original context, positively invites one to go down paths of interpretation that are clearly unjustified. Often when remarks are excised from a larger manuscript crucial features of context are lost. In what follows, although we have relied upon Anscombe's translation, we have often departed from it, and we shall not note down every point where we do. What we have done is to work to carefully reconstruct the context of manuscript remarks in terms of philosophical argumentation. Here the annotations to CPM were an aid and also a subject of clarification in their own right.

the 1970s: Joachim Schulte has reported to Juliet Floyd in informal correspondence that he saw the book at that time.

We saw that, like all of us, Wittgenstein had good days and bad days, days where he was tired and cranky, and days when he felt stronger, days when he saw a line of thought through to a new level, and days when he abandoned a line of thought. On the whole, the tête-à-têtes he had with himself, out of the gaze of students and colleagues, tell us a great deal about not only his method of working but also his underlying conception of philosophy and his intellectual aims.

He was far more concerned with philosophical progress, more diffident and exploratory about his investigations, than are many portraits of him that continue to be bandied about. We are especially concerned to deny that he was aiming to put an end to philosophy as such, dogmatically, on his own. Instead, his concrete engagements with mathematics, including its philosophy and its exposition, mattered to him very much. Our method of exposition is designed to give the reader an intimate feel for the kind of thinker he was, as well as to explore how philosophy of mathematics, if not philosophy as such, might be practiced in an exploratory, undogmatic vein.

In 1944 Wittgenstein himself stated, in a proposed biographical entry, that his "chief contribution has been in the philosophy of mathematics."[7] Whether or not this is true, only portions of his manuscripts from this period were published in RFM, often with chains of reasoning broken off and rearranged in the condensations of remarks made by the editors. In reconstructing how Wittgenstein stepped off from his 1942–1943 annotations to CPM into some of these further writings and reflections, we give specific rationales for many of his remarks about actual examples in real analysis and set theory, thereby shedding light on the potentials and pitfalls of his broader approach.

It is crucial to appreciate that *different* problems preoccupied Wittgenstein at different points during 1937–1944, the period during which his focus turned toward the foundations of mathematics. His treatment of Cantor's Diagonal Argument (RFM II, 1937/1938) is the subject of Mühlhölzer's essay in Part III. The issues raised here contrast with those in the later RFM III (1939–1940), where Wittgenstein focused on logicism and *Principia Mathematica* as a foundation for mathematics; Mühlhölzer (2010) is a lengthy commentary on this manuscript.[8] RFM II and III raise certain distinctive issues, including those concerning Wittgenstein's relation to Turing. Floyd's essay in Chap. 8 will discuss this relation.

The 1942–1943 annotations to CPM have their own special purview. In particular, they help to correct certain characteristic misinterpretations of Wittgenstein's philosophy of mathematics prevailing in the literature. Above all we shall argue against the idea that Wittgenstein was a constructivist or Brouwerian, or even a finitist or a verificationist.[9] Such philosophical positions reject, or should reject, Dedekind cuts outright. This, we believe, was no part of Wittgenstein's philo-

[7] Monk (1990), p. 466.

[8] See Floyd (2015) for a discussion.

[9] For a critique of the idea that Wittgenstein was a follower of Brouwer see Marion (1998) and Marion (2003).

sophical line of thought. Not because he was a "non-revisionist" philosopher of mathematics as a matter of principle, but, rather, because of certain specific examples, procedures, and lines of argumentation with which he was deeply and rightly preoccupied for many years.

Boston, MA, USA — Juliet Floyd
Göttingen, Germany — Felix Mühlhölzer

Acknowledgments

We are indebted to Michael Nedo, Ilyas Khan, and Jonathan Smith for providing us with the relevant images of Wittgenstein's copy of Hardy's *Course of Pure Mathematics* from the Cambridge Wittgenstein Archives and for supporting our effort to publish them. Without the work of the Archives, the annotations themselves may well have been forever lost.

The scholarly efforts of Alois Pichler and all those who have contributed over the years to the Bergen Wittgenstein Archives must also be acknowledged. The Bergen electronic edition of Wittgenstein's works (Wittgenstein 2015) enables open source use of images of Wittgenstein's manuscripts and typescripts, allowing scholars to cooperate and answer questions with lightning speed. We were especially helped by the guides through the manuscripts and published editions of Wittgenstein's works found in Biggs and Pichler (1993) and Pichler et al. (2011), the former based on the initial compilation (von Wright 1969).

Stefan Brandt, Eike Düvel, Arthur Gibson, Akihiro Kanamori, Wilfried Keller, Ilyas Khan, Juliette Kennedy, Ulrich Majer, Abdul Raffert, Björn Steinhoff, Norbert Schappacher, Sanford Shieh, and Jouko Väänänen provided us with comments and conversations for which we are also grateful. To Hans Biesenbach we owe thanks for sending us compilations of Wittgenstein's references to Dedekind and Hardy which were very useful (see his 2014).

We are grateful for Paul Goldberg's help in proofreading the entire manuscript, and the funding which he received from the Boston University Department of Philosophy, which also funded Merve Tapinc to begin indexing the book.

Juliet Floyd's research was supported by the Lichtenberg-Kolleg, Georg August Universität Göttingen, October 2009 to July 2010, under the auspices of the Deutsche Forschungsgemeinschaft. Felix Mühlhölzer's research was supported by the Opus magnum program of the Fritz Thyssen Stiftung and the Volkswagen Stiftung from October 2012 to September 2014.

Contents

Abbreviations

The following abbreviations, listed in alphabetical order, are used to refer to Hardy's and Wittgenstein's published works that are frequently cited. Parenthetical dates refer to the full entry in the Bibliography.

AWL	*Wittgenstein's Lectures, Cambridge, 1932–1935: From the Notes of Alice Ambrose and Margaret Macdonald* (Wittgenstein 1979)
B	*Gesamtbriefwechsel/Complete Correspondence* (Wittgenstein 2005b)
BB	*Preliminary Studies for the "Philosophical Investigations", Generally Known as "The Blue and Brown Books"* (Wittgenstein 1965b)
BFF	"*Bearbeitete Frühfassung*" of PI, in PUK
BGM$_1$	*Remarks on the Foundations of Mathematics/Bemerkungen über die Grundlagen der Mathematik*, first German-English edition (Wittgenstein 1956)
BGM	*Bemerkungen über die Grundlagen der Mathematik*, second German edition (Wittgenstein 1974a)
BT	*The Big Typescript, TS 213* (Wittgenstein 2005a)
CE	"Cause and Effect: Intuitive Awareness" (Klagge and Nordmann 2003)
CPM$_n$	*A Course of Pure Mathematics*, *n*th edition; "CPM" (Hardy 2008: is CPM$_8$
EPB	*Eine Philosophische Betrachtung*, MS 115 (Wittgenstein 1969b)
FF	"*Frühfassung*" of PI, in PUK
LWL	*Wittgenstein's Lectures, Cambridge 1930–1932, From the Notes of John King and Desmond Lee* (Wittgenstein 1980d)
LA	*Lectures and Conversations on Aesthetics, Psychology, and Religious Belief* (Wittgenstein 2007)
LFM	*Wittgenstein's Lectures on the Foundations of Mathematics: Cambridge, 1939, from the Notes of R.G. Bosanquet, Norman Malcolm, Rush Rhees and Yorick Smythies* (Wittgenstein 1976)
LPS	*Wittgenstein's Lectures on Philosophical Psychology, 1946–1947, from notes by P.T. Geach, K.J. Shah, and A.C. Jackson* (Wittgenstein 1988)
MS	Reference to manuscript number in Wittgenstein (2015)

MWL	*Wittgenstein: Lectures, Cambridge 1930–1933, From the Notes of G.E. Moore* (Moore 2016)
OC	*On Certainty/Über Gewissheit* (Wittgenstein 1974b)
PG	*Philosophical Grammar* (Wittgenstein 1974c)
PI	*Philosophische Untersuchungen = Philosophical Investigations* (Wittgenstein 2009)
PO	*Ludwig Wittgenstein: Philosophical Occasions, 1912–1951* (Klagge and Nordmann 1993)
PPF	Formerly "Part II" of *Philosophical Investigations*, based on (the now lost) TS 234, named by the editors "Philosophy of Psychology – A Fragment", in PI
PR	*Philosophical Remarks* (Wittgenstein 1980a)
PUK	*Philosophische Untersuchungen: Kritisch-Genetische Edition* (Wittgenstein 2001)
RFM	*Remarks on the Foundations of Mathematics* (Wittgenstein 1978)
RFM_1	*Remarks on the Foundations of Mathematics/Bemerkungen über die Grundlagen der Mathematik*, first German-English Edition (Wittgenstein 1956)
RLF	*Some Remarks on Logical Form* (Wittgenstein 1929)
RPP I	*Remarks on the Philosophy of Psychology, Vol. 1* (Wittgenstein 1980b)
RPP II	*Remarks on the Philosophy of Psychology, Vol. 2* (Wittgenstein 1980c)
SF	"*Spätfassung*" of PI, in PUK
SWL	*Wittgenstein's Whewell's Court Lectures, Cambridge, 1938–1941: from the Notes by Yorick Smythies* (Wittgenstein 2017)
TLP	*Tractatus Logico-Philosophicus* (Wittgenstein 1981a)
TS	Reference to typescript number in Wittgenstein (2015)
UF	"*Urfassung*" of PI, in PUK
VW	*The Voices of Wittgenstein, the Vienna Circle: Original German Texts and English Translations* (Wittgenstein and Waismann 2003)
WC	*Wittgenstein in Cambridge: Letters and Documents, 1911–1951* (McGuinness 2008)
WVC	*Wittgenstein and the Vienna Circle: Conversations recorded by Friedrich Waismann* (Wittgenstein and Waismann 1979)
Z	*Zettel* (Wittgenstein 1981b)
ZF	"*Zwischenfassung*" of PI, in PUK

Part I
Overview

Chapter 1
Introduction: The Context of the Annotations and Our Method of Approach

Abstract An overview of our approach to Wittgenstein's development and writings on mathematics, including his annotations to Hardy's *Course of Pure Mathematics*. Wittgenstein's development is divided into four stages: (1) The *Tractatus*, (2) The Middle Period (1929–1933), (3) *The Blue and Brown Books* (1933–1935) and (4) *Philosophical Investigations* (1937–1944). Wittgenstein's relation to Hardy is characterized, and the overall structure of this book is surveyed.

As stated in the editorial introduction to RFM, the posthumously published excerpts from Wittgenstein's writings on mathematics spanned many years of work, comprising manuscripts stemming from 1937 (RFM I) through the spring of 1944 (RFM VII). The method of arrangement of excerpts used by the editors was mainly historical: with a few exceptions[1] they were published in chronological order within each part.

In contrast to a comprehensive interpretation of RFM section by section, we offer here a more sharply focused, day-by-day account of how Wittgenstein's thoughts unfolded in his original manuscripts. We interpret the material by connecting these remarks to their future embeddings in more polished environments of manuscripts and typescripts, stepping off, initially, from the annotations to CPM. A particular portrait of Wittgenstein appears when we proceed in this way. We see the philosopher learning and teaching himself. Sometimes, as we shall note in what follows, our method allows us to explain that and why some of Wittgenstein's weaker remarks – widely known from their prior publication in RFM – were actually abandoned and never returned to again.

Our treatment analyzes Wittgenstein at work during a significant period in his philosophical life. The annotations to Hardy's textbook were written in 1942–1943, just as Wittgenstein was putting together the "Bearbeitete Frühfassung"

[1]At certain places the editors took the liberty of rearranging Wittgenstein's remarks. See their Preface to RFM, p. 33: "With a few exceptions we did not want to interfere with the order of the sections. Nevertheless we have sometimes (especially at the end of Parts IV and V) brought together material belonging to the same subject matter from different places".

J. Floyd, F. Mühlhölzer, *Wittgenstein's Annotations to Hardy's Course of Pure Mathematics*, Nordic Wittgenstein Studies 7,
https://doi.org/10.1007/978-3-030-48481-1_1

(hereafter BF), the third, "reworked early version" of his *Philosophical Investigations* (Wittgenstein 2009, hereafter PI). He submitted this typescript (TS 239) to the Cambridge University Press in September 1943, but never allowed it to be published. What follows is a brief outline of this context in terms of Wittgenstein's overall development on questions about the foundations of mathematics.

1.1 Foundations of Mathematics in the *Tractatus*

Wittgenstein's sole published philosophical book during his lifetime, the *Tractatus Logico-Philosophicus* (Wittgenstein 1981a, hereafter TLP) offered an account of the nature of logic that was deeply shaped by the work of Frege and Russell.[2] His treatment of the fundamental autonomy of logic contained a sketchy account of the concepts of number and arithmetic. But it was not clear – as Russell pointed out in his Introduction to the book – exactly how Wittgenstein's perspective could be developed to handle the transfinite.

As he wrote in his paper "The Foundations of Mathematics" (Ramsey 1926, p. 350), Ramsey thought it "obvious" that without further development, the *Tractatus* left us with "a ridiculously narrow view of mathematics, and confines it to simple arithmetic". He suggested extensionalizing propositional functions, postulating arbitrary correlations between propositions and individuals so as to allow for propositional functions "in extension" that we might never be able to define. Equivalence of descriptions in extension he construed in terms of potentially infinite disjunctions that were, he said, "tautologies".

Wittgenstein rejected this approach altogether – an early sign of his suspicions about a purely extensional point of view on mathematics. His debate with Ramsey focussed ostensibly on the issue of defining identity in terms of extensional equivalence, but this reflected a broader philosophical divide between them. Wittgenstein rejected the idea that equations of mathematics are logical identities, much less tautologies. Logical identities he regarded as theoretically eliminable *via* analysis or application of the theory of descriptions, whereas equations expressed rules of substitution for equational forms. Tautologies and contradictions constitute the realm of the so-called propositions of logic, limiting cases of genuine propositions without content. He regarded mathematical equations as rules of substitution to be used in calculations and inferences among mixed numeral sentences.[3]

---

[2]Moore was also a teacher of Wittgenstein's during his years at Cambridge 1911–1913. It was he who suggested the title for Wittgenstein's book, which reminded him, he said, of reading Spinoza. Wittgenstein himself used his own, original German title in referring to his book, "Logisch-Philosophische Abhandlung" (see Monk 1990, p. 206).

[3]The arguments Wittgenstein had in 1927 with Ramsey over these issues are analyzed in Sullivan (1995). As is explained in Floyd (2001a), n. 14, the notion of "calculus" must have been emphasized by Wittgenstein in conversations he had with Ramsey during March, May and July 1924.

The *Tractatus*'s characterization of mathematics was, as a matter of fact, not accepted by the book's most enthusiastic followers, including those in the Vienna Circle. They held that mathematics was – as the *Tractatus* had said logic was – *tautological* or, more broadly *analytic*, whereas Wittgenstein never construed mathematics in this way.[4] Instead, he took mathematics to consist essentially of equations and our operations with them, calculations carried out in the course of setting out possible formulations of natural laws. An equation expresses a "standpoint" on the expressions themselves (TLP 6.2323), rather than an assertion of the co-referentiality of names. Because of its connection with "mixed" statements and formulations of propositions, Wittgenstein's *Tractatus* view of number may be regarded as a variant of "logicism". But it is an idiosyncratic one. He construed the number words as markers of "form", rather than names, indicators of possibility and necessity rather than objecthood – something Frege and Russell, with their antipathies to modality, did not do.[5]

Of course, when it came to arithmetic, like Frege and Russell Wittgenstein never believed that we possess a faculty of mathematical "intuition". Instead, he argued (TLP 6.233),

> To the question whether we need intuition [*Anschauung*] for the solution of mathematical problems it must be answered that language itself here supplies the necessary intuition. The process of calculation brings about just this intuition.

A distinctive earmark of his view of generality in logic was that Wittgenstein took to be fundamental to logic our ability to follow out, step-by-step, "form series", or formal rules of expansion. These were rules specified, not extensionally, but in terms of procedural, formally specified recipes in language. The generality in our use of such well-founded series of signs and variables containing "and so on" reflected, according to the *Tractatus*, something essential to the very nature of logic, the very idea of a "logical syntax". Here Wittgenstein elaborated Poincaré's objection to Russell's logicism, and applied Russell's adoption of the "vicious circle principle" even to Frege and Russell's analysis of the ancestral (4.1273).

As we shall see, in his annotations to CPM Wittgenstein is still concerned to contrast an extensional point of view on real numbers with an intensional, procedural approach. As we shall also see, by 1942–1943 he had embedded the earlier debates with Frege, Russell and Ramsey in a more sophisticated, flexible, open-ended philosophical setting. Our interpretation of his annotations to CPM concerns just this question: What kind of *philosophical* progress had he made on the foundations of mathematics between the writing of the *Tractatus* and the writing of *Philosophical Investigations*?

In the *Tractatus* Wittgenstein offered an analysis of the ancestral of a relation, but it was not given in full second-order logic. Accusing Frege and Russell of engaging in a vicious circle, he wrote (4.1273):

---

[4]See Dreben and Floyd (1991), Floyd (2001a).

[5]On the importance of modality to the TLP, see Shieh (2014), Floyd (2018a).

If we want to express in logical symbolism the general proposition "*b* is a successor of *a*" we need for this an expression for the general term of the formal series:

$$aRb,$$
$$(\exists x)\colon aRx.xRb,$$
$$(\exists x, y)\colon aRx.xRy.yRb,$$
$$\ldots\ .$$

Utilizing variables for such well-founded formal "operations" – which he distinguished from (propositional) "functions" – he was analyzing the ancestral of a relation from below, in terms of possible, well-founded recursive procedures of calculation, and not from above, in terms of infinite systems (Dedekind) or concepts (Frege). The ellipsis "..." remained a primitive feature of his logic, however, not something analyzed in terms of a general notion such as "effective" or "computable". Weiss has shown that this kind of system is capable of expressing all inductive definitions, although the complexity of the consequence relation, if the universe is assumed to be greater than $\aleph_0$, is very high.[6]

More generally he expressed the notion of "..." in terms of variables ranging over all such operations (TLP 5.2522f.):

The general term of a formal series a, $O'a$, $O'O'a$, ... I write thus: "$[a, x, O'x]$". This expression in brackets is a variable. The first term of the expression is the beginning of the formal series, the second the form of an arbitrary term x of the series, and the third the form of that term of the series which immediately follows x.

The concept of the successive application of an operation is equivalent to the concept "and so on".

Envisioning (predicative) serial intersubstitution of such variables, he then defined *number* as an index or "exponent" of any such operation (TLP 6.021).

Although mathematical induction was never explicitly mentioned, clearly Wittgenstein was taking that mode of proof to reflect our general logical capacity to "go on" in a formal series with signs. The general form of cardinal number he wrote this way:

$$[0, \xi, \xi + 1]$$

with "ξ" standing in, formally, for a number attaching to one or another property or series of exponents of formal operations.

The general philosophical framework of these variables was, however, problematic. For in the *Tractatus* scheme the well-foundedness of the logical operations – their ultimate base cases expressing a totality of mutually independent elementary propositions involving only "simple" objects – was envisioned to be, at least ideally, *absolute*. Wittgenstein's ideal of the unity of logic was that there can be only one logical "space", or system. Given the independence of the elementary sentences, there would *have* to be, he argued, absolute "simples" – that is, ultimate

[6]Weiss (2013, 2017).

starting points for all collections of form series. These would be the "substance" encompassing the "form" of all possible things to be said or thought.

In analyzing the "general form of proposition" as that of the general form of truth-function, and asserting that all propositions are truth-functions of elementary, basic propositions, Wittgenstein committed himself to the idea that a determination of what follows from what, what is logically independent of what, and so on, must yield sharp answers, Yes or No, on the basis of operations founded on the initial set of elementary propositions alone.

And yet Wittgenstein's view of logic was schematic and procedural. The discovery of what these elementary propositions actually *are* was a job he left to empirical science, something irrelevant to his task as a logician. Only in this way, it appeared, could logic be articulated universally, regardless of any particular proposition at issue or any state of affairs the world might find itself in. Only in this way, it appeared, could logic be conceived to be what it is independently of any particular claim about how things are. According to the *Tractatus*, logic does not represent reality, but arises as an activity settling the very idea of what it is to possibly so represent.

1.2 Wittgenstein's Middle Period, 1929–1933; *The Blue and Brown Books 1933–1935*

Wittgenstein came to see this wholly schematic ideal of necessary absoluteness and simplicity as dogmatic and mistaken. He made his retraction of this view explicit in a lecture he wrote for the 1929 Joint Session of the Aristotelian Society and the Mind Association in which he renounced the *Tractatus* claim that elementary sentences must all be logically independent of one another.[7] In the lecture he used the grammar of primary colors as a counterexample, and it is convincing, in that it revealed a kind of dogmatic promissory note at the heart of the TLP. That an object *a* cannot be both red all over and green all over at the same time is not an accident. In TLP 6.3751 Wittgenstein had remarked on this, but suggested, with a vague gesture, that this *must* be a necessity hidden deep in the laws of physics or psychology, and that in such a way logic might some day produce a contradiction from supposing otherwise.[8] But this, he came to believe, was an ideal there was no good reason to think could be cashed in.

[7] Wittgenstein (1929), hereafter RLF. In the end this paper was not publicly presented, and in the mentioned Session of the Aristotelian Society Wittgenstein instead spoke on generality and infinity in mathematics. Anscombe reported that Wittgenstein thought the paper he had written was "quite worthless"; see Klagge and Nordmann (1993), hereafter PO, p. 28.

[8] Ricketts (2014) gives a charitable reconstruction of what Wittgenstein *might* have had in mind.

So he surrendered his ideal of a "final analysis" in which the simplicity of objects would be shown in the structure of the totality of mutually independent elementary sentences. He rejected that idea that there *must* be a "general form of proposition". Instead, he initially argued (1929–1933), simplicity must be taken to be *relative* to particular systems of grammar in which there are simples. For example, in one system we might take the three primary colors as "simple", and in another, based on the color octahedron, a larger group of colors. Because he still subscribed to the idea that there must be simples at work in any full analysis, some of what he had worked out in the *Tractatus* about analysis and logic remained in place, though now relativized. But he insisted now that a broad investigation of different *Satzsysteme* was required to see logic rightly.

This relativization of simplicity posed grave problems, however, for his account of the foundations of mathematics. Once he had given up his "general form of proposition", he realized that he could not attach his account of number to logic directly *via* the theory of operations, which had been based on the general propositional form.

And so the foundations of mathematics – which he had given a "nebulous" introduction to anyway in the *Tractatus*[9] –rose to the fore of his attention. He had discussed the issue of "calculation" with Ramsey in 1923, and stimulated him to develop his ideas into a genuine foundational position.[10] By 1928 Wittgenstein was discussing Ramsey's paper "The Foundations of Mathematics" (1926) with the Vienna Circle.

After his return to Cambridge in 1929, Wittgenstein frequently held discussions with Ramsey, whom he later credited as having helped him formulate his ideas.[11] He began to work hard at drafting a second book. Ramsey's death in January of 1930 came as an intellectual blow, because from him he had received the most trenchant criticisms of the *Tractatus* and been challenged to develop a clearer sense of his own perspective. From Ramsey he also learned something about the flavor, and some of the challenges, of empiricism and pragmatism in logic.[12] We do not take traces of pragmatism as a theory of truth or verifiability to be directly embedded in Wittgenstein's mature philosophy of mathematics, though we do see a kind of realism, in the sense of a realistic spirit, there.[13] For his part, Ramsey appears to have himself converted to some form of finitism in the final year of his life, perhaps

[9]As he wrote in Wittgenstein (1980a), hereafter PR, §109.

[10]See Nedo and Ranchetti (1983), pp. 192–3, for an image of the remarks Wittgenstein copied by hand into Ramsey's copy of the TLP when Ramsey visited him in Austria, and compare Lewy (1967) for a discussion of this, as well as McGuinness (2008), pp. 141ff., Letters 100–103. Ramsey had already published his review of TLP, Ramsey (1923), before he visited Wittgenstein.

[11]See Wittgenstein (2009), Preface; compare the discussion of Wittgenstein and Ramsey and Ramsey's notes on their discussions in Sullivan (1995) and Ramsey (1991a).

[12]On Wittgenstein and pragmatism in relation to Ramsey, see Misak (2016).

[13]On which see Diamond (1991), Floyd (2020, in press).

reacting to discussions with Wittgenstein.[14] And Wittgenstein alludes to Ramsey's view of analysis in his last papers in a notebook from 1939–1940.[15]

Meanwhile, at Schlick's request Wittgenstein had begun in 1929–1930 a collaboration with Friedrich Waismann in Vienna. The initial idea was to write a book that would provide a simplified, more accessible presentation of the *Tractatus*. In addition to participating in discussions with the Vienna Circle 1929–1932 (Wittgenstein and Waismann (1979), hereafter WVC), Wittgenstein also gave to the Heretics Society "A Lecture on Ethics", distancing himself from the Vienna Circle's emotivist ideas about ethics.[16] Finally, he wrote up a new manuscript, *Philosophical Remarks* ((1980a), hereafter PR), which he submitted to Cambridge as part of his successful application for a fellowship (1930–1936). He continued to try to write a book with Waismann through 1934. However, as his thinking rapidly matured, a "rehash of theses" – filled as it would be with ideas that he by now regarded as fundamental errors – came to him to seem not at all what was needed.[17] The collaboration with Waismann turned by 1932 toward a new book, initially to be co-authored. By 1933 Wittgenstein had put together, *The Big Typescript* (Wittgenstein 2005a, TS 213, hereafter BT), a massive book manuscript of over 500 pages drawn from reworkings of PR and new remarks.[18] BT was divided into two parts, the first devoted to an analysis of the notions of proposition and sense, and the second applying this basic conception to logic and the foundations of mathematics. But the collaboration became increasingly difficult. Waismann found it impossible to keep up with Wittgenstein's rapidly evolving thoughts, and his numerous efforts to extract a straightforward position from Wittgenstein's discussions were never satisfactory. While Wittgenstein was at Cambridge he continued lecturing and discussing with others, leaving Waismann in Vienna writing up conversations that were later regarded by Wittgenstein as misleading. This effort to write a joint work ended by 1934, Waismann utilizing some of the material in his own writings, Wittgenstein turning back to his own manuscripts and to working primarily with his students and colleagues at Cambridge.

Wittgenstein's fellowship at Cambridge was gained largely on the basis of Littlewood's testimony, which was based on reading PR and having several conversations with Wittgenstein about his new ideas. Littlewood wrote to the Council of Trinity College June 1, 1930:

> Wittgenstein has explained some of his ideas to me verbally in some 6 or 8 sessions of 1 hour to one hour and a half. This has been inadequate for a full understanding, but enough to give me the impression that his work is of the highest importance. Revolutionary as his

[14] See Ramsey (1991a,b).

[15] ms. 162b, 9v.

[16] Wittgenstein (1965a), hereafter LE.

[17] Wittgenstein and Waismann (1979), p. 104. For an account of the Waismann collaboration, see Wittgenstein and Waismann (2003) (hereafter VW), Preface.

[18] The volume *Philosophical Grammar* (Wittgenstein 1974c, hereafter PG), was compiled by Rush Rhees and published as a selection from this fuller typescript.

ideas are, some of them seem to me clearly destined to become eventually part of logical thought. There are cases of new absolutely first class work when even an amateur can feel confident, and justifiably, in its importance. This seems to me to be such a case, and as the Council has asked for my opinion, here it is. But I am only an amateur in logic, even in mathematical logic.

I began, for some or no reason, with the idea that W. might be living on old capital, or unable to communicate his ideas. I was agreeably surprised, and it is true to say that I always became interested when once the session started, and enjoyed it. His illustrations struck me as highly illuminating, he often guessed what was at the back of my hesitations, and so on. In fact whatever his idiosyncracies in other matters may be, on his own ground he is simply the first rate mind.

The idea about old capital is entirely groundless. W. wrote a book once before, and I mean it literally when I say that I see no reason whatever why he should not write another, and perhaps a more important one.

J. E. Littlewood[19]

The Big Typescript was far too long to be published. In subsequent years it was used by Wittgenstein as a kind of "quarry"[20] for his future manuscripts, and he often mined and reacted to it in his further writings. Its two part overall structure – the first part devoted to sense and the nature of the proposition, the second an application to logic and the foundations of mathematics – always remained the one he envisioned for his second book. However, the second part was never finalized, despite the great deal of time he spent on it.

One of us has argued that it was in 1936–1937, as a result of reading Turing's epochal paper "On computable numbers, with an application to the *Entscheidungsproblem*", that Wittgenstein's mature philosophical method and style of writing solidified, and he turned toward his latest concerted work on logic and the foundations of mathematics.[21] This concerted effort became less focussed by 1944, when topics in the philosophy of psychology and other areas began to preoccupy Wittgenstein more and more, and he seemingly surrendered his primary focus on the application to foundations of mathematics and logic.[22] Be this as it may, the annotations to CPM from 1942–1943 represent one of the latest attempts he made to work comprehensively on these subjects. Properly speaking, these writings form a kind of "application" of his mature thoughts about the nature of the proposition and meaning to the foundations of mathematics.

Wittgenstein was constantly revising his thoughts throughout this time, and he had generated an enormous number of remarks in large notebooks, as well as a series of typescripts: the enormous number of pages had to be cut down. Here an important stimulus was his road-testing of his ideas in the classroom and in meetings with groups such as the Moral Sciences Club and the Trinity Mathematics Society.

---

[19]See McGuinness (2008), p. 187, Letter 133 (June 1, 1930).

[20]Kienzler (1997, 2001).

[21]Floyd (2016, 2017), and compare the chronology in Monk (1990).

[22]Monk (1990), Chap. 22 reports that in 1944 Wittgenstein apparently wrote a dictionary entry in which he said that his "chief" contribution had been to philosophy of mathematics, but then 2 or 3 months later said to Rhees that "someone else can do that".

Fairly complete student notes from the period 1930–1947 have been published: he offered lectures at Cambridge throughout the period of his fellowship 1930–1936, continuing after his return as Professor 1939–1947.[23] It is clear that he gained, and felt that he gained, a great deal from being forced to explain his ideas to an audience.

Beginning in the fall of 1932 Wittgenstein taught two separate courses, one called "Philosophy", the other "Philosophy for Mathematicians". These corresponded to the two main parts of the book manuscript he wanted to write. It was in his 1932–1933 "Philosophy for Mathematicians" that he began publicly to respond to Hardy's work in a critical vein, initially discussing Hardy's paper "Mathematical Proof" 1929.[24]

This method had obvious disadvantages. Wittgenstein's Cambridge fellowship was due to expire in the summer of 1936, and he was under pressure to publish his ideas, especially given his large reputation in the philosophical world. Some of his views were getting partially and sketchily discussed in public, creating tension and conflict with his student note-takers. In the fall of 1933 "Philosophy for Mathematicians" grew so large and contentious that Wittgenstein deemed useful discussion impossible.[25] He surprised everyone by cancelling the lectures and replacing them with dictations that he made to a small group of his mathematics and philosophy of mathematics students.

The resulting typescripts of the dictations were turned into typescripts and circulated among the students and certain close friends and colleagues. They became known for the colors of their covers: the so-called "Yellow Book" (1933–1934)

---

[23]Klagge and Nordmann (2003) lists the occasions on which we know that Wittgenstein lectured and discussed philosophy at Cambridge with mathematicians and surveys the published transcripts of lectures through 2003. The published volumes of transcribed notes from Wittgenstein's Cambridge lectures are these: Wittgenstein (1980d), hereafter LWL, covering the course "Philosophy" in the years 1930–1932, Moore (2016), hereafter MWL, covering "Philosophy" in the years 1930–1933, Wittgenstein (1979), hereafter AWL, covering the years 1932–1935, Wittgenstein (2017), hereafter SWL, covering 1938–1941, Wittgenstein (2007), hereafter LA, covering 1938, 1942, 1946, Wittgenstein (1976), hereafter LFM, covering spring 1939, and Wittgenstein (1988), hereafter LPS, covering 1946–1947.

[24]See AWL pp. 205–225. In addition, the Francis Skinner archives, currently housed at the Wren Library, Trinity College Cambridge, contain other lecture transcriptions and dictations, more extensive than those published, including a longer version of notes from "Philosophy for Mathematicians" and a more extensive version of *The Brown Book*. On these see Gibson (2010).

[25]A public spat flared up in May 1933 in *Mind* between Wittgenstein and Braithwaite over the issue of Wittgenstein's timeliness in publishing his ideas (reprinted in PO pp. 156–7, discussed in Floyd (2017)). Wittgenstein fell out with H.M.S. Coxeter when he set up a mimeograph machine to run off the course dictations from "Philosophy for Mathematicians" in the fall of 1933 (WC, p. 219). Alice Ambrose also ran into irreconcilable differences with Wittgenstein. Though he helped her to obtain her Cambridge PhD without an oral examination, Ambrose annoyed Wittgenstein intensely when she published, against his advice, two articles in *Mind* (1935a; 1935b) attributing finitistic views to him (See WC, pp. 241–2). Wittgenstein insisted that Ambrose publicly retract her attributions and resigned from Ambrose's thesis committee, dubious that he could work with her. She shared the feeling. Russell (1936) explicitly cited and argued against Ambrose's views, treating them as Wittgenstein's; she responded in her Ambrose (1937).

consists of notes taken down by Alice Ambrose and Margaret Macdonald (later Braithwaite) before Wittgenstein cancelled the class "Philosophy for Mathematicians" (a version is published in AWL). In 1934–1935 Wittgenstein dictated the so-called "Blue Book" (1933–1934), and the so-called "Brown Book" (1934–1935), and these were published together in Wittgenstein (1965b), hereafter BB. They were dictated to Ambrose, Macdonald, and the mathematics PhD students H.M.S. Coexter, R.L. Goodstein, and Francis Skinner (see Klagge and Nordmann (2003, p. 345–6).[26] The primary intent was to generate a record so that (as Wittgenstein later wrote to Russell enclosing for him a copy of *The Blue Book*) his pupils "might have something to carry home with them, in their hands if not in their brains".[27] But, in reality, the dictations produced remarks that helped impose order on his own thinking and writing.

In *The Blue Book* the notion of a "language-game" appears, and the language-games themselves are now regarded both as "simple" and as objects of comparison. In *The Brown Book* Wittgenstein's comparative, sometimes "anthropological" stance toward language as an evolving, nebulous mix of procedures among different "tribes" rises to the fore as a central feature of his mature philosophical method. It is here, in *The Brown Book*, that his later approach to logic, or "grammar", makes itself clearly recognizeable.[28]

1.3 The Emergence of *Philosophical Investigations*, 1937–1944

For a time Wittgenstein thought of *The Brown Book* as something he could publish. He made final revisions to it as late as the fall of 1936, when, after his fellowship was over, he travelled to Norway in order to finish his book. However, by November 1936 he gave up, writing in heavy strokes across the page that "This whole attempt at a revision, from the start right up to this point, is worthless."[29] He came to find "it all, or nearly all, boring and artificial" and the English version "cramped" his thinking.[30]

Two factors indicate that Wittgenstein was already undergoing a shift in his conception of philosophical method during his attempted revision of *The Brown Book*. First, he began to thematize the notion of simplicity differently, adding a

[26] As noted, the Skinner Archives, currently at the Wren Library, Trinity College Cambridge, contain alternative, more expansive notes of the Skinner materials.

[27] See the Editorial Preface to BB; the letter to Russell is in McGuinness (2008), p. 197.

[28] Engelmann (2013), compare Floyd (2017).

[29] See the Editorial Preface to BB. The manuscript he was working on has been published in German as Wittgenstein (1969b), hereafter EPB, and his remark occurs on p. 118.

[30] See Wittgenstein's letter to Moore of 20 November 1936, WC Letter 204, p. 257, discussed and analyzed in Engelmann (2013, Chapter 4) and in Floyd (2016).

reference to Plato's *Theaetetus* and applying to it a language-game with colored squares, explicitly juxtaposing his perspective with the idea of absolute simples in the *Tractatus*, presenting his ideas in terms of an overarchig revision of his thought.[31] Second, he brought in the notion of *Lebensform*, a kind of replacement for an anthropological notion of *culture* or *tribe*.[32] In *The Blue and Brown Books* to imagine a language is to imagine a culture [*Kultur*], but after the fall of 1936 the notion of culture is abolished from Wittgenstein's manuscript, never occurring again.

As one of us has argued, these remarks begin to solidify what would become a different attitude toward the idea of simplicity as it occurs in logical analysis and philosophy.[33] Wittgenstein moved from a cultural, anthropological approach toward a more fluid, multifaceted way of thinking about what foundations and simples are. In his mature thought touchstones of argumentation in living procedures, parametrizations, and practices are made a central feature of his philosophical method, but also the whole idea of a useable *technique* becomes of central importance to his conception of logic. Language-games must be embedded in forms of life, rather than simply accepted.

The overarching evolution of his thought may thus be summarized as follows. In the TLP, simplicity had been absolute. In his middle years, simplicity had been regarded as something relative to a *Satzsystem*, hence purely conventional. In BB simplicity was lodged in language-games, which were presented in a serial, step-by-step manner. Finally, on his mature view, simplicity is not simply relative, nor a figment of particular language-games or analyses. Instead, it is more fluid: a feature of language-games that, once described, may be contested, altered, shifted, and changed as one amalgamates, compares, and reconfigures examples of routines. This is why the notion of culture is eliminated, and replaced with that of the more flexible, less historicized notion of *Lebensform*.

This newfound attitude toward simplicity necessitates a new style, one that would become characteristic of Wittgenstein's mature writings beginning in late 1936 through the spring of 1937. This style is no longer serial and linear, in the manner of BB. Instead it is composed in multilogue form, with shifting dimensions of analysis, conceived explicitly as an "album" or collection of snapshots of a variety of interlocutory exchanges, rather than a quasi-evolutionary, controlled step-by-step presentation of a series of language-games.[34]

It may well be that this new conception was stimulated by reading Turing's "On computable numbers: with an application to the *Entscheidungsproblem*" (Turing 1936), a paper in which the *Entscheidungsproblem*, intuitively framed

[31]These remarks on *Theaetetus* form the basis for PI §46. On the novelty of juxtaposing his new thoughts with his old, see Engelmann (2013).

[32]Floyd (2018b).

[33]Floyd (2016).

[34]On the "album" structure, see PI Preface and Pichler (2004). On the necessity of Wittgenstein's later style, see Floyd (2016).

by Wittgenstein himself in 1913, was resolved in the negative.[35] For this paper involves a distinctive analysis of the nature of formalisms in the context Hilbert's mathematization of the logical elements of the axiomatic method, one that may well have been influenced by Turing's exposure to Wittgenstein's BB and the notion of a language-game as an undergraduate.[36] Turing's analysis involves a precisely targeted *object of comparison*: an analogy between a human calculating according to a rule with paper and pencil and a machine calculating the digits of a real number.[37] It is likely Wittgenstein had heard about Turing's result during the summer of 1936, when Turing was in Cambridge revising it for publication; we know for sure that Turing personally sent him an offprint before the end of February 1937.[38]

From this point onward, Wittgenstein regarded simples, not as relative to a particular system, but fluid, finding their place within ongoing activities of proceduralization and clarification. He ended a notebook taken down during the spring of 1937 – this way:

> The simple as a sublime term and the simple as an important form of the representation but with home-spun [*hausbackener*] application (MS 152, 96).[39]

This fits Turing's "domesticated" conception of the application of the notion of calculation-in-a-logic: with a Turing Machine, procedures are taken to operate from a given set of commands formulated with symbolic configurations that are taken to be simple ("taken in at a glance") in the context of a form of life. (We shall discuss the vexed notion of "application" later on, in Sects. 3.1 and 3.2 and the connection with Turing in Sects. 8.5 and 8.8.) Moreover, the fact that with Turing's analysis of a step-in-a-formal-system we must take the notion of a partially, but not totally defined function to be basic implies that at its foundations logic cannot be presented in terms of the "gap-free" ideal of proof that Frege espoused, as Wittgenstein acknowledges explicitly in the Preface to PI.[40]

---

[35]For this argument, again see Floyd (2016), and compare Dreben and Floyd (1991).

[36]For this argument see Floyd (2017). Turing explicitly notes that his version of the theory of types in his (2001b) was indebted to attending Wittgenstein's 1939 Lectures on the Foundations of Mathematics [LFM]. See Floyd (2013b) for further discussion of this paper.

[37]For discussions of the conceptual precision in Turing's analysis (a feature also of the nearly simultaneous analysis of Emil Post), see Sieg (2009), Davis and Sieg (2015), Mundici and Sieg (2017), Floyd (2012c, 2017).

[38]Floyd (2017).

[39]Compare Wittgenstein's use of "*ganz hausbackene*" at BGM V §9, quoted above on p. 69.

[40]See Floyd (2016); after Turing, it is clear that the basic notion of proceeding step-by-step in a formalized logic is that of a partial, not a universal function (see Sect. 8.5 below). The importance of Wittgenstein's having surrendered the ideal of a "gap free" [*lückenlose*] presentation of his thoughts that would draw out a "natural" order of ideas, is twice made explicit in PI: in the Preface and in §67. The English translations of these remarks – Hacker-Schulte use "smooth" and "continuous", respectively and Anscombe uses "with breaks" and "overlapping" – fail to bring out Wittgenstein's explicit allusion to Frege's ideal of "gap-free" proof in logic, the *Lückenlosigkeit der Schlusskette*. See Frege's Preface to his *Begriffsschrift* (Frege (1879)). Wittgenstein's mature

Between late November 1936 and May 1937 Wittgenstein put together the so-called "Original Draft" (*Urfassung*) of his *Philosophical Investigations*.[41] This material comprises the first draft of the first 188 remarks of the final version of PI, the final 102 remarks of which were written between January and May 1937.

Returning to Cambridge for the summer of 1937, Wittgenstein continued his earlier practice of fine-tuning his arguments by presenting them in discussion with his students and close colleagues. He had a typescript made of the whole (TS 220), which he showed to and discussed with G.E. Moore and some other colleagues, including Turing and his friend (and close student of Wittgenstein's) Alister Watson, who had originally introduced Turing to Wittgenstein.[42] Moore would later say to Rhees that Wittgenstein told him at this time that *The Brown Book* had applied a wrong or "false" method [*falsche Methode*], but that now he had come to apply the "right" or "correct" method [*die richtige Methode*].[43] Our view is that the idea of the fluidity of the notion of simplicity embedded in his treatment of logical analysis earmarks this mature stage of his thought.

The challenge of applying this conception to the foundations of mathematics was considerable. As we have already said, Wittgenstein's concerted interest in developing an application of his mature philosophical conception to the foundations of mathematics began in earnest in 1937, beginning with exchanges he had with Watson and Turing that summer.[44] Over the course of 1937–1938, again in Norway and occasionally lecturing and meeting with students at Cambridge, he generated the first full-length, so-called "early version" (*Frühfassung*) of PI (TS 225, TS

philosophy, with its "broken", partial, interlocutory style draws out instead the more artificial and "gappy" character of human-fashioned techiques, which evolve (see Sect. 8.3 below).

[41]The "Original Draft", or "*Urfassung*" [UF] (MS 142, 1936–1937) is published in Wittgenstein (2001), hereafter PUK, along with four other different versions of PI, accompanied by editorial commentary that allows us an evolutionary analysis of PI's construction: the "Early Draft" "*Frühfassung*" [FF] (TS 225, 220, 221, 1937–1938, containing a first draft of RFM I), the "Reworked Early Draft" "*Bearbeitete Frühfassung* [BFF] (TS 239, 1939–1943), the "Intermediate Draft" or "*Zwischenfassung*" [ZF] (TS 221, 239, 1944–1945), and the final or "Late Draft", "*Spätfassung*" [SF] (TS 227, 1945–1946). In PI is published, a separate part, "Teil II" ("Part II", from the missing MS 144, 1946–1949), but now with a new title [PPF]. Compare PI pp. xviii-xxiii.

[42]Watson (1938) discusses the recent undecidability results, acknowledging both Turing and Wittgenstein for discussions. Compare the discussion in Floyd and Putnam (2000, 2012), Floyd (2001b).

[43]Moore said he did not know what Wittgenstein meant; see the remarks by Rhees in the Preface to EPB, pp. 12–13. Rhees also says that the manuscript included TS 221, but the editors of PI disagree, dating *that* typescript to work done during 1937–1938 in Norway.

[44]Floyd (2016).

220 and TS 221).[45] In FF he attempts to apply the passages on rule-following directly to discussion of the foundations of logic and mathematics, as well as to the nature of proof, mentioning Gödel explicitly.[46] He also drafted remarks on Cantor's diagonal argument; these will be the focus of Mühlhölzer's Chap. 7 below. By the end of the summer of 1938 he had generated a manuscript which he submitted to the Cambridge University Press. By January 1939 he held a Professorship, having initially not expected to do so, but compelled, as a result of the *Anschluss* of Austria in March 1938, to seek British Citizenship.[47]

This initial application of his mature perspective to the foundations of mathematics in the period 1937–1939 was ultimately unsatisfactory to Wittgenstein. Once he became a Professor, he continued attempting to develop a better one, initially in his Cambridge seminar from 1939, LFM, that Turing attended. Manuscripts from this period through 1944 are characterized by an exploratory, hesitant effort to incorporate specific mathematical examples into his perspective. After Wittgenstein's death, selections from these manuscripts were published as RFM. On the whole, we may say that in the end Wittgenstein did not aim to put forward a foundation of mathematics of his own, but, rather, to work through the whole idea of what it would be to do this: a philosophical task. It is a *Klarlegung* instead of a *Grundlegung*.[48]

The sections after 188 were wholly excised from the PI manuscript sometime before Wittgenstein finished his third version, the *Bearbeitete Frühfassung* (BF, "reworked early version", TS 239) in 1943 (see PUK). These excised remarks were posthumously published as Part I of RFM, and were never reinstated in the later versions of PI, including the final one.[49] There are occasional notes concerning mathematics and philosophy of mathematics that remained in PI, and asides on these topics occur throughout Wittgenstein's latest writing, including *On Certainty* (hereafter OC Wittgenstein (1974b)). But such remarks remain short detours, rather than themes treated at length. It is clear that Wittgenstein was less than confident about the status of his remarks.

But though these manuscripts remained unfinished, there is no doubt that Wittgenstein engaged in a sustained period of his thinking about how to apply his mature conception of philosophical analysis to mathematics from 1937–1944. And as we have already said, it was during the central part of this period that he annotated his volume of CPM_8. Thus the annotations to Hardy's textbook stem from a relatively late part of this period, something interesting in and of itself. And we shall show that they belong to a tentative period of his thought, not a settled one.

Wittgenstein presumably wrote down his annotations to CPM in 1942, perhaps continuing into 1943. We conclude this from the fact that they were carried forward

[45]The *Frühfassung* is published in PUK. Lectures by Wittgenstein at Cambridge 1938, several on the notion of certainty, and one on Gödel, may be found in SWL.

[46]See *Frühfassung* §383 (in PUK), discussed in Floyd (1995, 2001b); Floyd and Putnam (2012).

[47]On this see Monk (1990) and the letters from this time in WC.

[48]On this see Mühlhölzer (2010), critically analyzed in Floyd (2015).

[49]See Floyd (1991).

and considerably expanded in his manuscripts, MSS 126/127, in remarks written from 11 December 1942 to 4 April 1943. It is plausible to think that the manuscript notes occurred subsequently to the annotations, though they may have been penned at the same time. We take these notes and manuscript writings to be elaborations of the annotations, for, being more expansive, they contain important clues as to what Wittgenstein had in mind. Some of these clues found their way into RFM V, itself an extract of MSS 126 and 127. We shall weave our discussion of the annotations into this larger framework, using these manuscripts to shed light on the annotations.

Of course far more is contained in these manuscripts of 1942–1943 than just the remarks directly relevant to Wittgenstein's CPM_8 annotations. Our analysis will deal only with portions of them, enough to give life to our extensive discussion of the annotations themselves in light of Wittgenstein's aims in discussing the foundations of mathematics. At the same time, we will be characterizing the significance of the annotations by embedding them in a reconstruction of the argumentation contained in the larger, continuous flow of thoughts to be found in the manuscripts.

1.4 Wittgenstein and Hardy

When did Wittgenstein first get to know the text of Hardy's CPM? Brian McGuinness tells us that when Wittgenstein studied engineering in Manchester from 1908 to 1911, the mathematician J. E. Littlewood was there and Wittgenstein attended some of his lectures on analysis.[50] The first edition of CPM appeared in 1908, and it is very probable that Wittgenstein at least knew about CPM's existence then.

The first paragraph of the Preface to the first edition of CPM describes the aim of the whole book and was carried over by Hardy to all subsequent editions. It reads as follows:

> This book has been designed primarily for the use of first year students at the Universities whose abilities reach or approach something like what is usually described as "scholarship standard". I hope that it may be useful to other classes of readers, but it is this class whose wants I have considered first. It is in any case a book for mathematicians: I have nowhere made any attempt to meet the needs of students of engineering or indeed any class of students whose interests are not primarily mathematical.

For Wittgenstein, who came from engineering but wanted from at least 1909 onward to investigate, as a philosopher, the nature and philosophy of pure and modern mathematics, this would certainly have seemed an inviting beginning. Hardy's book was in fact designed to present a decidedly modern approach to the English speaking world. Immediately after the passage just quoted, Hardy explicitly writes that "the general range of the book is a good deal wider than is usual in English treatises of the Calculus". These treatises he calls "orthodox": they were geared toward training students to solve problems on the so-called 'Tripos'

[50]McGuinness (1988), p. 62.

examinations in Cambridge. Problems of this sort survive in the exercises of CPM, but the main aim of this book was broader: to provide a systematic and conceptually rigorous presentation of analysis.

Hardy's model was the second edition of Camille Jordan's *Cours d'analyse* (1893), published in three volumes from 1893 to 1896. In Section 29 of his *A Mathematician's Apology* (1940) Hardy revealed how important this book was for his own mathematical development, which had begun in Cambridge:

> I was really quite ignorant, even when I took the Tripos, of the subjects on which I have spent the rest of my life; and I still thought of mathematics as essentially a "competitive" subject. My eyes were first opened by Professor Love, who taught me for a few terms and gave me my first serious conception of analysis. But the great debt which I owe to him – he was, after all, primarily an applied mathematician – was his advice to read Jordan's *Course d'analyse*; and I shall never forget the astonishment with which I read that remarkable work, the first inspiration for so many mathematicians of my generation, and learnt for the first time as I read it what mathematics really meant.

With CPM he wanted to present a comparable book, of smaller size, in English. In the first edition of 1908, Hardy was a bit half-hearted in his efforts to achieve systematicity and rigor: he avoided the rigorous definition of real numbers as it was known after Dedekind, and, taking the 'linear continuum' for granted, simply assumed, as he wrote, "the existence of a definite number corresponding to each of its points" (CPM_1, p. v). But from the second edition of 1914 onwards he presented the Dedekindian theory of real numbers as cuts of the rationals. It is above all this approach that Wittgenstein concentrated on in his philosophical investigation of CPM.

From the outset, there is one question concerning CPM that immediately suggests itself: Why does the book have this strange title, *A Course of Pure Mathematics*? For it is without doubt a textbook on analysis. Why, then, didn't Hardy choose the title: *A Course of Analysis*, or even better, *A Course of Modern Analysis*? We do not know the answer for sure. But one reason might be that in 1902 there had appeared E. T. Whittaker's book *A Course of Modern Analysis* (1902), a text with an aim similar to Hardy's. Hardy certainly would have wanted another title for his own book.[51] There is, however, another way to understand Hardy's title in light of the earlier impact of Jordan's work on him in his student days. If we place the emphasis on the words "A" and "Pure", the book may be regarded as a course of pure mathematics because there were (and are) other kinds of "pure" mathematics courses, for example, ones in algebra. But Hardy's is a course of "pure" mathematics in the sense that analysis is presented now in a really pure way, inspired by Jordan's *Cours d'analyse*.

This presentation, as Hardy himself wrote in his reminiscence, was quite different from the ways in which analysis had been presented in Britain before this time. As we have said, one of the main topics of Wittgenstein annotations to CPM was the

[51] Another book with such an aim, and appearing before CPM, was E. W. Hobson's *The Theory of Functions of a Real Variable and The Theory of Fourier's Series*, which went into three editions (1907, 1921, 1926 1927). Neither Whittaker's nor Hobson's books ever had the success of CPM, although, as we shall see in Chap. 8, Hobson forms an important link in the Cambridge context.

definition of real numbers via Dedekind cuts. Hardy doesn't call them "cuts" but "sections". One should not think, however, that Wittgenstein first learned of "cuts", or "sections", from CPM. He certainly got to know about this material at least by the time of his student days in Manchester (1908–1911) or shortly afterwards, either through Littlewood or through his reading of Russell's *The Principles of Mathematics* (1903), as well as numerous other writings of Russell's over the years in which Dedekind cuts, the issue of "theft over honest toil", i.e., Dedekind's idea of "free creation", were analyzed.[52]

It is also important to bear in mind that Wittgenstein was personally acquainted with Hardy himself, at the latest from 1913 on; in fact Hardy was one of his most important friends in Cambridge for a very long time.[53] Later on, in 1929, Hardy directed criticisms at Wittgenstein's *Tractatus* philosophical views in his article "Mathematical Proof".[54] Wittgenstein replied to these in his Cambridge lectures on philosophy for mathematicians in 1932–1933 and 1939.[55] It would be strange if Wittgenstein and Hardy had not talked about mathematics, and about analysis and Dedekind cuts in particular.

Given this background, it is surprising that the topic of irrational numbers and Dedekind cuts nowhere turns up in the *Tractatus*, even if the book is targeting the nature of logic, rather than any specific mathematics. In fact, in his early work the view of mathematics that Wittgenstein presents is utterly "nebulous" with respect to such crucial issues concerning the infinite, as Russell pointed out in his Introduction and Wittgenstein later admitted (PR, p. 131). This reticence abruptly changes, however, when Wittgenstein returns to Cambridge in January 1929. His very first note in his very first manuscript, written on 2 February 1929, commences with a remark about irrational numbers:

> Is a space conceivable that contains only all rational points but not the irrational points?
>
> And that only means: Aren't the irrational numbers already prejudged in the rational ones?[56]

From then until 1944 Wittgenstein gave mathematics a considerable weight in his philosophizing, and thoughts about irrational numbers, Dedekind cuts, the notion of a function, a limit, etc., regularly turned up.

One cannot be sure about the reasons for this new preoccupation, beyond the obvious point that Wittgenstein knew that the *Tractatus* had glaring lacunae with respect to the theory of numbers, as Russell had pointed out. As reported by Menger,

---

[52]For example, see Russell's presentation of Dedekind cuts in his (1903), as well as his (1920), which Russell sent to Wittgenstein in 1918, while he was still a prisoner in Monte Cassino, concerned with the publication of his *Tractatus*. We do not know whether Wittgenstein ever read Dedekind's classic *Stetigkeit und irrationale Zahlen* (1872). The idea of "free creation" is used in Dedekind (1872), §§3–4, and (1888), Preface to the first edition, §73.

[53]This may be gathered from WC, especially on pp. 53f. and 194.

[54]Hardy (1929), p. 24.

[55]AWL, pp. 205–225, and LFM, pp. 123, 130ff., 138ff., 238ff.

[56]MS 105, p. 1; we shall come back to this passage in Chap. 2 below.

Wittgenstein attended Brouwer's lecture "Mathematik, Wissenschaft und Sprache" in Vienna on March 10, 1928 (Brouwer 1929); this seems to have stimulated him to new philosophical activity.[57] In this lecture Brouwer sketched his idiosyncratic view of irrational numbers and presented his famous oscillatory numbers (*Pendelzahlen*), and Wittgenstein may have been impressed by the novelty and the radicalness of Brouwer's approach. This does not mean, however, that Wittgenstein adopted Brouwer's intuitionism – later on he explicitly rejected it[58] – but the encounter with Brouwer's ideas may have inspired him to have a new look at irrational numbers.

CPM is not mentioned explicitly in his manuscripts of 1929 and at the beginning of the 1930s. Instead, Wittgenstein wrote down some reactions to Hardy's (1929) article on the notion of mathematical proof.[59] Explicit discussion of CPM began in Wittgenstein's 1932–1933 Cambridge lecture course entitled "Philosophy for Mathematicians", where Wittgenstein read out extracts from the 5th edition of CPM (1928) and critically discussed them.[60] Hardy had returned to Cambridge in 1931 from Oxford (where he had been since 1919). Moreover in the early 1930s Wittgenstein and Hardy shared some graduate students, as did Wittgenstein and Littlewood: for example, Mary Cartwright, later a co-author with Littlewood, came and presented a paper on "What is three?" to Wittgenstein's seminar, at his request, in early 1933.[61]

Littlewood explicitly counted himself a personal friend of Wittgenstein, writing, as we have seen, that "on his own ground he is simply the first rate mind" (WC, p. 187). In fact it was Littlewood who, more than Russell, wrote a report that got Wittgenstein his fellowship in 1930. As we have seen, his report was based on at least eight discussions he had with Wittgenstein to talk about his proposed ideas in May 1930 (WC, p. 187). Whereas Russell worried that Wittgenstein's ideas were not correct, Littlewood stressed that Wittgenstein was not simply working out "old" *Tractatus* ideas, but thinking through deep and novel ones.

[57] See Menger (1994, pp. 130–131), Monk (1990, pp. 249–251), Hesseling (2003, pp. 190–198), and Marion (2003).

[58] See PG, p. 295, BT, p. 537, LFM, pp. 30–31, 82, 127, 184, 237–8.

[59] What is alluded to at this early stage is a remark in Hardy (1929), p. 5, that Wittgenstein would revisit several times: "that the finite cannot understand the infinite should surely be a theological and not a mathematical war cry". See TS 233a, p. 56 (1930; 1948), MS 119, p. 84v (1937), MS 116, p. 54 (1937–1938), TS 230, p. 100 (1945), TS 228, p. 11 (1945–1946). CPM provided Wittgenstein with an opportunity to look at what Hardy actually does with proofs.

[60] As we have said, a transcription of these lectures is published in AWL, pp. 205–225; See Monk (1990), p. 328, and somewhat more detailed notes are in the Francis Skinner archive now held at the Wren Library, Trinity College, University of Cambridge.

[61] See WC, p. 207 for Wittgenstein's letter to Cartwright thanking her for her paper on "number" and requesting that it be discussed in class. A footnote in this edition contains reminiscences of several of the mathematics students. Cartwright expressed disappointment that Wittgenstein never reached infinite numbers, and, like the others, notes how unconfident and slow he was at articulating his ideas. See the entry for Littlewood on the St. Andrews website http://www-history.mcs.st-andrews.ac.uk/Biographies/Littlewood.html, which discusses the collaboration with Cartwright.

Between his 1931 discussion and the 1942–1943 annotations to CMP with which we are concerned, there are only sporadic references to CPM; one, for example, on June 12, 1938, where Wittgenstein mentions "a textbook like Hardy's" in discussing Cantor's diagonal argument.[62] In the 1939 lectures that Turing attended, though Hardy is mentioned several times, only his (1929) on mathematical proof is treated. Wittgenstein objected at length to the ways in which Hardy's view of propositions makes ordinary mathematical phraseology irrelevant "gas" or mere "decoration".[63]

But then why did Wittgenstein resume commenting on CPM in 1942, with his annotations to the 8th edition and the elaboration in the MSS 126–127? As usual, we shall never know for sure. Nevertheless, with regard to the substance of his reflections Wittgenstein surely had good reasons to take up the topic of irrational numbers and related topics again.

First of all, his discussion in his 1932–1933 lectures had been rather lame, leading, as we have seen, to the contretemps with Ambrose over her publication of ideas associating Wittgenstein's views with finitism and with Russell's direct criticisms of Wittgenstein in his (1936). Wittgenstein might have felt determined to do it better now. Russell's comeback to philosophy in the late 1930s, advanced with an attack on what Russell (following Ambrose) described as Wittgenstein's "finitism", evidently piqued Wittgenstein and may have stimulated him to work more on these topics.[64] It is not to be ruled out that his discussions with Turing, and/or knowledge of Turing's famous paper "On Computable Numbers" had an effect between the fall of 1937 and 1940, as we have already argued. Finally, Nedo writes that conversations with Georg Kreisel in 1940–1941 may have been instrumental as well.[65]

A further reason might have been the publication of Hardy's book *A Mathematician's Apology* (1940). Wittgenstein loathed this book because of the philosophical dilettantism he saw in it. In MS 124, pp. 35f., dated 13 June 1941, he wrote down a draft of what would later appear in PI with Hardy's name removed, placing the phrase "the mathematician" in its place (see PI §254):

> The sentences which Hardy presents in his "miserable [*elenden*]" book, "Apology of a Math.", as the expression of his philosophy of mathematics are not yet philosophy, but may, as all similar effusions [*Ergüsse*], serve as raw material for philosophizing, and then should not be enunciated in the form of opinions, statements, or axioms, but in the form: "I am inclined to say: ...", "I always want to say: ...".

This "miserable" book might have incited him to have a second look at Hardy's textbook: not his philosophical effusions now, but instead what Hardy says and does as a competent mathematician. And in fact, Wittgenstein's annotations to CPM and his reflections in MSS 126/27 avoid discussion of Hardy's philosophical remarks altogether, addressing themselves to the mathematical writings alone. Wittgen-

[62]See Felix Mühlhölzer on Cantor's diagonal method in Chap. 7.

[63]For discussion of the significance of this see Floyd (2017).

[64]See Monk (2007), p. 283.

[65]Nedo (2008), p. 87.

stein diagnoses important defects in Hardy's writing. But Hardy's mathematical statements, unlike his philosophical remarks, cannot be dismissed as "effusions": Wittgenstein's criticisms require quite different philosophical moves from those needed to treat Hardy's philosophical pronouncements.

One must admit at the outset that Wittgenstein's criticisms of CPM are, unfortunately, pervaded by a subliminal resentment that is characteristic of many of his investigations of mathematics.[66] He is suspicious of the characteristically abstract traits of mathematics as they arose through the work of Cantor and Dedekind in the nineteenth century and as they show themselves within CPM in the notion of a Dedekind cut. He does not go so far as to reject these traits. But he does regard them as relying on a sort of "phraseology" that should not be adopted at face value, and he emphasizes that there are severe philosophical dangers in its use. We shall discuss this notion of "phraseology" below, in Chap. 2.

Above all, Wittgenstein is strong-minded, and at times quite defensive, about what philosophical investigations are and should be. We have seen that he is, in the large, exploring how he might apply his own, hard-won mature perspective on logic to the very idea of a foundation of mathematics. Within the context of this enterprise, deference to "effusions" of mathematicians just because they are great practitioners of their art is wholly unacceptable for a philosopher, and a constant danger – just as dangerous as confusing purely general philosophical questions about conceptual analysis with particular analyses that are entangled with mathematical questions directly.

Most of the time Wittgenstein's suspicions are kept in check, working only subliminally, and evincing themselves in clipped and brief phrases. But sometimes they break out and bubble up to the surface, as, for example, when in MS 127, p. 67, a central Dedekindian insight concerning the closedness of the continuum is called "crude" [*abgeschmackt*]. This may be read as a sort of tit-for-tat response to Hardy (1929, p. 2): "[philosophical] arguments for any answer must be [...] crude". Sometimes Wittgenstein simply lets himself get carried away in exploring his own insistence on fine nuances of language in the face of mathematicians' pronouncements, as in the following appalling remark of 28 February 1947 in MS 134, p. 1v, written in English and in Wittgenstein's code: "The vacant smile of the modern English mathematician." It is presumably Hardy whom Wittgenstein means here (a few pages later Hardy and his philosophical position are explicitly mentioned), and the alleged vacancy of his smile seems to correspond to the vacancy of the abstract phraseology of modern mathematics and the passing over in a superficial way of what are, for Wittgenstein, important, difficult and fundamental philosophical points.[67]

[66] See Menger (1994), p. 129 on this resentment.

[67] Ironically, Menger describes Wittgenstein as wearing "an abstract smile and eyes focused at infinity" when he recounts the story of his behavior at Brouwer's 1928 Vienna lecture (Menger 1994, p. 131). Wittgenstein himself used the word "vacant" in PI §412 where he writes: "Mein Blick war 'vacant'; oder ähnlich dem eines Menschen, der die Beleuchtung des Himmels bewundert und das Licht eintrinkt" (in the Hacker-Schulte translation: "My glance was vacant;

When reading Wittgenstein it is important not to be dragged into sharing and promulgating this resentment. We do not think, however, that it is all that difficult to guard against it and set it to one side. The thoughts that remain are, we think, mostly perceptive and worth taking seriously and in any case should not be set aside on the grounds of such *ressentiment*, especially given the working, private nature of the manuscripts that were never intended for publication as they stood.

Instead, we aim to interpret Wittgenstein's annotations to CPM charitably, while remaining aware of his own limitations. These limitations include, in addition, the evident fact that Wittgenstein's mathematical knowledge was rather restricted, even for his own time. This must be taken into account, of course. For example, when Wittgenstein complained about the vacant smile "of the modern English mathematician", we must remember that CPM had in fact been a relatively "modern" book for a long time, but that by the 1940s (when Wittgenstein wrote down this remark) CPM had already become old-fashioned, at least when compared with all that had recently happened in research mathematics. For example, the Bourbaki group originated in the 1930s, and its initial aim was to write an up-to-date textbook of mathematical analysis.[68] In the end, a bit paradoxically, they didn't really produce such a thing, but instead material oriented toward cutting-edge research. Their systematic treatise, *Élements de mathématique*, published in the early 1940s (Bourbaki 1942, 1949), presented analysis, including the theory of real numbers, in a style light-years beyond the presentation in CPM. In the 1930s and 40s the tendency to abstractness was dominating mathematics, crowned by the development of category theory beginning with Eilenberg and MacLane (1942, 1945).[69] All these developments were unknown to Wittgenstein, as it seems. We do not think, however, that this invalidates his remarks on CPM. For they forward a perspective belonging to a different territory altogether from the abstract developments just mentioned.

1.5 Structure of This Book

We have not seen the original copy of CPM_8 into which Wittgenstein wrote his annotations. Michael Nedo told us that it has been lost, but witnesses in the 1970s (e.g., Joachim Schulte) have told us that it did exist. What we have used are images

---

or again, like that of someone admiring the illumination of the sky and drinking in the light"). The context is Wittgenstein's discussion of someone's turning his attention to his own consciousness, which may also yield a characteristically "vacant" expression in such a way. Here, the word "vacant" is partly meant polemically, but also descriptively. When applied to 'the modern English mathematician', however, the polemical meaning is certainly the predominant one.

[68]The relevant information about Bourbaki may be found in Corry (2009).

[69]S. M. Ulam claims to have invented category theoretic notions already in 1932; see Ulam (1991), p. 46. Initially, mathematicians liked to call category theory "abstract nonsense", not in a derogatory sense, but humorously. We suspect Wittgenstein would not have liked this sort of humor. His own attitude to such "abstract nonsense" was very different, of course.

of at least parts of Wittgenstein's own copy of CPM_8. Nedo himself, under the auspices of the Cambridge Wittgenstein Archives, provided us with images of annotations Wittgenstein made from the beginning through p. 27, as well as on pp. 30–47 and 116–123. Pages 28–29 are reproduced in Nedo (2008), pp. 96f.[70] In Part IV we transcribe the annotations in colored gray boxes, juxtaposing them with those parts of Hardy's CPM on which Wittgenstein was commenting, and providing both German and English translations (some of Wittgenstein's annotations were made in English or partly in English).

There can be no reasonable doubt that these annotations are Wittgenstein's: they are written in his characteristic hand and their content is Wittgensteinian, with obvious connections to the elaborations in MSS 126–127. We cannot be certain at present that Wittgenstein's copy of CPM did not contain more than the annotations we have seen, but we have found enough to interest us in what we do have, and it seems plausible to think that what we have seen constitutes the totality of Wittgenstein's annotations to CPM.

In Part II, Chaps. 3, 5, and 6, below, we shall present and comment upon each of the annotations. Images as well as diagrammed transcriptions of each are provided in Part IV for the reader's use. Because the images we worked from are of fairly high quality, and Wittgenstein's handwriting is characteristically clear, in all but a few cases the annotations can be easily and unambiguously decrypted. To understand what they signify demands, of course, interpretive reconstruction, and this is what we provide, in addition to translation into English where necessary. While this reconstruction must remain partly speculative, in most cases, especially when we are able to connect the annotation to a relevant passage in MSS 126–127, we do not think the sense of Wittgenstein's annotations is all that unclear, given the outlook we have sketched above.

Chapter 4 stands as a bridge chapter, framing a broader account of Wittgenstein's attitude toward the law of excluded middle. This, we believe, should help to further ground our reconstruction, as well as opening it up to further philosophical discussion.

We intend this volume to be comprehensible in itself, not requiring any previous knowledge of CPM, let alone Wittgenstein's manuscripts and annotations. A certain familiarity with mathematical thinking is presupposed, as is necessary, but not more than what would be taught today at the very beginning of an upper-level undergraduate course in analysis. Wittgenstein's annotations are presented in English, while the original versions (with one exception all in German) appear not only in the images shown in Part III but also in footnotes directly below their English versions.

[70] In Nedo (2008), p. 91, p. 6 of Wittgenstein's copy of CPM_8 is also shown.

Quotations of other passages in Wittgenstein's work are presented in English and may be compared with their German originals by using the Bergen edition of Wittgenstein's *Nachlass* (2015). We alone are responsible for the translations.

When Wittgenstein's annotations to CPM are discussed, we present at length the passages and mathematical explications in CPM_8 to which he refers, for they must be carefully read in context in order to understand his remarks.

Chapter 2
Wittgenstein's Non-Extensionalist Point of View

> *The difference in kind between the two conceptions is represented, by a skew form of expression, as difference of extension.* (RFM II §22)

Abstract A treatment of what is meant by "non-extensionalism" in Wittgenstein's philosophy of the real numbers. The "extensionalist" point of view is treated as a perspective that is at points in tension with the non-extensionalist point of view, following Bernays. Numerous examples with Wittgenstein's remarks on them are canvassed.

In order to direct the reader's attention to what is really important in the annotations, before presenting and commenting on them we would like to say a few words about the general perspective under which they should be seen.

Let us begin by looking at how Wittgenstein went on from his very first manuscript-entry in Cambridge, 2 February 1929 (quoted above, p. 19). This remark, as we have seen, concerned what in the real numbers may and may not be "prejudged" in the rational numbers. It was extended on 16 July 1931 in MS 111 and then placed into Wittgenstein's largest book draft of the 1930s, the *Big Typescript* (p. 738):

> "The rational points lie close together on the number line": a misleading picture. [...] Is a space conceivable that contains only all rational points but not the irrational points? And that only means: Aren't the irrational numbers prejudged in the rational ones? No more than chess is prejudged in draughts. The irrational numbers do not fill a gap that the rational ones leave open.

Wittgenstein here starts with the usual geometrical illustration of the real numbers as points on a straight line ("the number line"), infinitely extended in both directions. It may seem strange that, after having mentioned this line, he then talks about a "space", but we think that this has its origin in his just mentioned first note of this sort, of 2 February 1929, where the context is the question about the continuity of the visual space [*Gesichtsraum*] and whether in this space one can

J. Floyd, F. Mühlhölzer, *Wittgenstein's Annotations to Hardy's Course of Pure Mathematics*, Nordic Wittgenstein Studies 7,
https://doi.org/10.1007/978-3-030-48481-1_2

practice coordinate geometry (MS 105, pp. 1–9). Couldn't it be that in this case only rational coordinates may be involved, or are the irrational ones then automatically prejudged? In the *Big Typescript* this sort of question is then carried over to the mathematical context of the number line – the text just quoted occurs in the Chapter "On Set Theory"! – and Wittgenstein seems to have simply failed to delete the reference to a 'space'. To mention 'space' is merely a remnant of the former context and should be ignored now. It is the straight line that must be meant here.

The illustration of the real numbers as a line is used in CPM from its beginning, and is invoked immediately after Hardy's presentation of the rational numbers, which he treats as already understood. This sort of beginning suggests that the rational numbers (rational points) already "prejudge" the irrational ones, in a way to be explained later on.

Hardy introduces the irrational numbers immediately after this illustration in both the standard geometrical way ($\sqrt{2}$, for example, as the length of the diagonal of the unit square), and in the standard arithmetical way ($\sqrt{2}$ as the positive solution of the equation $x^2 = 2$). Seen from this perspective, irrational numbers are filling up "gaps" that the rational ones leave open, and the need to fill these "gaps" appears, as Hardy says on p. 13, "almost forced". To be more precise, Hardy doesn't explicitly use the picture of "gaps". According to him what is "almost forced" on us is a "generalisation of our number system", i.e., the original system consisting only of rational numbers (p. 13). The "gap"-rhetoric was, however, quite common in many other texts, starting with Dedekind's inaugurating *Stetigkeit und irrationale Zahlen* (1872/1996). According to this text, a "gappiness" [*Lückenhaftigkeit*] is to be found in the domain of rational numbers (p. 771 [p. 322]) and a "filling up its gaps in thought" [*Ausfülling seiner Lücken in Gedanken*] is something to be desired (p. 772 [p. 232]). This picture pervades the relevant texts right down to the present day. It can even be found in texts about pure set theory when irrational numbers are introduced.[1]

We do not know where Wittgenstein first encountered the language of "gaps", but it appears in Russell (1903 §§272 and 343; 1919 Chapter X) when the theory of limits and continuity are treated.

The actual point of all these texts – as it is also of CPM, even CPM_1 – is to leave behind the geometrical illustration, replacing it with a purely arithmetical conception of real numbers. In most of these texts, the terminology of the gap persists, whereas CPM is an exception in this respect. One may nevertheless say that despite its terminological asceticism, the image of a gap is suggested. Thus the idea that the irrational numbers are in a certain sense prejudged with the rational ones is present, in one way or another, in all the texts we have mentioned.

Wittgenstein is hostile to this attitude, or at least deeply suspicious of it, and his reaction is characteristically drastic: he reverts to a counter-analogy, comparing the relationship between the rational and the irrational numbers to that between draughts and chess. The point is that in draughts nothing about chess is prejudged,

[1]For example in Hrbacek and Jech (1999), pp. 86–88.

and when going from the one to the other no gaps are filled. This sort of analogy – the relationship of rational to irrational numbers in analogy to the relationship between draughts and chess – pervades a considerable number of Wittgenstein's annotations to CPM, and one should keep it in the back of one's mind when reading them.

It is important to use the analogy with discretion, because taken at face value it is of course simply ridiculous. The differences between the respective relationships are too obvious to take the analogy seriously, and Wittgenstein's comparison can only be understood as a therapeutic manoeuver. He is inviting the reader to see the relationship between rational and irrational numbers under a new perspective, not as one of *extending* a domain but as one of *embedding*, the structuring of one *kind* of number in terms of another.

This perspective is expressed in the motto of our present chapter. To elaborate and defend it forms the center of many of Wittgenstein's reflections on numbers, including his annotations to CPM, and we shall be attempting to articulate what may be said on its behalf in what follows.

Mathematicians, of course, have perfectly good reasons to go beyond the domain of rational numbers. For Wittgenstein this presents no problem at all in cases of *individual* numbers like $\sqrt{2}$, π and so on when these are determined by definite techniques of producing their decimal expansion or by similar means of identification. He balks, however, at the transition from the rational to "all" real numbers at one stroke, so to speak, which in CPM is accomplished by means of the general notion of a cut ("section" in Hardy's terminology) of the rational numbers.

As we shall see, it is not the case that Wittgenstein rejects this transition as such. But he insists on the necessity of investigating it in a much more sensitive way than is usual within mathematics.[2] The standard justification invokes important theorems which can be proved on the basis of the transition from the rational to the real numbers: Dedekind's theorem, above all (see Chap. 5 below), but also the intermediate value theorem, the Heine-Borel theorem, and so on. While never quarreling with these proofs, Wittgenstein nevertheless sees a lot of potentially confusing metaphors involved in their heuristics and motivations. It is these that are observed and explored at the pertinent places in his annotations to CPM.

It is important to emphasize that there is nothing at all extravagant in the effort to diagnose problems, and even very deep problems, in the transition from the

[2]Like Hardy, Wittgenstein is only concerned with the Dedekindian transition *via* cuts. But there are a lot of other possibilities, most prominently the Cantorian one that construes the transition *via* Cauchy sequences of rational numbers. These other possibilities are ignored in CPM and accordingly also in Wittgenstein's discussions of irrational numbers. There is one exception, however: irrational numbers can be introduced *via* familiar decimal expansions. Hardy dismisses this possibility as being "hardly of the precise character required by modern mathematics" (p. 8), but for Wittgenstein decimal expansions are standard representations of irrational numbers and he always uses these. Even from a really rigorous point of view, decimal expansions may be used to represent a perfectly respectable transition from the rational to the real numbers, as pointed out, e.g., in Gowers (2017b) and in Tao (2006), pp. 380–388.

rational to the real numbers. There is Weyl's criticism of the impredicativity of the Dedekindian conception (1918 and subsequent papers), and there is also, of course, Brouwer's intuitionism.[3] Wittgenstein knew at least some of Weyl's and Brouwer's works well enough to formulate his own reactions.[4]

But his own approach – especially as presented in his annotations to CPM and in MSS 126/27 – is very different. Unlike Weyl, he does not criticize impredicativity, at least not in these notes. And unlike Brouwer, he does not reject the law of excluded middle and other logical laws, though he does repeatedly question the sense in which a "law" of logic may be compared to an empirical or scientific law. There is no trace of the Brouwerian idea of the continuum as a "medium of free becoming" (as Weyl called it), and furthermore, Wittgenstein does not, like Weyl and Brouwer, make any attempt at constructing a new version of analysis, at least not in any serious sense.[5]

Wittgenstein's criticism is directed at the common view of "modern" mathematicians that mathematics is about *extensions*. We call this *the extensionalist point of view* and the contrary view proffered by Wittgenstein the *non-extensionalist* one. We do not use the term "intensionalist" because it would suggest that words get their meaning *via* certain entities individually attached to them, entities called "intensions", and this is a conception totally contradicting Wittgenstein's stance. According to him, words (linguistic expressions, signs) get their meaning – their "life", as he sometimes says – through the uses we make of them within language, and for him it is a sort of category mistake to base or ground this so-called meaning in immediate relations to certain entities accompanying the words.[6]

The contrast between the extensionalist and the non-extensionalist point of view is emphasized and explained by Wittgenstein long before his annotations to CPM.[7] In this chapter we will explicate it with the help of remarks he made from 1929 to

[3]Even mathematicians without a bent to predicativity or constructivism sometimes raise the question concerning the real justification of the transition from rational to real numbers; so Timothy Gowers in his beautiful "dialogue concerning the need for the real number system" (Gowers 2017a).

[4]On Wittgenstein and Brouwer see Marion (1998, 2003).

[5]There are passages on pp. 134–149 of MS 126 which may appear to be something like the beginning of such an attempt, but we think that they are developed only in order to show an imaginable alternative to our actual practice in analysis so as to shed light on it. These passages, however, are extremely dark and sketchy, and fortunately they do not seem to be relevant to his annotations to CPM.

[6]See on this point Mühlhölzer (2012), pp. 122–124, Mühlhölzer (2014b) and Grève and Mühlhölzer (2014).

[7]In particular, the *Tractatus* contains a reworking of Frege's notion of "sense" and Russell's intensional conception of a propositional function. Wittgenstein used "form-series" generality, i.e., an iterative recursive procedure, to specify his general form of proposition (see 5.501). So his variables for notations express recursive operations, or rules. As Russell noted in his introduction to the *Tractatus*, this idea is not developed in rigorous detail by Wittgenstein, especially with regard to the real and irrational numbers. A number of recent interpretations of the *Tractatus* have rigorously shown how Wittgenstein's approach might be developed to capture first-order logic and even some predicative or ramified portion of analysis; see Rogers and Wehmeier (2012), Ricketts (2014), Weiss (2013, 2017). Disillusionment with the unclarity of his treatment of "operational"

the end of the 1930s, reserving a discussion of some of his later remarks for the chapters in Part II below.

With the term "extension" Wittgenstein has two things in mind. First, he will strictly distinguish between sequences of numbers that the extensionalist considers to be, in Cantor's sense, "finished" [*fertig*] entities or sets – these are the "extensions" – from the techniques or rules by means of which such entities may be produced, assessed, or accessed. If there are such techniques, the extensionalist's interest is ultimately only in their results, the produced sequences, and not the possible processes or conceptual motifs or definitions leading to them.

For the non-extensionalist, on the other hand, it is the processes and structured conceptual motifs, the grammar or logic of the notions, we should be concerned with. The most important cases discussed in Wittgenstein's texts are given by the conception of a real number as a rule-governed calculational procedure for which we can see that for any given n it will generate n digits: for example, a recipe for generating more and more successive digits of $\sqrt{2}$. About the number $\sqrt{2}$ Wittgenstein explicitly says that "we [as extensionalists] have a tendency to think that there is one result produced by $\sqrt{2}$, viz., an infinite decimal fraction". But on his view $\sqrt{2}$ produces a series of results, but no single result: "$\sqrt{2}$ is a rule for producing a fraction, not an extension" (AWL, p. 221).

Statements of the extensionalist view abound in mathematics. For example, we see it said, most strikingly, that "it is possible to write down decimal expansions $0.a_1a_2a_3\ldots$ where "$\langle a_i\rangle_{i=1\ldots\infty}$" is an arbitrary sequence of integers between 0 and 9" (Hrbacek and Jech 1999, p. 86). In its most extreme form, this "it is possible to write down" is understood concretely as the idea that it is only our human weakness or finitude that prevents us from realizing the infinite process of writing down the infinitely long $0.a_1a_2a_3\ldots$. In fact an idea of just this kind is explicitly and approvingly expressed both by Hardy and by Russell.

Here is what Hardy says about infinite processes on the very same page where he mentions Wittgenstein by name:

> No one disputes that there are infinite processes which appear to be prohibited to us by the fact of the physical world. It is true, as Hilbert says, that no mathematician has completed an infinity of syllogisms. It is equally true that there is no mathematician who has never drunk a glass of water, and so far I can see, one of these facts has neither more nor less logical importance than the other. There is no more logical reason why a mathematician should not prove an infinity of theorems in this world than why he should not (as he has been so often encouraged to hope) emit an infinite sequence of musical notes in the next (Hardy 1929, p. 5).

In a similarly cheeky way, mentioning Wittgenstein, Russell called the impossibility of realizing infinite processes of the considered sort a mere "medical impossibility".[8] Earlier on, in a text that was formative for Wittgenstein's own

generality in connection with the idea of "the general form of proposition" in the *Tractatus* spurred Wittgenstein toward a more concrete, calculational approach in his later writings on the infinite.

[8]Russell (1936), p. 143.

philosophical development, Russell had also insisted that "there are positions intermediate between pure intension and pure extension, and it is in these intermediate regions that Symbolic Logic has its lair" (1903, §66). On this view (to take an example Wittgenstein himself explored in numerous remarks after 1929), it is no problem to imagine the production of an infinite decimal expansion by repeatedly throwing dice or by similar random means that lead to totally irregular distributions of the a_i in $0.a_1a_2a_3 \ldots$. This is a case which is a paradigm of the extensionalist view. But for a non-extensionalist like Wittgenstein, all this is nonsense insofar as it ignores – necessarily – our actual mathematical practice, i.e., what we do. In the next chapter we will explore how this form of resistance to extensionalism makes itself felt in Wittgenstein's annotations to CPM.

The idea of decimal expansions of real numbers as finished entities smoothly corresponds with the geometrical image of the straight line: the finished expansions represent the points on this line, and to consider only the rational ones obviously leaves open *gaps* – all the many expansions that are neither finite nor periodic from some place on – and the filling of these gaps with irrational numbers appears to be prejudged in a totally natural way: simply consider all possible expansions and not only the ones that are finite or periodic from some place on. In the passage quoted at the beginning of this chapter (BT, p. 738), Wittgenstein confronted this picture with the provocative reply that the irrational numbers are "no more prejudged" than is chess in draughts.

Immediately before this, on the same page and beginning with the same question as in this remark, he raises the further question of whether the "structure" or "net" (as he says) consisting of only the rational points wouldn't be "fine enough". He immediately gives the answer that in this case it is not the "finishing off" of an "extension" that we lack, but, rather, it is the laws that govern irrational numbers that we have, and the question of what to do with the idea of expansions that are not determined by laws is left open at this point. In this way, in accordance with the motto of this chapter, the view of extending a domain is replaced by the view of introducing new kinds of numbers through new systematizations, *via* new kinds of laws and conceptual orderings that are involved when we deal with irrational numbers.

The second context in which Wittgenstein speaks of "extensions" is the context of sets, paradigmatically sets of numbers: $\mathbb{N}$, $\mathbb{Q}$, $\mathbb{R}$, and subsets of them. Considered extensionally, the laws or rules or techniques through which we may approach them are taken as irrelevant, as their identity is only determined by the elements of which they consist. From Wittgenstein's non-extensionalist view, however, it is precisely these laws, rules or techniques we should take as primary. In the case of the prime numbers, for example, he expresses his point of view thus:

> "The position of all primes must somehow be predetermined. We just figure them out successively, but they are all already determined. God, as it were, knows them all. And yet for all that it seems possible that they have not been determined by a law." – Time and again there's this picture of the meaning of a word as a full box, whose contents are brought to us along with the box and packed up in it, and now all we have to do is examine them. – What do we know of the prime numbers, anyway? How has this concept been given to us at all?

Don't we ourselves specify it? And how odd that we then assume that specifications must have been made about it that we ourselves didn't make! But the mistake is understandable. For we use the expression "prime numbers", and it sounds similar to "cardinal numbers", "square numbers", "even numbers", etc. So we think it can be used in a similar way, but forget that for the expression "prime numbers" we have given quite different rules – rules *different in kind* – and now we get into a strange conflict with ourselves.[9]

There is, of course, an intimate connection between these two Wittgensteinian uses of the word "extension", because every sequence $<a_i>_{i=1\ldots\infty}$ can be identified with the set of all the ordered pairs $\langle i, a_i\rangle$, and in pure set theory everything is a set anyway. But we should not project set theoretical thinking into a Wittgensteinian text: the relevant context in which he is working derives from elsewhere, viz., Frege, Russell, Brouwer, Weyl, and so on.

Considered in Wittgenstein's way, the domains of numbers do not appear homogeneous and clear-cut in the ways suggested; in the case of real numbers, by the notion of *cuts of the rational numbers*, or by the idea of *decimal expansions unrestricted by rules or techniques of producing them*, or – above all – by the illustration of the real domain *via* points of the straight line. From Wittgenstein's perspective – which stresses the rules or techniques we actually follow in formulating and proving laws governing numbers – the differing domains are "the logically most different structures, without any clear limits" (BT, p. 751).[10] And the irrational numbers are in fact, among themselves, "different kinds of numbers – as different from each other as the rational numbers are different from each of these kinds" (BT, p. 761f.).[11]

This Wittgensteinian view is, at least with regard to the homogeneity of the domain of real numbers, not as wildly speculative as it might appear. For it points to a characteristic tension in our concept of the *continuum*, represented by the real numbers, a tension that has been beautifully described by Paul Bernays:

The conflicting aspects of the concepts to be determined [for analysis] are, on the one hand, the intended homogeneity of the idea of the continuum and, on the other hand, the requirement of conceptual distinctness of the measures of magnitudes. From an arithmetical point of view, every element of the number sequence is an individual with its very specific properties; from a geometric point of view we have here only the succession of repeating similar things. The task of formulating a theory of the continuum is not simply descriptive, but a reconciliation of two diverging tendencies. (Bernays 1957, p. 4)

This sort of tension is drawn out rather dramatically in Wittgenstein's treatment, because he not only emphasizes the numbers as individuals, but as individuals that may even be of different "kinds".[12]

---

[9]BT, pp. 763f. By "cardinal numbers", Wittgenstein always means "finite cardinal numbers", in the simple, non-technical sense of "natural numbers". –This passage about prime numbers goes on until p. 765, but the part quoted suffices to convey Wittgenstein's point of view.

[10]We have changed the Luckhardt/Aue translation of BT.

[11]Again, we have changed the Luckhardt/Aue translation of BT.

[12]In his review of RFM (1959), Bernays is rather critical: he finds Wittgenstein's remarks about an omniscient God "not convincing", and Wittgenstein's remarks about "calculation" ambiguous between the technical notion of "effectiveness" and a more general abstract idea of the result of

That irrational numbers in fact *can* be regarded and treated as being of different "kinds" is, again, not really all that extreme a thing to think. In fact this is just what is accepted, for example, when mathematicians distinguish "measures of irrationality" and "hierarchies of complexity" for real numbers.[13]

Hardy himself is sensitive to the sort of tension emphasized by Bernays. At the beginning of §16 of CPM he distinguishes two points of view from which real numbers can be considered: "We may think of them as an *aggregate* [...] or *individually*". In the next chapter we shall carefully discuss this remark.

Wittgenstein's investigation of the limitations of the extensionalist viewpoint is motivated by a diagnosis: that it all too easily leads to illusory thinking. It should therefore be refused as anything more than metaphor. In this sense he does not reject the extensionalist viewpoint altogether. Instead, he accepts it as a sort of specific phraseology, but one that should be assigned a relatively peripheral place, or rather, one that actually has a peripheral place already within our mathematical practice.[14] This is something a philosopher should acknowledge. In subsequent chapters we will try to show in detail how this comes out in his annotations to CPM.

There is one apparently severe defect of Wittgenstein's considerations that must be emphasized: he does not discuss axiomatic approaches to the extensionalist view. It is these, however, that are of course prominent in set theory. As a matter of fact, the whole issue of axiomatics is conspicuously absent in Wittgenstein's philosophy of mathematics overall. With respect to his annotations to CPM this absence may be justified, because CPM too does without axiomatics. But in regard to the extensionalism vs. non-extensionalism issue in general, it certainly should have been given a very central place.

We can, however, make two suggestions here. First, with respect to the more purely logical – as opposed to the mathematical – side of things, the axiomatic perspective is of course nowadays deeply intertwined with the computational approach to formalizing theories. A set of axioms produces a perspicuous way or manner in which formalized sentences of a theory may be generated in a step-by-step manner, algorithmically and, at least with respect to classical languages, machine-implementably. This is itself an instance of the non-extensionalist approach, for

a recursion; he also finds Wittgenstein's later philosophical methods too negative, tending toward the nihilistic. Yet even Bernays admits that Wittgenstein's "main argument" offers an applicable criticism in cases where the intensional and extensional approaches are mixed up, creating, as he says, "the impression of a stronger character of the procedure than is actually achieved".

[13]See the discussion in MathOverflow.net under the heading "Are some numbers more irrational than others?", at http://mathoverflow.net/questions/53724/are-some-numbers-more-irrational-than-others. As Kanamori (2020, in press) explains, Cantor (1872) dealt with irrationals of higher kinds.

[14]The term "phraseology" was used by Russell in his (1920) account of names as descriptions. There he emphasized that "there will, as a rule, be nothing in the phraseology to show whether [words such as 'Sir Walter' and 'Scott'] are being used ... [as descriptions] or *as* names" (Chapter 16, p. 175). Wittgenstein may be read to hold that it is *only* by focussing on use and our characterizations of it that we can say anything sensible about the distinction between names and descriptions. His point was later picked up on by Turing (2001b), on which see Floyd (2017).

there may be a variety of different ways of capturing the power of a mathematical theory in a logic; proof theorists have developed the idea that proofs involve "types".

By 1942–1943 Wittgenstein was aware of the analysis that Turing gave of the general notion of a formal system in his famous paper "On Computable Numbers, with an Application to the *Entscheidungsproblem*" (Turing 1936).[15] In analyzing the notion of *taking a step in a formal system*, Turing adopted in this paper a non-extensionalist perspective, treating "computable" real numbers on the basis of a "comparison" between a human calculating with pencil and paper and the step-by-step processes of a machine (see Sect. 8.5 below). We know that Wittgenstein discussed the paper and its relation to general limitative results in logic with his colleague Watson and with Turing in the summer of 1937; there are some small hints that Turing may have temporarily thought that a whole new foundation of mathematics could be erected on this view.[16] Wittgenstein himself returned to reformulate the argument of "On Computable Numbers" himself in remarks written in 1947, which we shall discuss below in Sect. 8.8.[17]

These discussions had an impact on Wittgenstein's thinking between 1937 and 1942–1943, among the most important of which was Wittgenstein's drawing in the notions of form of life [*Lebensform*] and "technique" [*Technik*] into his later philosophy, and applying these to topics in the foundations of mathematics, ultimately the problem of what it is to follow a rule.[18] Wittgenstein may have assumed that he could treat the issue of axiomatization under the broad heading of "following a rule", or "taking a step in a formal system" conceived of as computation, where the notion of computation, as analyzed by Turing, would be founded upon a "comparison" between a human being following a circumscribed recipe with symbols and a "step" in a formal system of logic. If so, then the importance of the everyday fact that we must take up or accept or take in or understand a rule would be deeply embedded in his approach, at the foundation, so to speak, lying closer to the heart of the matter than broader and more mathematical questions about the axiomatic method.

A second point is this. One reason to think Wittgenstein might have been so receptive to this way of approaching logic was his earlier insistence, in the *Tractatus*, that axiomatization or "postulation" in the usual sense is irrelevant in logic, because all so-called logical "propositions" share the same status: they may be "postulated" insofar as we postulate an "adequate" notation (see TLP 6.1223). In other words, it is the quite particular role and status of logic in general as *formalizing* that counts, and not any specific formulation or axiomatization of it. As his philosophy matured, moreover, Wittgenstein came to emphasize the lack of

[15]See Floyd (2017, 2016).

[16]Floyd (2001b, 2017).

[17]Floyd (2012c).

[18]See Sects. 8.2 and 8.8 below. For more detailed see argument Floyd (2001b, 2016, 2017, 2018b), Floyd and Putnam (2000, 2012).

any sharp conceptual dichotomy between step-by-step proceduralizing in formal and informal languages.

On this view, we should draw a distinction between the logical role of axiomatization and mathematical theories. For Wittgenstein all the so-called "propositions" of logic are "of equal status", and they are not to be individuated by their apparent subject matter, for none are "essentially derived propositions" (TLP 6.127). In other words "proof in logic is merely a mechanical expedient to facilitate the recognition of tautologies [i.e., logical 'propositions'] in complicated cases" (TLP 6.1262).

This perspective on logic subsequently became more explicitly relativized and systematically developed in a Hilbertian manner in Carnap's hands, as Wittgenstein knew. As he also knew by 1942–1943, issues of incompleteness are of fundamental importance in thinking through the limits of axiomatic formulations of portions of mathematics. Nevertheless, the view that the axiomatization of a theory does not necessarily show us anything intrinsic to the epistemological nature of its sentences and propositions, but only something about the power of logic itself, is not an impossible view to defend. Quine, for one, embraced it. For now we shall not speculate further about Wittgenstein's reasons for remaining so silent about axiomatics in his later writings, though we believe that one should bear it in mind as a general point. The question of whether and in what way this relatively dark chapter of his writings might somehow be brightened must be answered elsewhere.[19]

[19]Wittgenstein may in fact have profited from discussions of axiomatic approaches because there the axioms can be seen as "implicit definitions" and this has an obvious similarity to Wittgenstein's view of mathematical propositions as rules. This point is ingeniously elaborated in Friederich (2011, 2014).

Part II
Analysis of the Annotations

Chapter 3
Irrational Numbers: The Annotations on pp. 2–9, with Commentary

The irrational numbers are – so to speak – individual cases.
(RFM V §37)

Abstract An analysis of Wittgenstein's annotations to Hardy's *Course of Pure Mathematics* pp. 2–9, with commentary on each. Themes treated include the notion of an "application" of mathematics, Wittgenstein's remarks about Hardy's "nonsense", and Wittgenstein's remarks about proofs of the irrationality of $\sqrt{2}$, as well as the idea of a "phantasmagoric" application of mathematics.

3.1 Application is Essential, First Pass

Referring to §2 of Hardy's Chapter I (Real Variables), Wittgenstein's first annotation is on p. 2, where Hardy – having introduced the rational numbers in §1, with the presupposition that we are familiar with them and with the ordinary arithmetical rules for manipulating them – immediately presents the usual geometrical representation of the rationals by points on a line. Hardy emphasizes that this representation is merely an *illustration* and does not have any systematic significance for analysis. He writes: "the use of geometrical illustrations in this way does not, of course, imply that analysis has any sort of dependence upon geometry: they are illustrations and nothing more, and are employed merely for the sake of clearness of exposition."[1] To this, on the left margin of p. 2, Wittgenstein remarks:

`Illustration is indeed inessential, not, however, application.`[2]

[1] Already in the Preface of the first edition Hardy emphasizes: "I have, [...] throughout the book, kept geometry in a strictly subordinate position and used it merely for purposes of illustration" (CPM_1, p. vi).

[2] The German original reads:

`Illustration ist allerdings unwesentlich, nicht aber Application.`

The word "Application" is written with a "c".

J. Floyd, F. Mühlhölzer, *Wittgenstein's Annotations to Hardy's Course of Pure Mathematics*, Nordic Wittgenstein Studies 7,
https://doi.org/10.1007/978-3-030-48481-1_3

With this "illustration" Hardy is playing down the relationship between analysis and geometry. He is right to say that geometry helps us to understand analysis by giving vivid illustrations that are useful for exposition; he might also have mentioned the heuristic value of geometrical intuition for analysis, though he does not.

Conversely, however, analysis – including arithmetic – can at the same time be applied to geometry, and so help us to gain new geometrical insights: for example, it can be shown that the side and diagonal of a square are incommensurable, and that therefore the straight line has "irrational points", and so on. Wittgenstein is right in emphasizing this. But why does he say that the applications of analysis are *essential* to it?

They *were* essential in the history of mathematics. But beginning in 1858 an incipient, new point of view began to be developed by Dedekind, Weierstrass, and others.[3] Up to this time, real numbers had commonly been taken to represent quantities [*Grössen*], or, more precisely, *ratios* of quantities, and the notion of a quantity was explained in terms of measurement. Real numbers, therefore, had been regarded as essentially "measurement-numbers". Wittgenstein was certainly aware of this fact, not least through Frege, who stuck to this conception of real numbers until the end of his life.[4] On p. 127 of MS 126, dated 14 December 1942 (appearing in RFM V §29), Wittgenstein alludes to this, writing:

> Originally the geometrical illustrations [of analysis] were *applications of analysis*.

In 1858, with Dedekind's invention of his "cuts" in analyzing the concept of real number, there began a kind of conceptual development in analysis, and this, in the course of time, would totally sever the notion from that of quantity, as well as from any allusion to applications, geometrical or otherwise. Only after Dedekind's analysis was it possible to draw a proper distinction between analysis as such and its applications. Before then, these had been inseparably mingled.[5]

This development, which occurred when Dedekind showed that and how it was possible, conceptually speaking, to draw a systematic distinction between analysis and its applications, engendered the need to write new textbooks in analysis, and Hardy's CPM is precisely a book of this sort. Accordingly, the severing of analysis from its applications, and especially its geometrical applications, is one of the main points of Hardy's book! This was the explicit aspiration of the first edition CPM_1; but the really telling realization of Hardy's aspiration did not appear until CPM_2,

---

[3]The developments, which began with Dedekind's 1858 lectures at the Zürich Polytechnikum and rippled through Weierstrass's 1865 lectures on the real numbers, reached public fruition in the publication of Dedekind (1872) and Cantor (1872). They are instructively discussed in Epple (1999) and in Kanamori (2020, in press).

[4]See the informative description of Frege's view in Chapter 22 of Dummett (1991a); the happy phrase "measurement-numbers" just used comes from this text, p. 281.

[5]Epple (1999) gives a good outline of this development. An extensive description of it can also be found in Russell (1903), a book that Wittgenstein knew well. As emphasized in Petri and Schappacher (2007), p. 359, however, a thorough investigation of the complicated and varied history of it still remains to be written.

for there Hardy developed Dedekind's theory of real numbers. And this is kept in all later editions of CPM.

In the first edition Hardy was content to explain the real numbers via "measures of lengths" (see Section 7 in CPM_1); he even speaks of "quantities", but this is something which he explicitly retracts in the Preface to CPM_1, writing that

> The word "quantity" occurs occasionally in the earlier chapters. It should be in each case altered to "number". Unfortunately I arrived at the decision never to use the term "quantity" only after the earlier sheets had been passed for press.

So CPM_1 is the typical product of a transition between old and new thinking, and only with the second edition and thereafter did the new thinking really prevail.

Of course, Wittgenstein must have been aware of this.[6] But then, for precisely what reasons did he remark in his annotation to p. 2 that application is "necessary"? In MS 126, he is a bit more explicit. The quotation above from p. 127 is only a short fragment of an entire passage relevant to our question. This passage (published at RFM V §29) is as follows:

> The geometrical illustration of analysis is indeed inessential, not, however, the geometrical application. Originally the geometrical illustrations were *applications of analysis*. Where they cease to be this they can be wholly misleading.[7]

This suggests that the essentiality of geometrical applications for analysis lies in their potential to put a brake on certain misleading ideas that otherwise geometrical illustrations may induce.

Wittgenstein does not say here what these misleading ideas are, and we postpone this issue to his later annotations. In any case, however, this particular idea of the essentiality of geometrical applications is rather lame, and we doubt that it is what Wittgenstein meant, or at least that it is everything that he meant. Couldn't Hardy easily reply that in the end all this geometrical stuff is not necessary, that in principle analysis takes care of itself (to use one of those phrases which Wittgenstein himself is so fond of)?

So the question remains: what does Wittgenstein really have in mind here? We will come back to this at the end of Sect. 5.5.

[6]The final chapter XVII of Russell (1920), which we know Wittgenstein read (and where he is credited by Russell with the idea of a purely formal or "tautological" conception of logical form), emphasizes just this point about "quantity".

[7]In Elizabeth Anscombe's version of this passage, in RFM V §29, the word "analysis" is written with a capital "A". We don't know what her reasons were. Hardy always writes "analysis", and this is the usual way the word is written. Of course, in Wittgenstein's own formulation "Analysis" occurs with a capital "A", but this is the word in German where all nouns are capitalized.

3.2 Mathematics and Its Applications, Second Pass

Wittgenstein is rather often represented as claiming that mathematics in its entirety is essentially based on its applications: after all, the reasoning goes, he was an engineer. In this vein, one of the regularly quoted remarks is the "mufti"-passage in RFM V §2:

> I want to say: it is essential to mathematics that its signs are also employed in *mufti* [im *Zivil*].
> It is the use outside of mathematics, and so the meaning of the signs, that makes the sign-game into mathematics.[8]

This passage is extremely unclear and it should not be taken at face value. *Why* "essential"? Why *outside* mathematics? Why does Wittgenstein write "so the *meaning* of the signs"? And what sort of "*making*" is at issue? Above all: does Wittgenstein really rely on the typical *formalist* attitude to see the difference between mathematics and a pure "sign-game" in the applications of mathematics to things *outside* mathematics?

To this last question, the answer is an emphatic No. First of all, this remark as a whole is a response to the question, "Does a calculating machine calculate?". Wittgenstein pursued this question throughout the 1930s as he investigated the comparison between language and a mechanism, and between mathematics and a mechanism, stimulated partly by Russell's raising the issue in his (1921).[9] He would continue to revisit these comparisons later on. After his discussions with Turing in 1937–1939, this gets connected with his investigation of the distinction between calculation and experiment (his conception of a machine "symbolizing its own action" derives from this period).[10]

From the underlying manuscript MS 126 we see that this remark and its attending metaphor – "*in mufti*" [*im Zivil*], that is, in a dress worn in civilian life, when a military or official person is off-duty – are not the result of Wittgenstein's investigations, but their starting point. Wittgenstein seeks in the manuscript to clarify what might be right in it, and what might be wrong. Very clearly Wittgenstein struggles with what he "wants" to say here (MS 126 pp. 26ff). This struggle has no definite result. And this is so not only in MS 126, but within Wittgenstein's philosophy of mathematics in its entirety.

Why is there so much uncertainty on Wittgenstein's part? Why is the question about the applicability of mathematics so ticklish and difficult for him? Why does he take no final stand?

---

[8]This is the German original: "Ich will sagen: Es ist der Mathematik wesentlich, daß ihre Zeichen auch im *Zivil* gebraucht werden. | Es ist der Gebrauch außerhalb der Mathematik, also die *Bedeutung* der Zeichen, was das Zeichenspiel zur Mathematik macht."

[9]See e.g. WVC pp. 106,164.

[10]See RFM I §122; PI §193, and Floyd (2017) for discussion.

The simple reason is that this question touches the center of his later philosophy, in particular his advice that we should not fail to look to the *uses* we make of our language, including our symbolic activities. If we are after an understanding of how to regard the "sense" or "meaning" of our symbols, it is to their uses and possible uses we should look, especially if we want to distinguish between sense and nonsense, and better and worse ways of discussing what we really mean.[11] There are the words, and their possible ways of saying what there is to say; but there is also what we *can* do with the words given how we *do* use them. And there is also what we do quite apart from words and symbols – for example, with images or pictures or models. There can be slippage, a divergence, and so there is a need to reconstruct very carefully precisely *how* it is that these aspects of meaning interpenetrate and work with one another.

At this point it is important to understand that Wittgenstein does not aim at a *theory* of sense and/or meaning. He simply tries to call attention, in the course of his investigations, to the uses we make of our symbols. Some of these we are normally unaware of, although they lie right before our eyes. In *Philosophical Grammar* (p. 126f.) he expresses it thus:

> I would like to say: "I must *begin* with the distinction between sense and nonsense. Nothing is possible prior to that. I can't give it a foundation [*Ich kann sie nicht begründen*]."

To begin with, this distinction means taking into account our elementary understanding of what is nonsense and what not in particular cases of our using sentences or phrases to accomplish something. Wittgenstein gives many examples of the sort of nonsense he has in mind here. To mention only one, occurring in PI §265: in everyday life it is certainly nonsense to buy several copies of today's morning newspaper in order to assure oneself that what it said was true. It is not simply that it is pointless or uneconomical to do so, or an unsuccessful sort of a habit in an evolutionary societal sense, where higher probabilities for getting things right all around might be modeled. Rather, if the first copy didn't do the job of convincing me about how things are, then buying a second and third copy will not achieve this. This does not mean I could not buy several copies to do something else entirely: I might, in my doubting – and even for that very reason – plan to distribute the copies widely to mount a revolution, or make propaganda, or disseminate falsehood.

Re-assurance in the face of doubt is not a matter of playing the same card over and over again, determined to make it mean something (as it were, shuffling symbols until one hits upon one that means). As Wittgenstein elsewhere remarks, in everyday life – i.e., without special reasons or a context for doubt in which opening one's eyes would be a good idea – it is not that I am purposefully shutting my eyes at each step to keep doubt at bay and looking for a public verification to reassure myself that a symbol means something. Rather, there is no issue of opening my eyes in the first place: I am not "in action", on-duty, always in official dress in seeking justifications

[11]Throughout his life Wittgenstein associates "sense" [*Sinn*] with sentences [*Sätze*], true or false, and "meaning" [*Bedeutung*] with names, refusing Frege's assimilation of the two categories of linguistic expression.

for what I do. The ways and manners of certainty are the way and manner of the particular "language-game" (see PPF, §331–2).

Of course it also belongs to traditional philosophy to try to explore what happens when words that in everyday life do have civilian uses are taken out of their ordinary roles, with no point in particular at hand, and are drawn into battle. But it is one of Wittgenstein's most important discoveries – as well as one of his most important themes – that skepticism and other philosophical positions turn on an insistence that no civilian uses are capable of getting at what the philosopher is trying to say.

What is at issue is not whether and how such a position can or cannot be meaningfully pursued, but instead, whether Wittgenstein can make philosophical hay out of his quite different investigation of civilian uses. Within the lines of his – often analogical – investigations, philosophy may not dismiss altogether our usual understandings. This is not matter of "going native" with fancy dress in a foreign language-game. Rather, it is a matter of relaxing into the ordinary ways of proceeding we humans actually have – and there are many – and exploring speaking, thinking and how we do and might use words to refer to and mean things.

Of course ordinary discourse, with its contingencies, comparisons, and analogies, holds a variety of kinds of pitfalls: it would be nonsense to think of it as sacrosanct, because there are so many ways in which we often get confused, even we expert users of language. There are also many ways in which it can be stretched. Certainly in the context of proofs, mathematicians have a great many indices, definitions, and parameters to keep track of with notations that they devise, and hints and pointers play important roles too. Among the many issues here is one particularly important one: in textbooks such as Hardy's, there are challenges of exposition – and this is one of Wittgenstein's main concerns with Hardy.

We stress that this concern is not only a preoccupation on Wittgenstein's part with "civilian" uses, imposed on Hardy from outside by idiosyncratic philosophical prejudice or decree. It is also a genuine concern when we are talking about the exposition of mathematics in which outsiders or newcomers are drawn in, and in which foundational first steps are discussed. The pedagogical challenges are well-known to mathematicians, especially when they introduce modern, unfamiliar, and/or abstract methods into an area.

A couple of quotations from the Preface to Littlewood's *The Elements of the Theory of Real Functions* (1926), notes of lectures delivered at Cambridge in 1925 – a text also known to Wittgenstein (as well as to Turing[12]) – will illustrate this point.

[12]In October 1932 Turing annotated his copy of Littlewood (1926) with a panoply of logical symbols (see the Turing Digital Archive AMT/B/46, at http://www.turingarchive.org/browse.php/b/46). Turing had certainly read Russell (1920) by the spring of 1933, his second undergraduate year, following Littlewood's advice to do so in the Preface of his (1926); see Floyd (2017) for discussion of this.

Littlewood's book has an extremely misleading title. It is not at all a book on "real functions" but one on set theory.[13] The first edition was written only for those who had attended his 1925 lectures; the second was an expansion of that. The aim of the lectures, he wrote in the Preface, was

> to inculcate the proper attitude of enlightened simple-mindedness by concentrating attention on matters which are abstract but not complicated.

He immediately added that

> I have aimed at excluding as far as possible anything that could be called philosophy

and pointed out that students who have read Russell's *Introduction to Mathematical Philosophy* (1919) will have sufficient background to handle his course. Finally, he addressed in his Preface the issue of lectures versus written textbooks, and "nonsense":

> I hope [...] that my possible pupils in the University of Cambridge will not too hastily assume that the existence of the notes makes attending the lectures themselves entirely valueless. I am one of those who believe that lectures can have great value, and particularly at a certain moderately advanced stage of a mathematical education. The modern standard of conciseness and lucidity in original papers and advanced text-books is on the whole a high one, but the style is one for the expert only. We may demand two things of an original paper, a complete and accurate exposition on the one hand, and on the other that it should convey what is the real "point" of the subject-matter. For various reasons, among which a sufficient one nowadays is sheer lack of space, the second demand is inevitably sacrificed to the first. A lecture, however, more particularly when it is supported by a complete exposition in print, is the very place for the provisional nonsense that the second generally calls for. This would appear ridiculous if enshrined permanently in print, and its real function is to disappear when it has served its turn. To take a specific example, it is, I think, possible to approach the rather difficult Prop. 5 of Chapter I in such a way as to make both the underlying ideas seem intuitive and the official proof natural. The infinitely greater flexibility of speech enables me here to do without a blush what I shrink from doing in print.

Thus Wittgenstein's own teacher Littlewood himself placed a high premium on the manner in which the theory that is the subject of his book would be introduced, emphasizing the importance of making underlying ideas intuitive, and getting across the "point" of a particular subject matter.

Now *this* talk about "nonsense" in no way implies that everyone must always and everywhere stand on guard, attending to meaning. We can, so to speak, drive off the bridge of confusion when get there. To think otherwise – to think that nonsense is to be enshrined or placed off to the side, or at a limit of some kind – is to construct a false guardhouse that no one can actually inhabit. Even the image of a place where the safety of meaning itself could be lodged, once and for all, is the image of such a guardhouse: intrinsic justification as such, perceived as such.

From Wittgenstein's point of view, philosophy doesn't *need* a general account of that and why something is nonsense, even if a satisfactory account of such a thing in

[13]Wittgenstein refers to its way of presenting set theory in his MS 162a, pp. 15–18, which will be discussed below in Felix Mühlhölzer's Chap. 7.

general were available to be had (something that is very far from obvious). We can stick to working things through. Of course, drawing out the particular message with an example such as that of Wittgenstein's morning newspaper requires delicacy and discernment, a sense of its point, and a certain amount of conversation. Inevitably in the course of this working through there will be shiftings and missings of points. Wittgenstein's philosophy is full of this.

As a result, one of Wittgenstein's characteristic philosophical methods consists in what he calls passing "from unobvious nonsense to obvious nonsense" (PI §464; see also §524, where the reverse process is also commended). This method works very well in the case of the series of investigations that have come to be known as the "private language argument": a variety of routes are explored in which the relevant "can't understand" is unpacked, none of them satisfying the interlocutor that she has gotten the point across. That failure is designed to elicit from us a sense that the whole idea of "private" language can't be made sense of in a number of ways that one might think, at first blush, it can be.[14]

As for of the morning newspaper remark just mentioned, in PI §265 Wittgenstein produces it in reaction to an alleged justification that appeals to one memory, and then to another, and yet does so when the memory is supposed to involve routes among objects that exist only in our imagination. The sort of nonsense involved here is not produced by Wittgenstein's insisting on a public-verificationist conception, or a form of behaviorism, or an attack on the very idea of a memory or inner imaginative experience. Instead, he is aiming to show how it is that in the course of the interlocutory justifications and moves that the different voices make, apparent "subjective justifications", are playing the role of what he elsewhere calls "idle wheels" [*leerlaufende Räder*].

A wheel is "idle" if it can be fully turned though nothing else moves with it, so that it "is not part of the mechanism", i.e., "has no point", although it may at first appear so (PI §271).[15] We know how to grease the wheels of language, and we also know how to find friction for them by imagining a variety of contexts in which we may imagine meaning particular things with our words – this is, after all, part of what we imbibe when we learn a language. But Wittgenstein shows us far more sophisticated ways in which the wheels can be greased and detached from all friction – cases where we might not imagine we are greasing them at all.

In PI §265 it is the wheel of thinking that one can justify one "subjective" memory by another one, when one at the same time isn't convinced by any one such memory, so that one piles imagined objects one by one upon one another, trying to

[14]An excellent survey of some of the interpretive issues here can be found in Stern (2011).

[15]In §271 of the PI the expression "idle wheel" is actually not used, but one can find it within precisely the same context in MS 179, p. 19r, where Wittgenstein writes in brackets: "That must be an idle wheel/The wheel does not belong to the machine/that can be turned though nothing else moves with it" ["Das muß ein leerlaufendes Rad sein, was man drehen/Das Rad gehört nicht zur Maschine das man drehen/kann, ohne daß sich etwas anderes mitbewegt"]. In Wittgenstein's manuscripts the expression "idle wheel" turns up rather often, but in the PI – a bit surprisingly – it actually does not appear.

give constitutive sense to one's claims about oneself and one's inner states. This is as idle as buying several copies of today's morning-paper to assure oneself that what it said was true – assuming one is operating within everyday life, and not in some other radically differing situation.[16]

Wittgenstein's orientation is often expressed in an analogy between language and a mechanism: in PI §132 he explains that "[t]he confusions which occupy us arise when language is, as it were, idling, not when it is doing work". Wittgenstein brings out the image of language as disengaged from its ordinary tasks; he wants us to attend to the difference between this situation and one in which language is doing real work.

In his earlier writings on mathematics Wittgenstein seems to have thought of giving a similar kind of diagnosis. However, here the situation is rather different. For the relation of mathematical language to the notion of "doing real work" is more contestable and much less clear. As working drafts, his manuscripts register this unclarity. They lack the finely-tuned orchestration of remarks such as those in the first parts of PI. However, the original context for many of his analogies and metaphors are framed initially in the manuscripts. In particular, his "idle wheel" metaphor was framed as part of an extended investigation overlapping his annotations to Hardy's CPM.

In MS 120, p. 60v, Wittgenstein raises the beautiful question: "How does the working wheel pass into an idle one?".[17] In the context of the private language scenario, it seems that one can find a good answer. Such a passing occurs when the public background and public surrounding that gives our ordinary practice of talking about sensations its point is gradually removed and/or suppressed as irrelevant. The result, if one attempts to remove *all* such surrounding and background, is that what remains appears in the end as only the illusion of real content, like the illusion of work done by wheels that are actually running idle, having lost all friction.[18]

But what might be the parallel to this process in the case of mathematics? If we understand him correctly, Wittgenstein's idea of a parallel is as follows. First one may think of those parts of mathematics that possess empirical applications. Gradually one imagines setting various parts of these aside. Eventually one sets aside also anything having to do with our own, characteristic human mathematical activities, interests, conceptualizations, and experiences of mathematics as well. The removals imagined express the idea of getting to something we think of as "pure mathematics", or mathematics as it is regardless of us or our uses or conceptions of it. At this point, idle wheels begin to appear.

[16]An interpretation along these lines of PI §263 and its surrounding is given in Mühlhölzer (2014c). See also Schroeder (2006), pp. 206ff, where one thread in the private language argument is called the "idle-wheel argument".

[17]There is no question mark in the original. – We find this question beautiful although Wittgenstein himself marked it with a squiggly line at the margin, indicating (if we correctly interpret this) that he is not quite content with his formulation. We do not know why.

[18]See the detailed exposition of this argumentative structure in Mühlhölzer (2014c), Section 4.

This at least partly explains what he is up to in his remark about "civilian" roles in the above-quoted RFM V §2. He "wants to say" that in the process of attempting to set aside or suppress all elements of the applications of mathematics the signs that we ordinarily use in mathematics lose what is essential for them to figure *in* mathematics. The situation becomes even worse when from the very outset one assumes a priori that there exist no applications outside of mathematics.

In all these cases Wittgenstein tends to want to say that we have given to mathematics the character of a collection of idle wheels. At the end of §38 of RFM II this is expressed almost explicitly, when he writes:

> I can of course form the expression: "class of all classes which are equinumerous with the class 'infinite series'" (as also: "class of all angels that can get on to a needlepoint") but this expression is idle so long as there is no employment for it.[19]

What he has in mind here is set theory. But we will see that his suspicion extends also to the sort of "modern" analysis as presented in CPM, and especially to the notion of a Dedekind cut as Hardy presents it (which in CPM does *not* occur in a set theoretic surrounding). His suspicion also is directed toward foundationalist reconstructions of mathematics such as the logicist ones, at least when they envision themselves wholly abstracting from ordinary procedures and articulations in mathematics. In the *Tractatus*, as we have seen, Wittgenstein invoked Poincaré's objection to logicism, accusing Frege and Russell of falling into a *circulus vitiosus* in their second-order definition of the ancestral (see Sect. 1.1 above). In his *Tractatus* terminology, he accused them of leaning in their analyses on the idea of a "form series", or well-founded, step-by-step iterated operation with signs, when they were claiming to have analyzed this away in terms of concepts or sets. Here, in the early 1940s, he might have said to them – as he wrote in another context – "You describe an imaginary (idle) language-game behind the real one" (MS 121, p. 20v).[20]

Yet there is a major difficulty with this philosophical move. In the case of pure mathematics there remains enough "detached", operational, symbolic practice even when the empirical applications and purely logical foundations are omitted or non-existent. This is simply the usual practice of pure mathematics. Here we have a complicated symbolic practice, overwhelmingly rich in ideas and concepts, that is pursued quite independently of any imagined empirical applications.

So what might be respectable reasons to regard *this* practice with suspicion, as consisting of "idle wheels" that do not constitute properly meaningful activities because they do not belong to a "working mechanism"?

---

[19]We have corrected Anscombe's translation here. In Wittgenstein's German the end of this passage reads: "dieser Ausdruck ist leer, solange es keine Verwendung für ihn gibt", where the word "leer", of course, comes from the picture of "*leerlaufende Räder*", that is, of "idle wheels". Consequently, this word should be translated here as "idle". Anscombe's translation of "dieser Ausdruck ist leer" as "this expression is empty" doesn't convey the point.

[20]In Wittgenstein's German original: "Du beschreibst ein imaginäres (leerlaufendes) Sprachspiel hinter dem wirklichen." This remark actually refers to the philosophical preconception that it must be inner pictures that give words their meaning, in contrast to the public use we make of them. But the remark is appropriate, and very beautiful, with regard to our present context as well.

For modern mathematics, Wittgenstein's view as we have so far described it has much less persuasive power than in the case of the private language considerations, and Wittgenstein has a hard time substantiating it. In PI §107 he formulates the slogan: "We want to walk: so we need *friction*. Back to the rough ground!" But in mathematics the question precisely is where and how to see this rough ground, where and how to locate the "friction". Why should we demand "applications"? What are "employments" that count? Must there be applications "outside" of mathematics?

We must ask what precisely is and should be meant by "application" here. In his *Big Typescript* Wittgenstein emphasizes the quite different uses of the word "application" (BT, p. 556). Moreover, in the §§166–7 of RFM I he shows an astonishingly liberal attitude toward the question of what might count as one:

> 166. [Mathematics] forms ever new rules: is always building new roads for traffic; by extending the network of the old ones.
> 167. But then doesn't it need a sanction for this? Can it extend the network *arbitrarily*? Well, I could say: a mathematician is always inventing new forms of description. Some, stimulated by practical needs, others, from aesthetic needs – and yet others in a variety of ways. And here I imagine a landscape gardener designing paths for the layout of a garden; it may well be that he draws them on a drawing-board merely as ornamental strips without the slightest thought of someone's sometime walking on them.

However, what is remarkable in this passage is above all the fact that Wittgenstein does not even mention the idea that there might be purely intra-mathematical "sanctions" for the mathematicians' building of new roads in their network. The "new forms of description" he has in mind here concern, as the example with the gardener shows, only descriptions of non-mathematical states of affairs, and other functions of the mathematical network than aiming at such descriptions are not envisaged. Those states of affairs may not be realized, but according to Wittgenstein's view it is nevertheless they to which mathematics is, or should be, oriented.

The *Big Typescript* was finished in 1933 and the typescript TS 222, underlying RFM I, in 1938. Now, in 1942, in MS 126 (as we have emphasized, the basis of the first half of RFM V), Wittgenstein revisits the question concerning the role of applications for mathematics. He again commences his investigation with a remark expressed in the dogmatically worded statement, published in §2 of RFM V, that it is the use outside mathematics and "therefore the meanings [*Bedeutungen*] of the signs" that makes "the sign-game" [*Zeichenspiel*] into mathematics.

Since this statement is introduced with the words: "I want to say", it should be interpreted, we think, as marking the starting point of Wittgenstein's subsequent investigations and not as the expression of a Wittgensteinian result. This statement cannot be the expression of a result within Wittgenstein's later philosophy, for its formulation is much too general.

It is true that Wittgenstein sometimes lets himself get carried away into making statements of this sort. But then, in his mature philosophizing he often reproaches himself for doing so. We should not hesitate to reproach him either – as long as we are making an effort to understand him. Keeping track of his self-reproaches and qualifications of his own remarks is, after all, as important as – if not more important than – keeping track of his most extreme statements.

Consider, for example, §321 of *On Certainty* (1974b), written during the final weeks of his life (March–April 1951):

> Isn't what I am saying: any empirical proposition can be transformed into a postulate – and then becomes a norm of description. But I am suspicious even of this. The sentence is too general. One almost wants to say "any empirical proposition can, theoretically, be transformed...", but what does "theoretically" mean here? It sounds all too reminiscent of the *Tractatus*.

This methodological point about not over-generalizing is of fundamental importance for the mature Wittgenstein. As he had said just before this remark (OC §§318–320):

> 318. 'The question doesn't arise at all': Its answer would characterize a *method*. But there is no sharp boundary between methodological propositions and propositions within a method.
> 319. But wouldn't one have to say then, that there is no sharp boundary between propositions of logic and empirical propositions? The lack of sharpness *is* that of the boundary between *rule* and empirical proposition.
> 320. Here one must, I believe, remember that the concept 'proposition' itself is not a sharp one.

In MS 126 and RFM V Wittgenstein's self-admonishments take the form of an investigation that concerns itself with special cases, especially cases from within analysis that are taken from CPM, and in fact already commented on in his annotations to it. We shall go into this further at the end of Sect. 5.5.

3.3 "*Unsinn*"?

On the left margin of p. 4 of CPM Wittgenstein adds the remark "Unsinn", that is, "Nonsense". We take him to be referring to a rather perfunctory passage in which Hardy talks about a figure that would "look and behave very much like a line" and "properties which common sense attributes to the straight line" – a passage in which Hardy himself says that he puts "the matter roughly".

Wittgenstein does not explain what he means with his " Unsinn " here, but we presume that he is bothered both by Hardy's handwavy appeal to "common sense attributions" and by Hardy's mixing of arithmetic and geometry, application and illustration. It is reasonable for him to ask what can be meant by "look and behave". Even if we try to meet Hardy half-way with a pinch of salt, trying to take him to be on the way to saying what he wants to say, "look and behave" are words that work utterly differently in the two domains of analysis and geometry.

In arithmetic our familiar approach is based on techniques of calculation. In contrast, in geometry the calculations with numbers originally derive from a parametrization of unit lengths and ratios, so that here we are engaged in reasoning that involves pictures produced with ruler and compass (for example). From Wittgenstein's non-extensionalist perspective, Hardy's way of putting the matter "roughly" is much too crude because it misleads us, in its very wording, about deep differences within our mathematical practices. There is no *one* "matter" to be put. Students reading such a textbook should not, Wittgenstein thinks, be encouraged to accept such talk or to teach this way in the future. There is, after all, such a thing as being misleading.

3.4 First Proof of the Irrationality of $\sqrt{2}$

Wittgenstein's annotation to CPM p. 6 is a bit more interesting.[21] On this page, Hardy gives two proofs of the irrationality of $\sqrt{2}$. The first one is a generalizing variant of the classical proof of the Greeks, the one most often given in courses.[22] It generalizes to the square root of m (if m is not itself a square number) assuming prime factorization. Hardy begins it with the sentence: "For suppose, if possible, that $p^2/q^2 = m/n$, p having no factor in common with q, and m no factor in common with n"; he shows that then m must be equal to p^2 and n to q^2. This is the general result. In the special case when $p^2/q^2 = m$ (because $m = m/1$), $p^2 = m$ and $q^2 = 1$, so $q = 1$. Hence an integer cannot be the square of a rational number, unless that rational number is itself an integer. Therefore, there cannot exist a rational number the square of which is 2 because no square of an integer is 2.

This is Hardy's first proof of the irrationality of $\sqrt{2}$.

Wittgenstein comments upon this by putting a squiggly line under the phrase "if possible" at the beginning of Hardy's proof and writing on the left and upper margin of p. 6:

> "Suppose, if possible," means, I believe, nothing at all. It appears to me like a cowardly evasion. - It might mean, say, something like: try to suppose, and instead of this one might simply say "try". For example: "Check $(p/q))^2 = 2$ . . . ". [23]

[21] Its image is reproduced, with Wittgenstein's annotations, in Nedo (2008), p. 91.

[22] The contrapositive of what Hardy shows is "If m is not a square number, then the square root of m is irrational". For $m \leq 17$, this is known from antiquity as the "theorem of Theaetetus", a statement of which is alluded to in Plato's dialogue *Theaetetus* (p. 147e), though in a more down-to-earth way, and of course without any explicit appeal to modern algebra (see Kanamori 2018a). Setting $m = 2$ in Hardy's supposition, we have a kind of algebraical rendition of Plato's presentation of the Pythagorean theorem in the *Meno*, presumably a much older Babylonian argument which turns on the concepts of even and odd.

[23] "Suppose, if possible," heißt, glaube ich, gar nichts. Es scheint mir wie eine feige Ausflucht. – Es könnte etwa soviel heißen wie: try to suppose, und statt dessen könnte man dann einfach sagen "try". Etwa: "Prüfe $(p/q)^2 = 2 \ldots$"

The phrase "suppose, if possible", with which Hardy's proof begins, is in fact a strange one. What if the supposed state of affairs is *not* possible? How can one *suppose* what one takes to *be* impossible?[24] Yet this is the case with proofs by contradiction, by *reductio ad absurdum*. Wittgenstein enunciates precisely a trouble of this sort on pp. 120f. of MS 126, which is published as the middle part of RFM V §28:

> The difficulty which is felt in connection with *reductio ad absurdum* in mathematics is this: what goes on in this proof? Something mathematically absurd, and hence unmathematical? How – one would like to ask – can one so much as assume the mathematically absurd at all? That I can assume what is physically false and reduce it *ad absurdum* gives me no difficulty. But how to think the – so to speak – unthinkable?

As applied to the Hardy passage at hand, however, Wittgenstein's trouble is, strictly speaking, beside the point. For here the "suppose, if possible" does not commence a proof by *reductio ad absurdum*. Hardy says: "suppose, if possible, that $p^2/q^2 = m/n$, p having no factor in common with q, and m no factor in common with n", and this is trivially possible for $m = p^2$ and $n = q^2$. The *reductio* happens *afterwards*, when Hardy infers that then p^2/q^2 cannot be equal to 2 by trying out the values $n = 1$ and $m = 2$. He could have presented this (but actually did not) by saying: "suppose, if possible, that $p^2/q^2 = 2$, p having no factor in common with q", concluding that then p^2 must be equal to 2, which contradicts the already accepted fact that the square of no integer can be 2. And this would have been much simpler.

Despite being off-point, however, Wittgenstein has nevertheless good reasons to criticize Hardy's phrase "suppose, if possible". For Hardy uses it irrespective of the fact whether said possibility exists or not. As a matter of fact, all parties to the proof here know it does not, so the supposition is a staged, or sham one.[25]

Hardy simply leaves this issue open, however, proceeding along blithely about "possibility" as if the notion is quite clear given the grammar of a sentence. As we know, Wittgenstein had long struggled to bring the notions of possibility and necessity back into discussion of the nature of logic and the very notion of a proposition. Hardy's blithe use of modal notions reflected a semi-cultivated view of propositions that was, however, never fully defended (see Hardy 1929, 1940).

Hardy said that he believed, for example, that he could argue on the basis of an "invincible feeling" that there must be a way to resolve, e.g., the Goldbach conjecture.[26] This led him to explicitly align mathematical possibility with empirical possibility. As we know, Wittgenstein reacts allergically, here and in many other places, to the drawing in of "invincible feelings" to constitute a standard of

---

[24] Here there is a danger of confusing the giving of meaning to an assertion with having a strong "belief" or feeling. See Wittgenstein's remarks on Hardy's (1929) remarks about Wittgenstein and the Goldbach conjecture in TS 309, p. 16 (1933–4), i.e., BB, p. 11.

[25] We will see that Hardy also uses this way of speaking, for example, in the case of his second proof of the irrationality of $\sqrt{2}$, which is a *reductio* from the outset.

[26] Hardy (1929), p. 24.

correctness. Hardy has run roughshod over problems of the a priori that had long been debated by philosophers.

So what is Wittgenstein's strategy here? Instead of simply thundering out "*Unsinn!*", he presents a counterproposal for the presentation of the proof which is pleasantly down-to-earth. Simply begin such proofs with "Try" or "Check". This directs one's attention not – as with Hardy's phrase – to supposed possible states of affairs that are stated in the relevant sentences (e.g., the sentence "$p^2/q^2 = 2$"). Instead, one is directed toward possibly different ways of proceeding, and so can see Hardy's particular manner of checking in CPM in a clearer light.[27]

But *why* does Hardy's "suppose, if possible" appear to Wittgenstein like a "cowardly evasion"? *What* is evaded by Hardy? Why is this evasion a *cowardly* one? Because what Hardy doesn't face up to here are the many complications that accrue when one takes seriously the question of what gives a mathematical statement its *particular* sense on a *specific* occasion of use. The "cowardice" consists in the fact that Hardy represses this question, as he always did, having in fact a philosophical commitment to so doing.

For Hardy, there is the *one* statement to be made, e.g., "$\sqrt{2}$ is irrational", and this has but one specific sense, for he takes it to describe *one* particular mathematical fact, and express a unique proposition. The different proofs of it – the different ways of checking it – are only different ways to insure that what the statement says is the case. But for Wittgenstein this view is all too naïve. He struggles, therefore, with the question of whether different proofs, at least if they are sufficiently different, may articulate different senses even while the appearance of the sentence – and of attending senses articulated with it – remains unchanged.

Once again, we are confronted with a key underlying theme of Wittgenstein's later philosophy as a whole. In a given language like English or German, in everyday and everyday mathematical parlance, there are a range of possible things to be said, i.e., a range of senses. But the issue of which particular *one* among them is articulated on a given occasion requires elements of discernment, a shared understanding of *the point* of what is said, and why, and to whom: all the backdrop of the "language-games" that come into play.

In the case of mathematics, this overarching theme is especially complicated and challenging. To address it at all, one must be willing to explore an appreciation of context and its importance to the articulation and exposition of proofs in mathematics. This is no easy task: as Wittgenstein himself wrote to Schlick, warning him against drawing grand conclusions from the Gödel incompleteness proof, it is

> partly a matter of taste, that is, it is a matter of our judgment [*Ermessens*]

of how we weigh differing articulations of a theorem in prose-sentences, of whether we are inclined to apply a picture expressed in a prose-sentence or not.[28]

---

[27] See MS 126, p. 126 (entry dated 14.12.1942).

[28] See Wittgenstein to Schlick July 31, 1935 (in *Briefwechsel*, hereafter B); the letter is analyzed in Floyd (2012a).

A defense of this approach lies beyond the scope of our exposition of Wittgenstein's annotations to CPM – though we shall further explore this below, in our essays in Part III. For now it is enough to say that we believe that *this* is the basis of Wittgenstein's charge that there is, at this point in CPM, a "cowardly evasion" on Hardy's part in his presentation of this proof of $\sqrt{2}$.[29]

3.5 Second Proof of the Irrationality of $\sqrt{2}$

On p. 6 Hardy gives a second proof of the irrationality of $\sqrt{2}$. Here it is, in Hardy's own words:

> Suppose, if possible, that p/q is a positive fraction, in its lowest terms, such that $(p/q)^2 = 2$ or $p^2 = 2q^2$. It is easy to see that this involves $(2q - p)^2 = 2(p - q)^2$; and so $(2q - p)/(p - q)$ is another fraction having the same property. But clearly $q < p < 2q$, and so $p - q < q$. Hence there is another fraction equal to p/q and having a smaller denominator, which contradicts the assumption that p/q is in its lowest terms.

This less well-known proof can already be found, in an only slightly and inessentially different formulation, in Dedekind's *Vorlesungen über Differential- und Integralrechnung* of 1861/62 (Dedekind (1985), pp. 24ff).[30] In their Introduction, the editors of these *Vorlesungen*, Max-Albert Knus and Wilfried Scharlau, stress that this proof avoids prime factorization, as the usual proof does not. Although it is not clear that Dedekind was the first to frame it, they call the proof "extraordinary" [*ungewöhnlich*] and "inventive" [*originell*] (p. 25n). They also remark that it is the special case for $D = 2$ of Dedekind's proof of the irrationality of $\sqrt{D}$ on p. 12 in his (1872). While they state that this proof is rarely to be found in the literature, they do mention two exceptions: Hardy's CPM (the proof occurs already in its first edition of 1908 and in all the subsequent ones) and also Landau (1930), where it can be recognized in the proof of *Satz* 162 (albeit not in a very transparent form). The backdrop to Dedekind's proofs are the algebraic methods known to him through Dirichlet, and these, not primarily aimed at obtaining proofs of the irrationality of numbers *per se*, were well known to Hardy.

This second proof of the irrationality of $\sqrt{2}$, framed by Hardy purely algebraically, is a *reductio* in outward form. Before we consider this point, it is worth emphasizing that the proof can be rendered in a variety of ways, not only algebraically but also geometrically.

---

[29] In Schappacher (2008) it is shown that not only Wittgenstein but also sensitive historians of mathematics avoid this cowardice. Wittgenstein deals with the question of whether the sense of a mathematical statement depends on its proof in RFM III §§55–63, for example; see the commentary on this in Mühlhölzer (2010), pp. 351–389.

[30] We are grateful to Norbert Schappacher for calling our attention to these lectures.

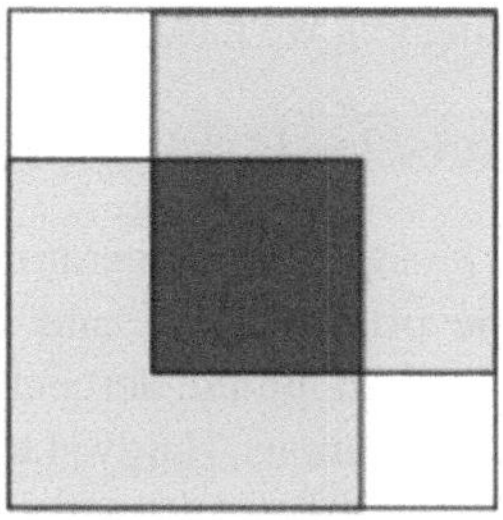

Fig. 3.1 Tennenbaum's Proof

Let us first choose one extremely simple geometrical version, due to Stanley Tennenbaum, who apparently discovered it in the early 1950s.[31]

Begin with a $p \times p$ square, p a positive integer (the enclosing square in the diagram above). Next, superimpose onto this square's corners two $q \times q$ squares such that the area of the original square is twice the area of each $q \times q$ square. In this way one has represented the equation $p^2 = 2q^2$ by a $p \times p$ square the area of which is the sum of the areas of the two $q \times q$ squares placed inside.

Figure 3.1 lets us see *geometrically* that the area of the dark square in the middle, where the two $q \times q$ squares overlap, must be the sum of the areas of the two remaining little white squares top left and down right. Because of the way they are built, the lengths of the sides of all these squares must be integers; and in fact the middle square is a $(2q - p) \times (2q - p)$ square and the little ones are $(p - q) \times (p - q)$ squares. This, however, contradicts the condition that "p/q is a positive fraction, in its lowest terms", which geometrically simply means that the $p \times p$ square and the two $q \times q$ squares are the smallest squares for which $p^2 = 2q^2$ holds.

To illustrate a kind of transition from the algebraical proofs to this geometric one, we can emphasize what Akihiro Kanamori has called the "scaling" character of these arguments, arguably invoking a much older form of reasoning.[32]

First, let $p/q = \sqrt{2}$. Then, by cross multiplying:

$$p = q\sqrt{2} \text{ and } p\sqrt{2} = 2q. \qquad (*)$$

Notice that since $p < 2q$ ($\sqrt{2}$ being assumed to be positive and less than 2 itself), we have $0 < p - q < q$. "Scaling" down (i.e., multiplying numerator and denominator

[31] This proof has been presented and discussed in Conway (2006), Miller and Montague (2009) and Väänänen (2011). It may be seen at http://www.cut-the-knot.org/proofs as proof number 8 on a list of many different proofs of the irrationality of $\sqrt{2}$. Kennedy and Kossak (2011), p. xiii, write:

> Finally we include *via* Väänänen (2011) Tennenbaum's entirely geometric proof of the irrationality of the square root of 2. It appears to be a genuinely new proof of that theorem, as well as the simplest ever found.

[32] For more examples, see http://www.cut-the-knot.org/proofs/sq_root.shtml and Kanamori (2018a).

in the first equation) and applying (*), we have

$$p(\sqrt{2}-1)/q(\sqrt{2}-1) = (2q-p)/(p-q).$$

$2q - p$ is still an integer, as is $p - q$. Now numerator and denominator are smaller than the terms p and q in the original ratio, and this process can be repeated indefinitely. In other words, we get an infinite descent, which is impossible.

Returning to Wittgenstein's annotations, Hardy starts his second proof again with the phrase "Suppose, if possible". Wittgenstein again places under this "if possible" a squiggly line and his annotation on the left and the upper margin of p. 6 (quoted above) is also relevant to it – even more so because now, in contrast to the first proof, the second proof is really one by *reductio ad absurdum*.

As we have just seen, this proof is quite various in its modes of possible presentation. And this makes Wittgenstein's next annotation quite striking. Hardy precedes his second proof with the remark (on p. 6):

> The following alternative proof that no rational number can have its square equal to 2 is interesting.

To which Wittgenstein adds in brackets – this time in English – a direct continuation of this sentence:

> But in what way <u>is</u> it interesting?

This is a good question that Hardy might have raised himself, and we have just considered a number of different kinds of proof of the irrationality of $\sqrt{2}$. Hardy, however, remains totally silent on these matters and this bothers Wittgenstein.

In MS 126, p. 122, he writes:

> What troubles me in an exposition like, e.g., Hardy's is the seemingly useless variety of proofs of one and the same sentence.
>
> I want to say: each of these proofs belongs to a particular occasion where it has to be precisely applied.

What Wittgenstein has in mind here is hinted at in the next annotation. He himself does not think of the sort of variety of proofs we just gave, but of something different. Let us see what he does.

3.6 A Strange Transformation of Hardy's Proof

This annotation is on the lower margin of p. 6, with a bent line pointing upwards from it to Hardy's second proof of the irrationality of $\sqrt{2}$:

> How [would it be] if one used here numerals instead of the letters? That is: From $5^2 = 2 \times 3^2$ follows $(2 \times 3 - 5)^2 = 2 \times (5 - 3)^2$. But there <u>should</u> be $3 < 5 < 2 \times 3$ and hence $5 - 3 < 3$, hence $5^2/3^2 = (< 5)^2/(< 3)^2$.[33]

[33]Wie, wenn man hier Ziffern statt der Buchstaben gebraucht? Also: Aus $5^2 = 2 \times 3^2$ folgt $(2 \times 3 - 5)^2 = 2 \times (5 - 3)^2$. Es <u>soll</u> aber $3 < 5 < 2 \times 3$ sein und also $5 - 3 < 3$, also $5^2/3^2 = (< 5)^2/(< 3)^2$

Wittgenstein here has simply replaced Hardy's letter "p" with the numeral "5" and the letter "q" with the numeral "3" (and we have tacitly changed Wittgenstein's erroneously written "$5 - 2 < 3$" to the correct "$5 - 3 < 3$").

One needs MS 126 in order to figure out the point of this *prima facie* strange and seemingly arbitrary transformation of Hardy's proof. There, on pp. 118ff. (published within RFM V §28), Wittgenstein proposes trying to take every proof by *reductio ad absurdum* to be directed at an opponent who makes a mathematically refutable claim.[34] On pp. 122ff. of MS 126 he applies this idea to our present case. The opponent claims that $2 = (1414/1000)^2$ – that is, Wittgenstein here replaces the ridiculous example of the annotation just quoted by a more realistic one – and a mathematician then shows that the train of thought in Hardy's second proof, now for the case $p = 1414$ and $q = 1000$, leads to smaller numbers, $2q - p$ and $p - q$, such that $2 = ((2q - p)/(p - q))^2$, and then the same train of thought leads to still smaller numbers with this property, and so on – which ultimately the opponent has to accept as being impossible.[35]

This seed of an idea, however, cannot of itself show *why* the devising of different articulations of proofs of the same (or apparently the same) mathematical sentence might be of any value, i.e., be "interesting". For it can be applied to all such proofs in an entirely mechanical way: to Hardy's first proof, to the classical proof in which one shows that $p/q = \sqrt{2}$ forces p and q to be divisible by 2, implying again that positive integers p and q can never be minimal with respect to the property $p/q = \sqrt{2}$, *and so on*.

At this point the *geometrical* version of Hardy's second proof – to which there doesn't seem to be a good analogue in the case of the first proof – can come to our aid. Simply imagine a person who has the feeling that in order to gain real understanding of a mathematical proof one has to see a geometrical idea in it and that all other proofs are only what Schopenhauer called mousetrap proofs, i.e., proofs that force us to accept the theorem at hand but without any real understanding of *why* it is true.[36] For such a person, geometrical or appropriate alternative algebraical versions of Hardy's second proof, with their content and variety, might be necessary in order to come to see the *point* of the proof, as well as to grasp it as a very simple one.

---

[34]Hardy himself makes use of the idea of an interaction with an opponent on p. 119 of CPM with respect to the (ϵ, δ)-definition of a limit (see our Sect. 5.4). But he doesn't rely on a similar approach in the present context.

[35]Wittgenstein's example with 1414 and 1000 is actually still a very bad one, because both numbers are obviously divisible by 2, immediately leading to the smaller numbers 707 and 500 so that $(707/500)^2 = (1414/1000)^2$. A better example, based on the same idea, would use 141, 421 and 100,000, say. (This is a typical example of Wittgenstein's carelessness when presenting formal issues.)

[36]See Schopenhauer (2012b), Book I, §15, and the end of Ch. VI of Schopenhauer (2012a).

3.7 "Everything Leans a Little"

Wittgenstein's next annotation is on the upper margin of p. 7:

> Here everything leans a little. Just enough to show that the whole thing is no organism.[37]

As usual, this remark is *prima facie* rather cryptic and we are not fully confident about how best to understand it. But pp. 121–127 in MS 126 let us presume that Wittgenstein here refers to p. 6 of CPM in its entirety, and in particular to Hardy's "suppose, if possible" interpreted as the introduction into a proof by *reductio ad absurdum*. For after Wittgenstein's critique of the phrase "suppose, if possible", as quoted above ("But how to think the, as it were, unthinkable?!"), he immediately adds on pp. 121f. (the end of §28 of RFM V):

> What the indirect proof says, however, is: "If you want this then you cannot assume *that*: for only the opposite of what you do not want to abandon would be combinable with *that*."

Here Wittgenstein opens our eyes to wider connections, to an "organism" that would, so to speak, be disrupted, "leaning over" from being centered, or standing straight up, if one tried to combine *this* with *that*. According to his analogy, if one thinks of a "system" of real numbers, then one must appreciate that its inner complexities distinguish it from other "systems" of numbers, and that this affects what we may *do* with it.

With respect to indirect proofs of the irrationality of $\sqrt{2}$ this is expressed quite explicitly on pp. 126f. of MS 126:

> Instead of "Let us suppose $(p/q)^2 = 2$" it would be better to say: "Check '$(p/q)^2 = 2$' ", or "Check: $(p/q)^2 = 2$". – For now you can check it in different ways.
>
> The question of whether (p^2/q^2) will ever be 2 is connected to a great number of other mathematical relationships and this proof shows the respective connections and which positions we would be forced to abandon if we permitted $(p/q)^2 = 2$.
>
> This is a particular proof-machinery, not the eternal form of a proof. (I think of Gödel's introductory casual line of proof.)

When in the annotation on p. 7 of CPM Wittgenstein writes: "Here everything leans a little", what he presumably means is that the supposition with which a proof

[37] Hier steht alles ein bißchen schief. Gerade genug um zu zeigen, daß das Ganze kein Organismus ist. – Wittgenstein's "Gerade genug" is actually ambiguous. It can mean "just enough", as in our translation, or "straight enough". The idea behind our translation is this: that "the whole thing is no organism" is shown by what is reasoned about leaning over, deviating from straight, like the leaning tower of Pisa. This, as it were, shows the intrinsic lack of inner organization of the irrational numbers as we make the transition to them from the rational numbers. The image of an organism leaning over, deviating from upright, may also suggest a potential movement toward the conceptual collapse of our original structure of concepts.

by contradiction starts does disrupt the "organism", but only a *little bit*, and this bit is then taken back by the proof in the end. But seeing precisely *how* this happens may take several different versions of the proof to be understood, several efforts at disrupting the system by a precise bit. It is as if one were testing a machine as a whole, altering its inner parts bit by bit, in order to see precisely what forces it can tolerate and still perform its proper functioning – and why.

So the contrasting kinds of proof we have considered, the second by scaling regress (whether it uses specific geometric or algebraical means), show us different sorts of disruption of the mathematical organism when the first step of the *reductio ad absurdum*, its assumption, is taken up.

As for Gödel's incompleteness theorem, mentioned by Wittgenstein in brackets, we do indeed confront an analogous situation. For the first proof in the introduction of Gödel (1931/1986), labeled by Wittgenstein "casual", is very different from the second, main proof with its sophisticated machinery. In this case too, from Wittgenstein's point of view, it would be inappropriate to consider one of these, whether the first or the second, as having "the eternal form" of proof.[38] Instead, Gödel shows us something dynamic about our notion of "provable", which similarly "leans a little" when we consider models of arithmetic and the notions of "definable" and "true".[39] In Chap. 8 we shall return to Wittgenstein's idea of a proof as a piece of technology, and this dynamism, placing the "contradiction" to which Gödel refers in his Introductory remarks, the Richard Paradox, into historical and philosophical perspective.

3.8 The "Phantasmagoric"

Pages 1–7 of CPM are propaedeutic to the systematic study of irrational numbers, which begins in Section 4 ("Irrational numbers (*continued*)"). Here Hardy introduces the idea of Dedekind cuts, calling them "sections" (see p. 11), a term hardly

---

[38]See Floyd (2001b, 2012a), p. 254. Note that Gödel added a footnote 48a to his 1931 in which he states that "the true reason for the incompleteness inherent in all formal systems of mathematics is that the formation of ever higher types can be continued into the transfinite" (1931/1986, p. 181). Wittgenstein was concerned to question this idea of a one "true" reason for the incompleteness, whereas Gödel, for his part, was very disturbed by Wittgenstein's apparent unwillingness to take on transfinite types. For further discussion see Wittgenstein's lecture on Gödel in Wittgenstein (2017) and Floyd and Kanamori (2016) on Gödel's attitude toward Wittgenstein.

[39]See Floyd and Putnam (2000, 2012) for arguments that this dynamic character of the notion of "provable" comes into play when we consider models of arithmetic, and compare Floyd (2020, in press).

ever used nowadays. Nevertheless we adopt it in what follows in order not to veer too far from Hardy's exposition.

On p. 8 Hardy describes how in the totality of the positive rational numbers[40] the irrational number $\sqrt{2}$ produces a section with the "lower class", or "left-hand class", L, and the "upper class", or "right-hand class", R, where $L = \{x|x$ is a positive rational number with $x^2 < 2\}$, and $R = \{x|x$ is a positive rational number with $x^2 > 2\}$. Immediately after this he writes:

> [...] it is easy to convince ourselves that we can find a member of the class L whose square, though less than 2, differs from 2 by as little as we please, and a member of R whose square, though greater than 2, also differs from 2 as little as we please.

And he concretizes this by means of the obvious sequence of decimal expansions 1, 1.4, 1.41, 1.414, 1.4142... (representing rational numbers approaching $\sqrt{2}$ from below), and 2, 1.5, 1.42, 1.415, 1.4143... (representing rational numbers approaching $\sqrt{2}$ from above), each of these figuring, of course, in the decimal expansion of $\sqrt{2}$.[41]

In Sect. 3.5 (again with the caption "Irrational numbers (*continued*)"), Hardy connects this with his standard geometrical illustration, the straight line, called by him "Λ".[42] As he writes at the beginning of this section (p. 9):

> Our common-sense notion of the attributes of a straight line, the requirement of our elementary geometry and our elementary algebra, alike demand *the existence of a number x greater than all the members of L and less than all the members of R, and of a corresponding point P of Λ such that P divides the points which correspond to members of L from those which correspond to members of R.*

At this point, on the lower and the right margin of p. 9 – alluding to Hardy's presentation on p. 8 of the processes of approaching $\sqrt{2}$ from both sides – Wittgenstein comments:

> These two approaching series,[43] one wants to say, determine a place – but not a rational one. I should have to say of this place that it is located left of all the right

[40] We use the term "totality" and not "set" in order to follow Hardy, for he himself is avoiding set theoretic talk throughout his book. This is one of its salient characteristics, and, as we shall argue below, one of its weaknesses as exposition. Presumably Hardy adopts this approach for pedagogical reasons. From a systematic point of view, Hardy did not reject set theory at all, for he had published papers in it himself (see Grattan-Guiness (2001), pp. 412f.). In our exposition of Hardy's text, we sometimes take the liberty of using set theoretic *notation* because of its lucidity, but in so doing we do not presuppose any set theory, let alone axiomatic set theory.

[41] On p. 8, Hardy remarks in small letters that the reasoning using this sort of representation "is hardly of the precise character required by modern mathematics". Yet one can very well introduce real numbers via decimal expansions in a rigorous and satisfying way. See Gowers (2017b) and Tao (2006), pp. 380–388, as already mentioned in Chap. 2 above.

[42] At this stage of the exposition Hardy refers with "Λ" merely to the points corresponding to positive numbers, but this is not an important restriction.

[43] Today one would call them "sequences", but on p. 8 Hardy himself uses the word "series" for them.

and right of all the left members. But this is a misleading form of expression, for the determination of the place does not come to an end. [44]

This seemingly trivial remark should be read in light of Wittgenstein's first annotation, concerning the "application" of analysis. What he has in mind here is the application of analysis – including arithmetic and what Hardy calls "algebra" – to geometry. The non-geometric result that $\sqrt{2}$ is irrational, for example, can be *applied* to geometry in such a way that we accept that the diagonal of the unit square, if transposed onto the straight line starting at its origin and extending to the right, does not end in a rational point. This non-rational point is determined by such a procedure, a procedure which consists of only a few, totally unproblematic geometrical steps, steps that are described by Hardy on pp. 5f.

We have a deeply different procedure in mind, however, when we think of determining this point by approaching it from below and above as described on p. 8 of CPM. This latter procedure consists of infinitely many steps and – when considered as an "application" of analysis – *obviously* does not come to an end.

When Wittgenstein speaks of a "misleading form of expression" what he means is the sweeping talk of the "*determination* of a specific point". This obliterates the deep difference between the many different finite ways (consisting in real geometric constructions) and the many different infinite ways to which we are led when bringing *sections* into play, and the accompanying idea of *approaching* points as described by Hardy on p. 8.

As already mentioned in Chap. 2, Hardy considers such a lack of determination – in effect, an impossibility – to be a mathematically inessential one, as inessential to the irrationality of $\sqrt{2}$ as the impossibility of a mathematician having drunk no water. Russell concurs in calling it a mere "medical impossibility" (1936, p. 143). But this point of view ignores our real mathematical practice, the "grammar", or specific logic of the situation.

Wittgenstein considers the sort of idealized, infinitely extended application that Hardy and Russell seem to be entertaining to be "phantasmagoric" or "imagined", as he expresses himself in MS 126, pp. 127f.:

Here we have then the phantasmagoric application. The imagined application.[45]

This remark occurs immediately after the passage quoted above in Sect. 3.1 about Hardy's illustrations being potentially utterly misleading if not seen as "applications" – where he means real applications, not phantasmagoric ones.

[44]Diese beiden sich nähernden Reihen, möchte man sagen, bestimmen einen Ort – aber keinen rationalen. Ich müsse von diesem Ort sagen, er liege links von allen rechten und rechts von allen linken Gliedern. Das aber ist eine irreführende Ausdrucksweise, denn die Ortsbestimmung kommt zu keinem Ende.

[45]The German reads: "Hier haben wir dann die phantastische Anwendung. Die eingebildete Anwendung."

The whole train of thought appears as RFM V §29, an important section that we now quote in its entirety:

> The geometrical illustration of analysis is indeed inessential, not, however, the geometrical application. Originally the geometrical illustrations were *applications of analysis*. Where they cease to be this they can be wholly misleading.
>
> Here we have then the phantasmagoric application. The imagined application.
>
> The idea of a 'cut' is one such dangerous illustration.
>
> Only in so far as the illustrations are also applications do they avoid producing that special feeling of dizziness which the illustration produces at the moment at which it ceases to be a possible application; when, that is, it becomes stupid.[46]

This distinction between "real" and "phantasmagoric" applications is a typical Wittgensteinian one, and the fact that Wittgenstein calls the "phantasmagoric" ones "stupid" shows how important the distinction is for him. It concerns the central and difficult point that we already elaborated upon in Sect. 3.2: the difference between, on the one hand, "idling" illustrations, i.e., ones failing to be engaged with any particular context of explanation or proof, and, on the other hand, "real" ones, i.e., *working* applications whose rationale and purpose are clear. The pictures or illustrations of the former sort are "idle wheels". When they show up parading as "applications" they are merely phantasmagoric, merely imagined. Accordingly, in RFM V §37 he writes of the "yawning gulf between illustration and application".[47]

It is worth stressing that this view distinguishes Wittgenstein's thinking deeply from intuitionism, let alone strict finitism. None of what he is saying is capable of raising any objections to usual mathematical practice as such. What interests Wittgenstein instead is how it is that what has appeared, naturally, *within* mathematics as "nonsense" – as a war-cry – bears the possibility of being drawn out of its real workings. How was it possible that mathematics – that subject in which conviction, agreement, justification and clarity seemed to be so nicely enshrined since before Plato – could have *produced* such war-cries? How might the wars, the argumentation, be resolved? What are we to say now? *These* are Wittgenstein's themes (compare RFM VII §22).

In this regard, Wittgenstein directs his remarks only toward what he calls the "prose". It is exemplified by that which Hardy cocoons around his discussions of the continuum – perhaps inevitably – thereby enshrining in a textbook what he would have "blushed" to put into print in a published article (recall Littlewood's 1926 prefatory remarks). On Wittgenstein's view, such blatant cocooning is nearly inevitable in the teaching of abstract modern mathematics, as perhaps it is not in other branches of the subject. That is one reason Wittgenstein interests himself in Hardy's textbook, and in Dedekind cuts.

The main theme – as so often in later Wittgenstein's philosophy generally – is the philosophical task of investigating dubious pictures that are parading as

[46] We have slightly changed Anscombe's translation.

[47] In German: "klaffende Spalt zwischen Illustration und Anwendung".

justifications when they are not really serving as such, and putting certainties and matters unquestioned into the right perspective.

As has often been pointed out, it is certainly true that a potential tension looms here.[48] Wittgenstein sometimes speaks as if he longs for, and is urging, a real change in mathematical practice, in just the same way as was Brouwer. For example, there is his famous reaction to Hilbert, who had written in "On the Infinite" (1926) that no one would drive mathematicians from Cantor's paradise. Wittgenstein's response was voiced in his 1939 Cambridge lectures on the foundations of mathematics – with Turing present in the audience:

> I would say: "I wouldn't dream of trying to drive anyone out of this paradise." – I would try to do something quite different: I would try to show you that it is not a paradise – so that you'll leave of your own accord. I would say, "You're welcome to this: just look about you" (LFM, p. 103).

Similarly, on pp. 129f. of MS 126, in a context parallel to the present one, he writes: "We fight now against a direction. But this direction will die, pushed aside by other directions. And then one will no longer understand our argumentation against it; not comprehend why one had to say all that."

Yet, characteristically, even here the image is of a *direction* moving off somewhere else, an ebb and flow of battles that emerged, skirmishes that have been fought, and afterwards mathematics being re-fashioned so as to continue its lava-like flow. It is not that something turns out not to have been part of mathematics at all. It is, rather, that we, in the practice of mathematics, have rejoined the onward flow of mathematics back to itself, have returned mathematics to civilian life, something of which it had always been capable, if we had allowed this.

Wittgenstein himself is often a generator of novel conceptions. As already mentioned in a footnote in Chap. 2 (page 30, note 5), there are for example some rather strange remarks on pp. 134–149 of MS 126, where he toys with designing a totally new sort of mathematics dealing with continuity. This, however, is not meant as a serious proposal to change our actual practice of doing analysis, but as something Wittgenstein rather often does: to present a conceivable alternative that sheds light on what we actually do. It does not contradict the numerous and systematic remarks from his later philosophy in which he emphasizes that he does not want to entangle himself directly in the doing of mathematics.[49]

For example, right at the beginning of his 1939 Cambridge lectures on the foundations of mathematics he explicitly said: "[It] will be most important not to interfere with the mathematicians" (LFM, p. 13). Since Turing was teaching a course on foundations of mathematics that very same term at Cambridge, it was important for Wittgenstein to make his approach clear to the students, as well as

[48] See the *locus classicus* on this, Wright (1980).

[49] Of course, earlier in his life Wittgenstein had pursued applied mathematics, as well as mathematical logic. He held a patent for his design for a propeller back in 1909, the result of research done while in Manchester. See McGuinness (1988), Chap. 3, Spelt and McGuinness (2001), Lemco (2007).

to Turing himself.[50] As one of has emphasized elsewhere,[51] Wittgenstein was not pursuing the *foundations* of mathematics, in the sense of Frege, Russell, or Hilbert (a *Grundlegung*). Instead, he was aiming to work through, investigate, and clarify what to say about mathematics (a *Klarlegung)*. This is a unique kind of task, and he should be credited with making the pursuit of it relevant to the heart of philosophy.

What Wittgenstein reacts to with distrust is not mathematics, but the "prose" in which mathematics is usually presented, and perhaps must be presented to students, other mathematicians, and also to philosophers. Here – whether it is a mathematician or a philosopher or anyone else speaking – a philosophical context of discussion may, wittingly or unwittingly, be broached before one realizes it.[52]

By "prose" Wittgenstein means the verbal statements of mathematical results considered apart from their serving as elements contributing significance to the activity of mathematics. His approach – taking into account what we are *doing* with words and symbols we use – was developed over a long period of time across many differing areas of our uses of language, especially those that have interested philosophers. It became crucial for him, in the period 1937–1944, to work out how this approach would work in connection with mathematics, as we have already explained in Chap. 1.

Wittgenstein's approach serves to question one prevalent danger. It consists in our tendency to lapse into conceiving of mathematician's "prose" statements as functioning in a way essentially similar to the ways statements made by empiricial scientists function – be they physicists, cosmologists, biologists, psychologists, or sociologists, and/or thinking that they are, in everyday language, capable of clarifying philosophical problems about mathematics generally by describing how things are, and this *apart* from any specific, usual ways of clarifying. One may hold, for example, that all these statements do the same work in that they all do or do not describe facts: the empirical scientists describing facts of nature, and the mathematician describing facts of the mathematical world, or not describing at all. But this parallelism and dichotomizing involves a misunderstanding of the way Wittgenstein is thinking about language everywhere, in mathematics, in science, and in everyday life.

Of course one can say that the prose-sentences in Hardy's CPM are true or are false, describing a "reality".[53] But one needs to be very careful about justifying in detail the point of saying this. There is no telling precisely what work the words are being put to before one takes oneself to have understood precisely which particular thing is being said, among a variety of possible things that are there to be

[50]See Floyd (2012c, 2017) for an emphasis on what Turing seems to have drawn from the 1939 seminar discussions: an interest in formulating "types" with ordinary, working concepts of mathematics to allow for easier applications of logic in mathematics.

[51]Mühlhölzer (2010).

[52]Floyd (2001b, 2012b), Floyd and Putnam (2000, 2012).

[53]On Wittgenstein's remarks about allowing us room to discuss the idea of "corresponding to a 'reality'" with regard to mathematical statements, see Diamond (2018).

said with the words. There are a variety of contexts in which mathematicians put words, proofs, particular numbers and theorems to work. This is what is at issue in Wittgenstein's annotations to CPM.

As we have repeatedly emphasized, it is central for Wittgenstein to carefully keep track of the difference between words, sentences, signs and pictures that are being used to a purpose of some kind or other, and those that we – in our treatment and conceptions of them – have *made* non-useful (as we would have made inner imaginary sensations non-useful, or multiple purchases of the morning newspaper non-useful, in looking to use them as unendingly-ever-more-helpful justifications). Given this point, we can see why a great deal of philosophical argumentation in the later Wittgenstein is devoted to criticizing the idea that words, symbols, actions, thoughts or signs must be seen to be immediately and intrinsically meaningful, *sans* context.

In particular, it was Russell, with his construal of singular thought in terms of the notion of "acquaintance", who ran afoul, from Wittgenstein's point of view, of the extrusion of psychology from the foundations of mathematics that Frege had called for. Russell believed that "the limits of empiricism" must be intrinsically justified mental contact with universals, or else directly grasped, yet unguaranteed assumptions. This struck Wittgenstein as ill-founded from the bottom up: we have seen how he shows that multiple appeals to multiple imaginary objects and acts fail to do the work one might think they can do.

Wittgenstein is actually trying to usher Russell's notion of "acquaintance" back to the everyday, to our usual understanding we have of what it is to be "acquainted" with something or someone (a proof, a person, a word's uses, and so on). As he wrote at RFM VII §21, alluding to Russell's invocation of our "medical" limitations with respect to the infinite (Russell 1936):

> The limits of empiricism are not assumptions unguaranteed, or intuitively known to be correct. They are ways in which we make comparisons and in which we act.[54]

Throughout his life Wittgenstein characterizes the contrast between mathematical prose and genuine mathematics as the contrast between *prose* and *calculus*. His explicit uses of the distinction become less pronounced after his turn to his mature method of philosophizing in 1937, although the idea of what he is trying to convey remains.[55]

What Wittgenstein means by "prose" is any sort of non-working cocoon that parades as an explanatory or descriptive use of language when it is not really explaining or describing what is going on. In his conversations with members of the Vienna Circle in the early 1930s, Wittgenstein explicitly contrasts the construction of mathematical calculi with the expression of "contentual insights", of which he

[54]Other references to Russell (1936) are to be found in RFM at III §72, IV §29, VII §17. See also Wittgenstein's letter to Russell in WC, p. 199, explaining his lack of attendance at the lecture Russell gave, as well as the analysis in Mühlhölzer (2010), pp. 391–469.

[55]On the middle period use of the idea, see Marion (1998); on the softening of the distinction, compare Kienzler and Grève (2016).

says that it is "the prose that accompanies the calculus" (WVC, p. 129). Evidently he is at this time influenced by Hilbert's "finitism" with its conception of "contentual" [*inhaltliche*] reasoning, vs. the template-like, schematic mode of proof expressed in arguments by mathematical induction.

As an example of such prose, he mentions the proposition "There are infinitely many prime numbers". According to Wittgenstein, this proposition cannot really convey the genuinely mathematical result, which only lies in the relevant proof. Such a proof, Wittgenstein says, "lets us see something. What it shows, however, cannot be expressed by a proposition" (WVC, p. 137). Wittgenstein here means that the proof shows something that can only be misleadingly expressed by this prose.[56] What is important to "there are infinitely many prime numbers" is that it reflects acceptance of a kind of schematic procedure (proof by mathematical induction). To hold that such a manner of proof may be justified independently of our accepting it as a procedure capable of supporting general claims is to misread its specific logical role and status, its "grammar".

Of course, mathematicians often talk about mathematical "theories", and they say that there *are* infinitely many prime numbers. But this should not mislead them into thinking that their so-called "theories" and their "propositions" are norm-free explanations or descriptions of what is the case. What is really mathematical in these "theories" and "propositions" does not, as the expression in prose suggests, go beyond the specific procedures (in the terminology of the middle Wittgenstein, the "calculus") to which they belong.

In Wittgenstein's middle phase, he draws sharp and apparently ultimate distinctions between what may be stated in a proposition and what may only be shown, or taken in. This is reminiscent, of course, of the show/say distinction as he handled it in the *Tractatus*. In his later work, by contrast, his language becomes more moderate, especially in his use of the term "calculus". But his warning about the intrusion of prose persists, and he still wants us to avoid looking at mathematical propositions as essentially descriptive of facts, at least in their characteristic uses.[57] Sometimes he suggests that we look at them as commands, initiating specific procedures (see RFM V §13f.).

In 1944 he talks about "the curse of prose [...] in mathematics" (RFM VII, §41), and in this context he specifically mentions "Russell's prose", by which he means formalized propositions in the language of *Principia Mathematica*. He suggests that we may regard these as nothing but sophisticated translations of ordinary prose (e.g.,

[56] See WVC, p. 137. In both cases (WVC, p. 129, as quoted above, and WVC, p. 137), the English translation is tendentious in replacing Wittgenstein's single word "*Prosa*" by "everyday prose". Wittgenstein does not contrast "everyday" prose with "mathematical" prose here, but prose with "calculi".

[57] See on this Mühlhölzer (2012) and especially Friederich (2011, 2014), where the *normative* character that Wittgenstein diagnosed in mathematical propositions is explicated in an ingenious way.

the phrase "there exists").[58] They are as potentially misleading as any other prose trying to give an ultimate clarification of the nature of mathematics.

Let us come back to Wittgenstein's distinction between "phantasmagoric" and "real" applications. With regard to the application of the decimal expansion of $\sqrt{2}$, RFM V §9 reads as follows:[59]

> I want to say: it *looks* as if a ground for a decision were already there; and it has yet to be invented.
>
> Would this come to the same thing as saying: in thinking about the learnt technique of expanding [$\sqrt{2}$, for example,] we use the false picture of a completed expansion (of what is ordinarily called a "series") and this forces us to ask unanswerable questions?
>
> For in the end every question about the expansion of $\sqrt{2}$ must be capable of being formulated as a practical question concerning the technique of expansion.

Wittgenstein is clearly trying to *avoid* "unanswerable questions" here.[60] "For in the end" means: when we think of *real* applications and not of phantasmagoric ones, we will have to see things another way. The phantasmagoric applications come along with the idea that it is as if "a decision were already there", and with this phrase Wittgenstein has in mind typical questions about the law of excluded middle – e.g., "Does 0123456789 occur in the decimal expansion of π or not"?

Such questions, when pressed, led the intuitionists to propose a change in the logic of mathematics. Wittgenstein, however, treats the issue in quite another way: he tries to uncover the wrong pictures that go along with the questions that are unanswerable. These he aims to dispel by offering a sober look at our actual mathematical practices. The topic of the law of excluded middle takes so much space in MS 126, in connection with discussions that concern Wittgenstein's annotations to CPM, that it seems advisable at this point to say a bit more about it.

---

[58] See BT, p. 629: "If you want to know what the expression 'continuity of a function' means, look at the proof of continuity; that will show what it proves. But don't look at the result as it is expressed in prose, or as it reads in Russellian notation, which is simply a translation of the prose expression; rather, direct your attention to the part of the proof where calculation is still going on. For the verbal expression of the allegedly proved proposition is in most cases misleading, because it veils the real goal of the proof, which can be seen completely clearly in the proof itself."

[59] We have modified Anscombe's translation.

[60] Floyd (2012b,c).

Chapter 4
The Law of the Excluded Middle: A Digression

"But don't you really understand what is meant?" – But may I not believe I understand it, and be wrong? (RFM V §12)

Abstract An excursus on the status of the law of excluded middle in mathematics, regarded both from Wittgenstein's point of view and our own.

4.1 Improper Pictures

The passage quoted at the end of the foregoing Chapter, with its charge of a "false picture" (RFM V §9), was explicitly prompted by the Brouwerian question whether a certain finite sequence ϕ of digits occurs in the decimal expansion of π; where it is presupposed, of course, that we – at present or maybe if we continue until the end of the world – have not found any means to decide it.[1] The remark RFM V §9 begins this way:

> We only see how strange [*seltsam*] the question is whether the pattern ϕ (a particular arrangement of digits, e.g. "770") will occur in the infinite expansion of π, when we try to formulate the question in a quite homemade, garden variety way [*ganz hausbackene Weise*]: humans have been trained to put down signs according to certain rules. Now they proceed according to this training and we say that it is a problem whether they will *ever* write down the pattern ϕ in following the given rule.

When understood in this way, what precisely makes the question about a particular arrangement of digits appearing in the infinite expansion of π "strange"? According to Wittgenstein, part of its strangeness is to be seen in the tension between, on the one hand, a very suggestive picture – the picture of a completed infinite series that

[1]The specific patterns ϕ that Brouwer and Wittgenstein actually gave as examples (e.g. "0123456789", "770", and so on) have become obsolete because modern computers have found them; but they can be replaced by other patterns so that the question remains open. In what follows we will retain Wittgenstein's examples in order not to complicate the presentation of his thoughts.

J. Floyd, F. Mühlhölzer, *Wittgenstein's Annotations to Hardy's Course of Pure Mathematics*, Nordic Wittgenstein Studies 7,
https://doi.org/10.1007/978-3-030-48481-1_4

leads us to say that either one will come upon ϕ in the series or one will not – and, on the other hand, our inability to make a respectable mathematical argument out of this picture.

In §352 of *Philosophical Investigations*, in the context of certain considerations about mental images, Wittgenstein expresses it thus:

> At this point, our thinking plays us a strange trick [*einen seltsamen Streich*]. That is, we want to quote the law of excluded middle and say: "Either such an image [i.e., a mental image] floats before his mind, or it does not; there is no third possibility!" – We encounter this strange [*seltsame*] argument also in other regions of philosophy. "In the infinite expansion of π either the group '7777' occurs, or it does not – there is no third possibility." That is to say: God sees – but we don't know. But what does that mean? – We use a picture: the picture of a visible series, the whole of which one person can survey and another can't. Here the law of excluded middle says: it must look either like *this* or like *that.* So really – and this is surely obvious – it says nothing at all, but gives us a picture.
>
> And the problem is now supposed to be: does reality accord with the picture or not? And this picture *seems* to determine what we have to do, what to look for, and how – but it does not, precisely because we do not know how it is to be applied. Here, saying "There is no third possibility" or "There really isn't a third possibility!" expresses our inability to turn our eyes away from this picture – a picture which looks as if it must already contain both the problem and its solution, while all the time we *feel* that it is not so.[2]

This remark was originally drafted and revised in the same year that Wittgenstein was working through Turing's epochal paper "On Computable Numbers, with an application to the *Entscheidungsproblem*".[3] But in RFM V §9 (1942–1943, contemporaneous with the Hardy annotations) Wittgenstein elaborates a bit more about these pictures.

He first observes, in §10 of RFM, that "[w]hen someone sets up the law of excluded middle, he [or she] is, as it were, putting two pictures before us to choose from, and saying that one must correspond to the fact". In §18 he further elaborates on this, emphasizing the asymmetry of the situation:

> [. . .] when I say it occurs ["it" = the pattern ϕ], a picture of the series from its beginning up to the pattern floats before my mind – but if I say that the pattern does not occur, then no such picture is of any use to me, and my supply of pictures gives out.

Indeed there is a logical asymmetry here. If a pattern *does* occur, then it can simply be produced by calculation (at least in principle). But to show that a pattern does *not* occur demands what one might call a "negative proof", that is an impossibility proof covering all possible continuations of the series. This in turn requires a mathematical

---

[2]Note that in the Hacker/Schulte translation of PI, which is the basis of our quote, the second occurrence of Wittgenstein's word "seltsam" is not translated as "strange", as in the case of its first occurrence, but as "curious". In view of Wittgenstein's linguistic sensitivity we do not think that this is a wise decision and have changed it accordingly.

[3]See Sect. 1.3 and Chap. 8. This passage goes back to an entry in MS 116, pp. 148–150, which, according to von Wright, stems from 1937/38 or at least contains Wittgenstein's editing of writings from 1937/38, that is, of writings that are much earlier than his annotations to CPM and his remarks in MS 126.

and/or conceptual analysis of the very concept of "all possible calculations (of real numbers) in accordance with the rules".

Now as Wittgenstein very well knew, such an analysis had been given by Turing in "On Computable Numbers": Turing personally sent off an offprint to Wittgenstein in the first round of distributing it in late 1936 or early 1937.[4] Turing's analysis, which served as an analysis of the general notion of taking a "step" in accordance with a "formal rule", was based on a comparison between a human calculating a real number with paper and pencil, on the one hand, and, on the other, the operations of a machine capable of only a finite number of configurations (1936 §1). This was a comparison which Wittgenstein appreciated and explored later on, in 1946–1947, just after he had taught a course on the philosophical foundations of psychology.[5]

The remark with which we began this section (RFM V §9) should remind us of Turing's perspective. For notably, that perspective – analyzing as it does "computable (real) numbers" by way of an analogy between a "step" in a routine of computation[6] and human calculation with paper and pencil – does conceive of humans as beings trained to write down signs in accordance with certain rules. Moreover, Turing's analysis of "computation in accordance with a formal rule", quite strikingly, in no way depends upon taking up any kind of special stance on the law of excluded middle, but, rather, artfully dodges the issue.

It is indeed not a part of our concept of "calculating in accordance with a (formal) rule", or "computing the digits of a real number" that one must accept the law, or even see it as a necessary law.[7] This, combined with the striking fact that, as Turing also showed, there is no single method of calculation for determining whether or not a particular calculating routine will or will not end with a specific result, brings us face to face with the "giving out" of the intuitive pictures if we try to construe them as if they, in and of themselves, relate determinately to calculations, proofs, and mathematical arguments in anything other than a pleasing, perhaps psychological way.

Wittgenstein's most important addition in RFM V, however, consists in his observation, stated in §29, that in analysis people often make use of certain pictures that naturally come along with calculations, and treat them as if they constituted *applications* of analysis, applications to geometry, say. In reality, however, these are merely *imagined* and *phantasmagoric* applications: they have no clear or genuine use as explanations or clarifications of the real numbers, whatever their psychological appeal. These pictures, as is said by Hardy, can of course be quite helpful as illustrations, and they can also be useful in a heuristic way. But they are

[4]See Turing's letter to his mother February 11, 1937 AMT/K/1/54 in the Turing Digital Archive at http://www.turingarchive.org/browse.php/K/1/54.

[5]See Wittgenstein's 1947 remark at RPP I §§1096ff. about Turing machines being "*humans* who calculate", discussed in Floyd (2012c, 2017) and in Sect. 8.8 below. Compare Gandy (1988), Mundici and Sieg (2017) and Davis and Sieg (2015), as well as LPS.

[6]In the context of the *Entscheidungsproblem*, the idea is to analyze, in general, the idea of a (finitistically comprehensible) "step" in a formal system of the relevant kind.

[7]See Floyd (2012c, 2013a) and Chap. 8.

not "applications" in any serious sense. In terms of Wittgenstein's image of language as a mechanism: here the pictures and words are doing no definite work.

As we have seen, in his very first annotation to CPM Wittgenstein says of the genuine applications that they, by contrast, are *essential* to analysis (see Sect. 3.1). We shall explain at the end of Sect. 5.3 precisely what he means by this.

4.2 The Status of the Law of Excluded Middle

If one wants to explicitly maintain the law of excluded middle come what may – even when no method of decision is to be seen – then, as Wittgenstein writes in §9 of RFM, it seems to him that one may be, in saying this, actually setting it up as a rule or a postulate. The law then has the character of a sort of methodological rule. But it is one that can only be supported by pragmatic considerations; for example, a wish to aim at a system of mathematics that is maximally rich in proofs and results and which, in turn, may be justified by the usefulness of such a system in physics. In the end, this may be the ultimate framework of reasoning which may be said to legitimize accepting the law of excluded middle without restriction.[8]

In any case, when this law is accepted without a respectable mathematical proof of one of the two alternatives, one should say that, strictly speaking, one does not bestow on it a validity "*inside* mathematics", as Wittgenstein writes in §18. This can be seen as another strange fact concerning the law of excluded middle: its acceptance in the case of the non-decided Brouwerian question leads us *outside* mathematics, if we are to make of it an explicit principle!

One might object that this cannot be right, because an important part of our normal mathematical practice consists in making *conjectures* about the correct answer to questions like the Brouwerian one and in trying to prove them. Must it not, therefore, be the case that such a question has a perfect sense *within* mathematics? But this is not necessarily so, if we attend, not merely to the compositional putting together of familiar words, but to the uses this particular combination may be put to in a given context: this, as we have said, is the perspective on mathematics – and language generally – that Wittgenstein is exploring in his later philosophy. The general issue is in fact one he confronted in many places in his writings on mathematics, often drawing examples from history.

Let us consider a prominent example. Does it make perfect sense to say, from the point of view of Euclid's *Elements*, that the regular heptagon is constructible or that it is not constructible – *tertium non datur*? Someone who has learned only Euclidean geometry in the style of Euclid's *Elements* and who asks whether the regular heptagon is constructible will presumably think that this construction, if possible, should be analogous to, say, the construction of the regular pentagon. But

[8] A sentiment like this is expressed by Hermann Weyl in his (1948), p. 332.

this, of course, overlooks the fundamental difference between the regular 5- and 7-gon, a difference revealed by analytic geometry.

This geometry was developed, of course, centuries later than Euclid (see the discussion at LFM pp. 46ff). If someone had conjectured back then that this construction is not possible, she could not have had any respectable mathematical idea – without a great deal of other mathematical theory – of what it could mean to claim this impossibility. The sentence gets *mathematical* content by bringing in the concepts of analytic geometry, which provides the needed conceptual edifice to prove it and draw it into the conceptual edifices of mathematics. The philosophical task, in such a case, is to better understand the nature of these "conjectures" in order not to fall prey to certain forms of philosophical naiveté.[9] And exactly the same is Wittgenstein's task, as he conceives it, in case of the Brouwerian question.

Given his perspective, Wittgenstein has good reasons to say that, strictly speaking, the acceptance of the law of excluded middle as a general law occurs "outside" mathematics – or (as we might better put it) at best at its margins. That is, it may be either primordially embedded in (what are recognized to be) ordinary procedures, or it may be made explicit by way of an axiom.[10] In RFM V §20 Wittgenstein expresses his attitude thus:

> But does this mean that there is no such problem as: "Does the pattern ϕ occur in this expansion?"? – To ask this is to ask for a rule regarding the occurrence of ϕ. And the alternative of the existence or non-existence of such a rule is at any rate not a mathematical one.
>
> Only within a mathematical edifice [*Gebäude*][11] which has yet to be erected does the question allow of a *mathematical* decision, and at the same time become a demand for such a decision.

With "such a rule" Wittgenstein here means a mathematically respectable way to decide this problem. The logical structure of the request for a rule does not provide this on its own, for there are a variety of different things that a request for such a rule might involve. This logical structure of the "law of excluded middle" is then, as Wittgenstein often put it throughout his life, a kind of *scaffolding* for the erection of an edifice. It is not part of the building blocks, elements, and materials that will be needed to structure the deciding of the problem and the assertion of a mathematical

[9]See Floyd (1995) and Mühlhölzer (2001) for attempts to clarify the nature of such conjectures. Compare also Säätelä (2011).

[10]The assumption that the law of excluded middle holds for all real numbers can be given different interpretations, in any case. Sometimes it is formulated semantically as a "principle of bivalence", though this can also be understood in stronger and weaker ways, e.g. (1) the schema $p \vee \neg p$ is accepted as part of the deductive apparatus of the theory of real numbers, or (2) every sentence of mathematics that concerns itself with real numbers is "determinately true or false." See Dummett (1991b) on this distinction; Wittgenstein's point applies either way.

[11]Anscombe's translation here is "structure" which sounds ontological – because of the association with "mathematical structure" – and this is certainly not what Wittgenstein meant.

result, but something more modular, which we will bring to bear on the construction to help build it as we wish.

As philosophers we should be sensitive to this: in mathematics one must wait until a new mathematical edifice is erected in order to make such a decision, assertion, rule, and so on. And there are criteria for this that are part of ordinary parlance within mathematics itself, for mathematicians are very good at spotting gaps in reasoning, places where trouble with a proof-sketch might arise, and so on. The picture of the law of excluded middle alone is not enough to get at our working decisions and proofs, or to clarify them very far.

If and when the question can be decided, i.e., if a new mathematical edifice has been found that makes the problem decidable, then the former rule or postulate changes its *status*, as Wittgenstein writes in §9, "[f]or a connection is made then, which formerly *was not there*." What had been merely a rule or postulate with a justification *outside* mathematics is now transferred into something that visibly resides *inside* mathematics. This new situation shows nicely the deep asymmetry we already mentioned in Sect. 4.1 and which is beautifully described at the end of §18 as follows:

> The strange thing [*Das Seltsame*] about the alternative "ϕ occurs in the infinite series or it does not", is that we have to imagine the two possibilities individually, that we look for a distinct idea of each, and that one is not adequate for the negative and for the positive case, as it is elsewhere.

Here we run across another strange thing. Recall that, first, it is not only strange how pictures that are very suggestive can be in tension with our actual impotence to decide a certain question; it is also, second, strange that the insistence in the general validity of the law of excluded middle leads us outside mathematics. It is, third, strange that the emerging possibilities for the variety of ways of deciding questions indicate that the characteristic sort of asymmetry described in §18 is of fundamental importance within mathematics itself.

4.3 Understanding in Mathematics

These considerations show that it is not helpful to claim that one simply *understands* the alternative "ϕ occurs in the infinite series or it does not" in the Brouwerian situation. Instead, in this situation a clarification of what we are doing in using this form of words to communicate, what we are doing in attempting to apply our words, is needed. In §516 of *Philosophical Investigations* Wittgenstein repudiates the idea of a blunt reference to understanding here as follows:

> It seems clear that we understand the meaning of the question "Does the sequence 7777 occur in the development of π?". It is an English sentence; it can be shown what it means

> for 415 to occur in the development of π; and similar things. Well, our understanding of that question reaches just so far, one may say, as such explanations reach.[12]

Thus Wittgenstein has a picture of understanding that admits of qualitative variance depending upon the explanatory context, that is, what understanding contributes depends upon how it is that we *take* understanding to serve meaning in a specific situation. This implies that then, as is said in PI §517,

> The question arises: Can't we be mistaken in thinking that we understand a question?

That is to say, if there is a difference between compositionally putting together a form of words that adheres to the form of the law of excluded middle, on the one hand, and on the other, understanding, on a given occasion of use, *what* application is being made of it, then this will affect our conception of what it is possible and impossible to do with those words. Wittgenstein reverts here to an example he often used[13]:

> For some mathematical proofs do lead us to say that we *cannot* imagine something which we believed we could imagine. (For example, the construction of a heptagon.) They lead us to revise what counts as the domain of the imaginable.[14]

Note that Wittgenstein does *not* write here what he is often taken to have held, and what he did have a great tendency to think during his middle period (before 1937), when he still believed that *Satzsysteme* constituted the senses of concepts and sentences within them: that the "meaning" or "sense" of "Does the sequence 7777 occur in the development of π" *changes* once we have a new kind of proof in hand. Instead, the emphasis in the first sentence of PI §516 (quoted above) is on "seems". And the context of these remarks concerns special cases: those in which one is all too easily mistaken when believing that one understands (without further ado or explanation) what is meant (see the quotation forming the motto of the present Chapter).

This is something that does happen often enough in everyday life, after all. In answer to the question, "Does God exist"? two people might both say "Yes!", but then act in ways that lead each to say of the other, "But you don't understand this answer in the ways I do". What is distinctive about Wittgenstein's approach is that he is investigating how and to what extent such phenomena locate themselves at the very foundations of our reasoning in mathematics and logic.

Thus when he writes, "our understanding of that question reaches just so far, one may say, as such explanations reach", he emphasizes the flexible character of what

---

[12]We do not know why Wittgenstein chose the pattern 7777 here, just as we do not know why he chose 770 in RFM V §9. Of course, the specific choice is not important as long as we cannot answer the respective question, which we always presuppose: see the first footnote of this Chapter.

[13]See Floyd (1991, 1995).

[14]These remarks stem from MS 163, pp. 52r–53r, written in summer 1941, i.e., more than one year before MS 126.

can be called "understanding". In RFM V §46 it leads him to say the following about understanding in mathematics:

> "Understanding a mathematical proposition" – that is a very vague concept.

This is to say that when we investigate the notion of "understanding", we see that it is vague. And this is so even when we are dealing with what might appear to be the sharpest, least vague case: mathematical propositions.

In RFM VI §13 Wittgenstein pursues a similar line of investigation, this time in the context of an investigation of our understanding of Fermat's Last Theorem.[15] Here he asks (in 1943/1944, when the mathematicians were far away from a proof of it):

> Don't [the mathematicians] *understand* it just as completely as one can possibly understand it?

He then repudiates this naïve question by noting:

> 'Understanding' is a vague concept.

This is the same philosophical response as in RFM V §46, and it is not meant as a complaint! It expresses instead an important insight concerning our mathematical practice. It implies that all positions in the philosophy of mathematics that rely wholly upon non-vague concepts of understanding must be inadequate.

One may think at this point of certain versions of logicism (Fregean or otherwise) or set-theoretic reductionism that are based on the idea that it makes sense to talk about the "full understanding" of a mathematical proposition, where this full understanding consists in recognizing the completely analyzed form of the proposition (its 'logical form' or 'set-theoretic form'). All positions of this sort can be rejected in one full sweep if Wittgenstein is right that "understanding a mathematical proposition" is a concept that spreads and shifts in response to what we know.[16]

4.4 God and the Law of Excluded Middle?

Wittgenstein discussed the law of excluded middle in many places. Perhaps his most famous remark about it is the following one in §41 of RFM VII, stemming from MS 124, pp. 175f., written on March 23, 1944:

> Suppose that people go on and on calculating the expansion of π. So God, who knows everything, knows whether they will have reached '777' by the end of the world. But can his *omniscience* decide whether they *would* have reached it after the end of the world? It cannot. I want to say: Even God can determine something mathematical only by mathematics. Even

[15] Frege shared this problem with Wittgenstein in a letter of 6 February 1916, a response to a request for clarification from Wittgenstein that has unfortunately been lost; see Frege (2011).

[16] For an account of this notion of understanding in PI, see Goldfarb (1992).

> for him the mere rule of expansion cannot decide anything that it does not decide for us.
> We might put it like this: if the rule for the expansion has been given us, a *calculation* can tell us that there is a '2' at the fifth place. Could God have known this, without the calculation, purely from the rule of expansion? I want to say: No.

In view of what we have now heard from Wittgenstein, this passage should be interpreted as follows: When Wittgenstein talks about "mathematics" what he refers to is *our* mathematical practice with *our* language-games. This is not something where – as Hardy and Russell put it – the impossibility of producing a complete infinite expansion is similar to the medical impossibility of a mathematician having drunk no water (see Sect. 3.6). Instead, when thinking about mathematics we should take seriously the idea that our actual practices – which are in many respects limited ones, having an end with the end of the world (at the latest) – are what they are. If one wishes to say that God can go beyond them, we are owed a story that we can understand from where we are about what this "going beyond" looks like. Until this is filled in, all we have any real grip on are our practices: mathematics, such as it is and might be envisioned to be from where we are. As Austin would say, epistemology should be practiced *sub specie humanitatis*. So also for Wittgenstein. But Wittgenstein is also concerned, unlike Austin, with thinking through how it is that our uses of the word "God" might actually be thought to serve various things said. God's omniscience may be able to achieve something, perhaps as a kind of idealized model. But when God is imagined to make pronouncements about (our) mathematics, he (or she, or it) must be imagined somehow to take into account at least some of the limitations which are characteristic of it. When God is imagined to go "beyond" these in altogether unspecified (perhaps unspecifiable) ways, we are dealing with something else, and no longer with mathematics.

To put it differently: when God uses the word "mathematics" he (or she or it) shouldn't be assumed to be able to mean by this something wholly and unaccountably different from what we mean. For we are imagining him (or her or it) to engage in something entangled with our concepts. At the very least, we must use our words in speaking about this, however metaphorical and however far they may stray from mathematics. Thus in one way or another, at some point or another – which Wittgenstein does not need to specify or characterize in general in advance – we must take into account the limitations, workings, and procedures characteristic of our practices.

But now suppose that, up until the very end of the world, we were not able to decide whether a certain pattern occurs in the decimal expansion of π. Shouldn't we say, then, that in such a case it would not be *determined* whether this pattern does or does not occur in it? To this question, one should react as does Wittgenstein in §189 of *Philosophical Investigations*: "The question contains a mistake"! The mistake lies in the assumption that it is *clear* what "determined" means here. That is certainly wrong. And Wittgenstein therefore immediately responds to the question as follows:

> We use the expression: "The steps are determined by the formula ... [e.g., a formula for the expansion of π]". *How* is the expression used?[17]

When we face up to this question, we certainly find appropriate contexts in which it makes perfect sense to hold that the occurrence of the pattern should be regarded as *not* determined (e.g., when "determined" is understood in the sense of "actually decided at some point before the end of the world") as well as appropriate contexts in which the occurrence *should* be regarded as determined. Within classical mathematics, e.g., it *is* regarded as determined, in the plain sense that one *accepts the law of excluded middle*. In other words, it is the particular use of this law itself which gives sense to the term "determined", when one has in the background all the techniques that classical mathematics provides.

As we said in Sect. 4.2, the unrestricted acceptance of the law of excluded middle may be strongly motivated by our wish to have at our disposal a powerful instrument for empirical applications. We might say, therefore, that it is ultimately this motivation which lets us claim that the occurrence of the pattern in question is determined. But of course then we would have to explore a variety of other considerations: the law of contradiction; the strength of assumptions in the framework of reasoning involved; a host of alternatives, more and less developed and applied, and so on.

As we press more and more on assumptions and alternatives, we might say that the law of excluded middle is "determined" as we manage to keep embedding the claim in ever richer contexts. But at each step, the "law" is determined in the sense that we accept it, in that context, as a sufficiently justified postulate, one which we allow ourselves to make thoroughgoing use of.

We are in this sense opportunistic about our picture. And we manage to keep it going. But the important point is that this "law" depends upon *our* responsibility in applying it reasonably, in ways we can discuss, argue over, and, ultimately, connect up in some way or another with mathematics as it is practiced by us. And the fact that there is a multiplicity of mathematical ways of deciding questions must be taken into account. Even by god.[18]

---

[17] See Floyd (1991).

[18] See Floyd (2020, in press) for a more detailed discussion of RFM VI §41 applying our perspective on Wittgenstein.

Chapter 5
The Continuum of Real Numbers: The Annotations on pp. 10–30, with Commentary

> *When the idea of a cut of the* real *numbers is now introduced by saying that we simply have to extend the concept of a cut of the rational numbers to the real numbers – all that we need is a property dividing the real numbers into two classes (etc.) – then* initially *it is not clear what is meant by such a property that thus divides* all *real numbers. (RFM V §34)*

Abstract An analysis of Wittgenstein's annotations to Hardy's *Course of Pure Mathematics* pp. 10–30, with commentary on each remark. The focus is on the notion of the continuum and the diversity of techniques for regarding the real numbers. Dedekind's notion of a cut is contrasted with ordinary mathematical techniques for working with real numbers.

5.1 Measuring and π?

Let us proceed further with Wittgenstein's annotations, having finished our digression on the law of excluded middle.

After the first paragraph of his Section 5, which we quoted at the end of our Sect. 3.8, Hardy supposes "for a moment" (as we analyzed this before) that there is such a number x [greater than all the members of L and less than all the members of R (given L and R as described on p. 8 of CPM)], and he shows on p. 10 that it follows then that

> $x^2 = 2$, and that x is the number which in algebra we denote by $\sqrt{2}$.

He also says that

> this number is not rational, for no rational number has its square equal to 2.

J. Floyd, F. Mühlhölzer, *Wittgenstein's Annotations to Hardy's Course of Pure Mathematics*, Nordic Wittgenstein Studies 7,
https://doi.org/10.1007/978-3-030-48481-1_5

Furthermore he remarks that, just as he showed in the case of $\sqrt{2}$, we can accept the existence of other irrational numbers that turn up as solutions of algebraic equations. He concludes:

> We are thus led to believe in the existence of irrational numbers x and points P on [the line] Λ such that x satisfies equations such as these, even when these lengths cannot (as $\sqrt{2}$) can be constructed by means of elementary geometrical methods.

Examples of lengths of the latter sort are the side of a regular heptagon inscribed in a circle with radius 1, the length corresponding to the cubic root of 2 (which gives the solution to the Delian Problem of doubling the cube), and so on.

Responding to this on p. 10, Wittgenstein writes an annotation on the left margin:

> **[A]** It would be strange to say of any piece of wood which is ca. 3 1/7 feet long: "it is about π feet long".[1]

On the upper margin he writes:

> **[B]** It makes no sense to say "the length of this rod is $\sqrt{2}$ cm" - except you might want to say with this that it is the hypotenuse of a triangle etc.[2]

And on the lower margin:

> **[C]** Bear in mind that a 'determinate length' need not be defined by a direct measurement.[3]

These simple notes do not need extensive commentary, for what they say is not in dispute between Wittgenstein and Hardy – or, really, anyone else.

[A] is obvious in ordinary cases, though there can be exceptions. That there can be such exceptions is the point of **[B]**, which refers to another case, a case where the geometry of triangles might come into play (e.g., when the piece of wood is the bent stave of a circular barrel). Such cases show what is meant in **[C]**: statements giving lengths need not rely on direct measurements. Instead, $\sqrt{2}$ and π are convenient mathematical constructs that may be used to talk about lengths of diagonals or circumferences in another, more economical way.

The main point of this train of annotations **[A]**, **[B]** and **[C]** is that in any of these kinds of cases, infinite sequences of numbers approaching the number one wants to determine are simply not thought of. The *real* determinations of lengths – and in particular the *real* applications of analysis to geometry – do not involve such things.

What is the point of these remarks?

---

[1] Es wäre seltsam von irgendeinem Stück Holz, das ca. 3 1/7 Fuß lang ist, zu sagen: "es ist rund π Fuß lang".

[2] Es hat keinen Sinn zu sagen "die Länge dieses Stabes ist $\sqrt{2}$ cm" - es sei denn, man wollte etwa damit sagen es sei die Hypothenuse in einem Dreieck etc.

[3] Bedenke daß eine 'bestimmte Länge' nicht durch direkte Messung definiert sein muß.

Presumably, Wittgenstein wants to say that up to now Hardy hasn't yet given good reasons for introducing irrational numbers *via* sections and standard approximating infinite sequences of rational numbers. Wittgenstein is waiting for such reasons. But it is not only this. Wittgenstein also finds questionable traits in the whole theory of real numbers as it is presented in CPM (beginning with its second edition), irrespective of the way this theory is motivated. And yet: this is the standard theory, or at least one form of the standard theory, right down to the present day.

This line of critical questioning starts with Wittgenstein's annotations from p. 24 on. But before we get to this pressing on the standard presentation of the theory, we should look briefly at the annotations on pp. 11 and 15.

5.2 Here It Is Unclear What "Interpretation" Means

Wittgenstein's annotations on pp. 11 and 15 are of a rather casual nature.

Continuing in his Section 5 with his consideration of irrational numbers, Hardy presents the following problem:

> Show that if m/n is a good approximation to $\sqrt{2}$, then $(m+2n)/(m+n)$ is a better one, and that the errors in the two cases are in opposite directions. Apply this result to continue the series of approximations in the preceding example[4] (p. 11).

Wittgenstein puts a squiggly line under the word "good" and writes "any" above it.

If we understand him correctly, this annotation simply refers to the mathematics involved, because the obvious proof doesn't use the presupposition that m/n is a good approximation to $\sqrt{2}$. It runs instead as follows: Suppose that $m, n > 0$ and $(m+2n)/(m+n)$ is left of $\sqrt{2}$. We then have:

$$0 < 2 - (m+2n)^2/(m+n)^2 = (m^2 - 2n^2)/(m+n)^2 < (m^2 - 2n^2)/n^2 = m^2/n^2 - 2.$$

And analogously for the case when $(m+2n)/(m+n)$ is right of $\sqrt{2}$. This is the essential core of the proof, and it is true for any m/n with $m, n > 0$.

When making his annotation here, Wittgenstein is simply mathematically attentive. One might additionally be tempted to say that he is bothered by Hardy's prose: "a good approximation". This prose, however, seems to be quite harmless, especially in philosophical respects. Although it hangs in the air, not clarifying things very much, it is gesturing at a whole variety of placeholders that might be filled in.

The next annotation is on p. 15, in the context of Hardy's Section 8 ("Real numbers"). This begins with Hardy's definition of real numbers as sections of the rational numbers in which the lower class has no greatest member. Immediately

[4]The preceding example is: Find the differences between 2 and the squares of $\frac{1}{1}$, $\frac{3}{2}$, $\frac{7}{3}$, $\frac{17}{12}$, $\frac{41}{29}$, $\frac{99}{70}$.

after this definition (pp. 14f.), Hardy points to a characteristic ambiguity that now turns up with respect to the term "rational number".

In his Section 1 rational numbers were simply presupposed as given, as our familiar arithmetical fractions. But now they also appear as real numbers, that is as sections of the given rational numbers. Hardy's comment on this situation is as follows (p. 14f.):

> Ambiguities of this kind are common in mathematics, and are perfectly harmless, since the relations between different propositions are exactly the same whichever interpretations are attached to the propositions themselves. From $\frac{1}{2} > \frac{1}{3}$ and $\frac{1}{3} > \frac{1}{4}$ we can infer $\frac{1}{2} > \frac{1}{4}$; the inference is in no way affected by any doubt whether $\frac{1}{2}$, $\frac{1}{3}$, and $\frac{1}{4}$ are arithmetical fractions or real numbers [as just defined]. Sometimes, of course, the context in which (e.g.) '$\frac{1}{2}$' occurs is sufficient to fix its interpretation. When we say (see §9) that $\frac{1}{2} < \sqrt{\frac{1}{3}}$, we *must* mean by '$\frac{1}{2}$' the real number $\frac{1}{2}$.

The end of this passage, beginning with the word "occurs", is on p. 15. Wittgenstein puts a squiggly line under the word "*must*" and also a vertical squiggly line at the right margin together with the following remark:

> Here it is unclear what 'interpretation' means.[5]

This is a completely legitimate criticism. And yet it can be met in a satisfying way. It is a pity that Hardy didn't speak to this issue.

It is certainly *prima facie* clear that in the first interpretation of "rational number" this term is simply understood in the ordinary way: it is taken to refer to our familiar fractions. In the second interpretation it is explicated *via* sections: it is taken to refer to those sections of the rational numbers – understood according to their first interpretation – in which the lower class has no greatest member (i.e., is a real number) but the upper class has a least member. The decisive point, then, is to see the *connection* between these two interpretations of "rational number", for it is this that guarantees what Hardy emphasizes about them (p. 14):

> [. . .] the relations between different propositions are exactly the same whichever interpretations are attached to the propositions themselves.

This is established mathematically through the notion of an *isomorphic embedding*, as we shall now explain.

Let us temporarily call the set of rational numbers $\mathbb{Q}$ when the rational numbers are understood as fractions, and the set of real numbers $\mathbb{R}$ when they are understood according to Hardy's definition. Let us furthermore suppose that $\mathbb{Q}$ and $\mathbb{R}$ have their ordinary mathematical structure. Then $\mathbb{Q}$ can be "isomorphically embedded" into $\mathbb{R}$ by assigning to any m/n of $\mathbb{Q}$ the section of $\mathbb{Q}$ of which the lower class consists of all the fractions lower than m/n and the upper class of all the fractions equal or greater than m/n. This mapping is obviously injective and it respects the pertinent structures of $\mathbb{Q}$ and $\mathbb{R}$. It allows us to identify fractions with the respective sections of fractions, and in this way we can say that we simply find $\mathbb{Q}$ and its structure

[5]Es ist hier unklar, was 'interpretation' heißt.

appearing once again in $\mathbb{R}$. As Hardy says, situations of this kind are quite common in mathematics.

The description of the way $\mathbb{Q}$ is isomorphically embedded in $\mathbb{R}$ that we just gave was formulated in the extensionalist vocabulary of set theory. This may suggest that the elements of $\mathbb{Q}$ are, as it were, waiting from the outset to be considered as part of a more extensive structure. But the situation need not be understood in such a way. It can in fact be seen as compatible with Wittgenstein's non-extensionalist perspective, because Wittgenstein can interpret this embedding as a sort of *translation* of the familiar language in which we describe $\mathbb{Q}$ in the new language of sections that is now used to describe $\mathbb{R}$. Dedekind wrote of a "remodeling" of our concepts, an attitude Wittgenstein himself might be regarded as developing: we develop a new picture, or model, of the real numbers.[6]

To discuss possibilities of giving translations from one domain of discourse to another is not foreign to Wittgenstein's later philosophy.[7] In the case at hand, the appropriate translation is utterly clear. The sign "m/n" is translated into a suitable sign for the real number in which the upper class begins with m/n, and the vocabulary for the respective structural components is to be translated in a similar obvious manner. In this way the isomorphic embedding of $\mathbb{Q}$ in $\mathbb{R}$ may be described in a rather homespun manner.

It is not only a pity that Hardy didn't really clarify this topic, but also that Wittgenstein did not present it in accordance with his own point of view. But in saying this, we have made sense of Wittgenstein's annotations. For Wittgenstein seeks both a comprehensive understanding and one that is sensitive to alternative aspects of our procedures and concepts. This pluralism, sensitively implemented without giving into relativism, is the hallmark of his later philosophy of mathematics.

5.3 "Creative Powers" and Pictures: "A Point Exists" vs. "A Number Exists"

Wittgenstein's next annotation is on p. 24. Here Hardy's Section 15, headed "The continuum", begins with the following words:

> The aggregate of all real numbers, rational and irrational, is called the *arithmetical continuum*.
>
> It is convenient to suppose that the straight line Λ of §2 is composed of points corresponding to all the numbers of the arithmetical continuum, and of no others.* [*Note by Hardy: This supposition is merely a hypothesis adopted (i) because it suffices for the purposes of our geometry and (ii) because it provides us with convenient geometrical illustrations of analytical processes. As we use geometrical language only for purposes of illustration, it is not part of our business to study the foundations of geometry.] The points of

---

[6]See Dedekind (1888), p. 795, for "remodeling" language; compare the discussion in Floyd (2013a) at p. 1033.

[7]See, for example, PI §§206–208 and the discussion in Mühlhölzer (2010), pp. 397ff.

the line, the aggregate of which may be said to constitute the *linear continuum*, then supply us with a convenient image of the arithmetical continuum.

Responding to all this, Wittgenstein remarks on the bottom of p. 24:

What is actually misleading about this picture is the difference between the use of "a point exists" and "a number exists".[8]

The purpose of this remark seems to be similar to Wittgenstein's annotation "Unsinn" on p. 4 (discussed above in Sect. 3.3). Wittgenstein emphasizes the deep gap between our criteria for saying that a point exists – understood as a proposition in geometry – and for saying that a number exists.

But is there really something "misleading" involved? Doesn't Hardy make sufficiently clear how mathematicians should not be misled by this illustration? Wittgenstein here seems to preach to the converted. As we see it, his remark at this point is not very interesting or important, for it is not directed at any kind of characteristic *mis*understanding.

The situation becomes more interesting, however, when on p. 27 Hardy refers back to his explanation of a "section" of the rational numbers as a mode of division of the aggregate of all rational numbers

into two classes, L and R, such that (i) every member of R is greater than every member of L, (ii) we can find a member of L and a member of R whose difference is as small as we please, (iii) L has no greatest and R no least number (p. 9),

and when he also refers back here to p. 14, where *real numbers* are understood as nothing but certain sections as thus explained. Here is the official definition on p. 14:

A section of the rational numbers, in which both classes exist and the lower class has no greatest member, is called a ***real number****, or simply a* ***number****.*

This *identification* of real numbers with sections of rational numbers may in a certain sense be considered to be "modern", but it is by no means self-evident. Constructivists reject it because they do not accept the great generality of the notion of a "section" that Hardy has in mind. Wittgenstein also balks at this. But, importantly, it was even rejected by Dedekind himself, the inventor of the sections!

In a famous letter to Heinrich Weber from January 1888 Dedekind expounded the following view:

[Y]ou say that the irrational number is nothing other than the cut [= section] itself, while I prefer to create something *new* (different from the cut) that corresponds to the cut and of which I say that it brings forth, creates the cut. We have the right to ascribe such a creative power to ourselves; and moreover, because of the similarity of all numbers, it is more expedient to proceed in this way. The rational numbers also produce cuts, but I would certainly not call the rational number identical to the cut it produces; and after the introduction of the irrational numbers one will often speak of cut-phenomena with such

[8] Das eigentlich irreleitende an diesem Bild ist der Unterschied zwischen dem Gebrauch von 'ein Punkt existiert' und 'eine Zahl existiert'.

> expressions, and ascribe to them such attributes, as would sound in the highest degree peculiar were they to be applied to the numbers themselves.[9]

It would lead us too far astray to try to explain Dedekind's writing about the *creation* of numbers and about our respective creative *power*, which certainly sounds mysterious.[10] But there are many statements by Dedekind about numbers that show a rather down-to-earth attitude which may always have been in the back of his mind from the very beginning.

For example, in contrast to the formal analysis he eventually presents in his classic text *Was sind und was sollen die Zahlen?*, in the Preface to it he explains:

> This memoir can be understood by any one possessing what is usually called good common sense; no technical philosophical, or mathematical, knowledge is in the least degree required. I know that, in the shadowy forms [*schattenhafte Gestalten*] which I bring before him [in the subsequent text], many a reader will scarcely recognize his numbers which all his life long have accompanied him as faithful and familiar friends.[11]

By the "shadowy forms" Dedekind of course means his newly invented proxies of the familiar numbers. But why did he produce these "shadowy forms"? And how can it be that the production was not to depend upon any special technical, mathematical, or philosophical knowledge?

The answer is that Dedekind did not want to present the actual, or genuine numbers, but instead to extend the theoretical possibilities of devising more proofs in the theory of numbers on the basis of more rigorous, systematic, and generally applicable concepts. This is already hinted at in the very first sentence of his first edition Preface just mentioned: "In science nothing capable of proof ought to be believed without proof".[12]

And yet there remains "common sense" understanding of our "old and familiar" friends, even after the conceptual analysis: and so, according to Dedekind, it must

---

[9]This is the English translation to be found in Hallett (2010), p. 446. The German original (Dedekind (1930–1932), vol. III, pp. 489–490) reads as follows:

> Du sagst, die Irrationalzahl sei überhaupt Nichts anderes als der Schnitt selbst, während ich es vorziehe, etwas *Neues* (vom Schnitte Verschiedenes) zu erschaffen, was dem Schnitte entspricht, und wovon ich sage, daß es den Schnitt hervorbringe, erzeuge. Wir haben das Recht, uns eine solche Schöpferkraft zuzusprechen, und außerdem ist es der Gleichartigkeit der Zahlen wegen viel zweckmäßiger, so zu verfahren. Die rationalen Zahlen erzeugen doch auch Schnitte, aber ich werde die rationale Zahl gewiß nicht für identisch ausgeben mit dem von ihr erzeugten Schnitte; und auch nach Einführung der irrationalen Zahlen wird man von Schnitt-Erscheinungen oft mit solchen Ausdrücken sprechen, ihnen solche Attribute zuerkennen, die auf die entsprechenden Zahlen selbst angewendet gar seltsam klingen würden.

[10]See, however, for a first approximation to an explanation, the beginning of Mühlhölzer (2014b).

[11]This is the English translation (1888), p. 791. A similar sentiment is expressed in a letter Dedekind wrote to Dedekind (1932). For discussion see Sieg and Schlimm (2005), p. 154 and note 66 on p. 166.

[12]See Dedekind (1888), p. 790.

be. This is, perhaps, more generally, an instance of what later became known as Moore's "paradox" of analysis: in order to *be* an analysis, the concepts to which others are reduced must be new. But in order to be an *analysis*, it must be something, not new, but an elaboration of what we already had.[13]

Dedekind writes about all this with respect to the natural numbers, but it is quite clear that he also sees the real numbers in such a way, at least ones with which we have been familiar for centuries, like $\sqrt{2}$, π, and so on. In the quoted letter, Dedekind's explicit reason for not accepting the identification of real numbers with sections is simply that we then would have to ascribe unwanted properties to the real numbers. But another good – perhaps even better – reason would be that there are attractive alternatives for constructing the reals. In particular, there is the Cantorian one, published in Cantor (1872), in the same year as Dedekind's *Stetigkeit und irrationale Zahlen*. Cantorian reals are developed as limits of Cauchy sequences of rational numbers. This is a conception very different from the Dedekindean one, though it has the advantage of much greater generalizability and less distance from what mathematicians do.[14] In view of such alternatives, a further approach suggests itself: a purely axiomatic one in the spirit of Hilbert. Here one refrains from constructing the reals from the outset, simply listing their characteristic properties through a series of axioms. One then considers the theorems of analysis to be derivations from the axioms.

We shall not advocate one or another construal of the real numbers here. Our point is that there are clearly a number of quite *different* ways of introducing, interpreting, and construing the notion of a real number. What has been called "indifference to identification" is a crucial point in mathematics.[15] Textbooks on analysis can be distinguished by their decision to choose one or another approach.[16] As is characteristic of his later philosophy generally, Wittgenstein insists on sensitivity to the variety of aspects of the real numbers, as they show up in our procedures, analyses, and methods of proof. Our point is that he is registering something quite real within the mathematical tradition.

Yet in CPM Hardy remains silent about such alternative approaches to real numbers, sticking throughout to his identification of them with sections of the

---

[13]Floyd (2013a) connects this issue of Dedekind's "naïve" perspective with a variety of philosophical issues about simplicity and analysis.

[14]In fact, Cantor himself criticized Dedekind's cuts, writing in his (1883) that they have "the great disadvantage that the numbers of analysis *never* occur as 'cuts', but must be brought into this form with a great deal of artificiality and effort". – This remark is quoted and beautifully analyzed in Kanamori (2020, in press).

[15]For more on this see discussions in Pettigrew (2008), Shapiro (2008, 2012) and Burgess (2012).

[16]This is true down to the present day. The Dedekindian approach can be found, e.g., in Spivak (1994), Chapter 29, the Cantorian in Körner (2004), pp. 348–356, and the axiomatic approach on pp. 32–39 of Heuser (2009) (an extremely successful textbook in Germany). The historical development of the notion of real number from 1860 to 1910 is illuminatingly described in Epple (1999). Deiser (2008) gives a systematic presentation of the many alternatives for conceiving of real numbers.

rational numbers. After the definition given in Section 8, he introduces the term *arithmetical continuum* for the aggregate of *all* real numbers in Section 15 ("The continuum"). And on p. 27 he starts his Section 16 ("The continuous real variable") with a sensitive description of two points of view from which those "real numbers" may be regarded:

> We may think of them *as an aggregate*, the 'arithmetical continuum' defined in the preceding section, or *individually*. And when we think of them individually, we may think either of a particular *specified* number (such as 1, $-\frac{1}{2}$, $\sqrt{2}$, or π) or we may think of *any* number, *an unspecified* number, *the number x*. This last is our point of view when we make such assertions as 'x is a number', 'x is the measure of a length', 'x may be rational or irrational'. The x which occurs in propositions such as these is called *the continuous real variable*: and the individual numbers are called the *values* of the variable.

In the next paragraph, Hardy emphasizes that variables need not be continuous, as in the case the variable n with the positive integers as values. The paragraph after this begins with the assertion:

> Naturally 'x' and 'n' are only examples of variables, the variable whose "field of variation" is formed by all the real numbers and that whose field is formed by the positive integers.

Here, on p. 27, Wittgenstein places squiggles under the phrase "all the real numbers". And then, as if this were not enough, he puts a further vertical squiggly line on the right margin.

These squiggles are only a hint, of course. But they – and Hardy's own words – make clear enough what sort of criticism Wittgenstein has in mind. It is a sort that pervades the whole history of analysis since Dedekind's conception of what Hardy calls the "aggregate" of all real numbers. It also reflects Wittgenstein's own struggles with the general notion of a "variable". The difficulty resides in the attempt to develop a *general* point of view on what a real number *is*.

Criticism of the standard perspective within the history of analysis has typically been leveled from a constructivist perspective. Constructivists balk at Dedekind's unscrupulously general approach, in which without qualification *all* sections of the set of rational numbers are considered and quantified over.[17] Constructivists propose appropriate restrictions. Naturally, they have to *change* the Dedekindean mode of analysis. But it has become the standard one accepted by most mathematicians and is also presented in CPM (from the second edition onwards).

For all his questions about Hardy's presentation of the extensionalist point of view, Wittgenstein should not be called a "constructivist". Although he accepts many constructivist points of critique, sympathizing with the idea of considering alternative analyses and perspectives, his aim is *not* to change standard analysis. This might be the aim of a mathematician, but it is not the aim of a philosopher like Wittgenstein who, as he himself emphasizes, "leaves everything as it is" and in particular "also leaves mathematics as it is" (PI §124). Wittgenstein is not a constructivist but, rather, a *non-extensionalist*, as we called him in Chap. 2 above.

---

[17] Similarly with regard to Cantor's analysis, which is based on talking about *all* Cauchy sequences of rational numbers.

So far we have explicated the contrast between extensionalism and Wittgenstein's non-extensionalism in terms that were used by Wittgenstein in his middle phase. There, we recall, he stressed that the "all" of all numbers is connected with our inductive proofs: the schematic idea of recursion, implemented in a particular (finitistic) proof is what shows us infinitary generality. In his later work Wittgenstein takes much of this conception over, but now with a more sophisticated philosophical understanding of the best way to explore and investigate alternative conceptualizations. In the later period, his philosophy is more sensitive to particularities of both specific proofs and practices of generalization. As we have explained above in Sect. 1.3, Wittgenstein by this time had become more sophisticated about simplicity, more sophisticated about grammar and *Satzsysteme*. The issue of translating and embedding one conceptualization in another – the tos and fros of conceptual construction [*Begriffssbildung*] – are investigated with much greater care.

With regard to the real numbers, e.g., Wittgenstein writes in RFM V §37 that the extensionalist view consists of

> the idea that the real numbers are there spread out in the number line. They may be known or not; that does not matter.

This is a coarse way of putting his point, in comparison to the sophisticated attitudes held by both Dedekind and Cantor. As modern mathematicians, they took the real numbers to be autonomous, and also numbers that may, as it happens, be *correlated* with points on the straight line.[18] Whereas Wittgenstein's own position, the *non*-extensionalist view, highlights our specific access to particular numbers, which from case to case can be of very different kinds, distinguished in overlapping ways, and according to different analyses of procedures, including those of generalization.

The characteristic mistake in the extensionalist's manner of expressing herself, as Wittgenstein sees it, is that a "fiction of a procedure" is created through a manner of describing the real numbers (RFM II §22). Where certain descriptions of procedures and analogies (e.g., "remodeling") would have "a good and honest sense", the extensionalist, by contrast, "pretends to compare" sets as if they are of the same kind, but then speaks as if there is no "as if", and no differences to be accounted for at all. As Wittgenstein nicely puts it in the motto of our Chap. 2:

> The difference in kind [between the concepts of natural and real number] is represented, by a skew form of expression, as difference of extension.

This characteristic "taking over" and totalizing of the extensionalist's perspective is what Wittgenstein wishes to warn us against. And this explains why, in the passage from RFM V §37 (quoted above), Wittgenstein resisted one of the most misleading aspects of the geometric number line when presented as an "illustration" in Hardy's way: it immediately suggests, visually and directly, the extensionalist viewpoint.

As Wittgenstein sees things, this is one of those pictures, as he says in PI §115, that hold us "captive". What he has in mind are pictures that lie "in our language",

---

[18]See Kanamori (2020, in press) for a discussion of the differing attitudes between Cantor and Dedekind even here.

as he puts it: in misleading analogies, characterizations of starting points and steps, arguments, concepts and so on. What his annotations to Hardy show us is that he included here the sort of pictures that we naturally associate with certain conceptions of the real numbers, such as the picture of the straight line offered up by Hardy, and most other mathematicians and philosophers as well.

In RFM V §34 Wittgenstein understands this picture as one that contributes to the extensionalist misconception of Dedekind cuts (i.e., Hardyan "sections"):

> That every rational number can be called a principle of division of the rational numbers is perfectly clear. Now we discover something else that we can call a principle of division, e.g. what corresponds to $\sqrt{2}$. Then other similar ones – and now we are already quite familiar with the possibility of such divisions, and see them modeled by the picture of a cut made somewhere along the straight line, hence extensionally. For if I cut, I can of course choose where I want to cut.[19]

Wittgenstein is bothered by Hardy's seemingly unruffled transition from cuts that are defined by specifically given singular numbers – unproblematic from the non-extensionalist point of view – to a general notion of "all cuts", imagined as utterly uncoupled from all specific ways of producing them.

It is true that at the beginning of his Section 16 (quoted above) Hardy carefully distinguishes between a way of regarding the real numbers as members of an "aggregate", and a way taking them as given merely individually. But Wittgenstein is not content with these qualifications. The next annotations give some suggestions of what sort of criticisms he may have had in mind.

Before we consider these, however, we should have a look at the vertical squiggle on the left of the second and third line of p. 28.[20]

5.4 Examples from "Ordinary Life"?

After the statement on p. 27 that the "x" for the real numbers and the "n" for the positive integers are only "examples" of variables, Hardy stresses the great generality of the very notion of a variable itself – a logical issue. On the top of pp. 27f. he goes so far as to let "x" stand for an unlimited series of examples of objects in ordinary life, apparently struggling to make the general notion intuitive:

> Naturally 'x' and 'n' are only examples of variables, the variable whose 'field of variation' is formed by all the real numbers, and that whose field is formed by the positive

[19]The phrase "see them modeled by the picture" is our translation of Wittgenstein's "sehen sie unter dem Bild". Elizabeth Anscombe has translated this as "see them under the aspect", which is, strictly speaking, a mistranslation – and a surprising one, because to emphasize the misleading character of pictures is one of the most important themes in Wittgenstein's later philosophy.

[20]This squiggle is produced, together with p. 29, including Wittgenstein's annotations, in Nedo (2008), pp. 96f.

> integers. These are the most important examples, but we have often to consider other cases. In the theory of decimals, for instance, we may denote by x any figure in the expression of any number as a decimal. Then x is a variable, but a variable which has only ten different values, viz. 0, 1, 2, 3, 4, 5, 6, 7, 8, 9. The reader should think of other examples of variables with different fields of variation. He will find interesting examples in ordinary life: policeman x, the driver of cab x, the year x, the xth day of the week. The values of these variables are naturally not numbers.

Wittgenstein's vertical squiggles on the left of the second and third line of p. 28 refer to the assertions in the final two sentences, Hardy's directive to "the reader". As mere squiggles they provide no direct or explicit criticism. Yet it seems obvious that Wittgenstein takes offense at this mixture of mathematical and non-mathematical contexts, and clear to us why. To confront difficult conceptual differences between notions, to explore the notions of "variable" and "object", was always one of the most important traits of Wittgenstein's philosophy of mathematics, early middle and late. It is no wonder that he marks passages that threaten to level out or obscure the difficulties and transitions here.

It was then remarkable for us to learn, in double-checking the actual original 1941 edition,[21] that the "examples in ordinary life" – which Hardy presented from the first to the eighth edition (the latter is the one where Wittgenstein made his annotations) – totally disappear in the ninth and all subsequent editions. Only purely mathematical examples remain.

We do not know Hardy's reason for this alteration. Did he react to a Wittgensteinian criticism? As far as we know, he never says anything about this.[22] Although the deletion of these examples from "ordinary life" is striking, we do not know whether it was Wittgenstein's criticism, or word of a certain kind of Wittgensteinian criticism, that led Hardy to alter his text. But, be this as it may, Hardy's aside to "the reader" does not help at all to clarify the notions of *variable* and *field of variation* in the mathematical context.

5.5 The Role of the Examples, That Is Our Problem

On the upper and the left margin of p. 28, respectively, referring to Hardy's whole Section 16, "The continuous real variable" (whose essential assertions we presented

[21] And not the 2008 reprint by Cambridge University Press of the tenth, 1952 edition, CPM_{10}.

[22] The very short "Preface to the Ninth Edition", for example, doesn't help. It reads thus:

> I have altered some passages in the first two chapters on account of criticisms made by Prof. H. Davenport [a number theorist who in 1943, when Hardy wrote this Preface, worked at the University of Manchester]. Otherwise the text is unchanged except for the correction of some minor errors and the insertion of a few additional references.

in our Sect. 5.3), Wittgenstein writes:

[A] Whoever teaches someone to plane planks doesn't teach him thereby to make tables and chairs, not even in general, but rather he teaches a technique of very general application.[23]

[B] A mode, or form, of reasoning is expounded which will later be used. One might say: considerations according to this schema will often be useful later on.[24]

We take these notes to refer to the general idea of a section.

In the first remark, **[A]**, Wittgenstein says that the teaching of this idea doesn't automatically grant us access to specific numbers like $\sqrt{2}$, π, etc. To teach the latter is something very different, analogous to the way in which teaching someone about mass-manufacture of tables and chairs is something very different from teaching someone about how to plane planks. And it is true: to teach students about specific real numbers is something very different from discussing real numbers in general. To invoke an analogy Wittgenstein framed elsewhere, in the context of his criticisms of logicism's foundational aspirations: teaching people about cabinet-making is not simply a matter of teaching them how in general to glue pieces of wood together.[25]

[B] enunciates this same point without reference to teaching, but instead to the subsequent uses of teaching after it is assimilated. With the general conception of a section, a mode or form of reasoning is expounded which will be used later on, but this form is far away from the familiar reasoning about specific numbers like $\sqrt{2}$, π, etc.

This sounds plausible enough. But how can it amount to a criticism of Hardy, as it is undoubtedly intended to be? At the beginning of his Section 16, Hardy himself emphasizes the contrast between the generality of the notion of real numbers (i.e., essentially sections) and the way we access specific numbers like $\sqrt{2}$, π, etc. Furthermore, in Wittgenstein's annotations (as we interpret them) there is nothing like a rejection of the general notion.[26] What he says is merely that the generality of the notion in terms of "sections" functions differently from the specific notions that underlie our reasoning about $\sqrt{2}$, π, etc. But then: what precisely *is* his critique?

[23]Wer einen Bretter hobeln lehrt, lehrt ihn damit nicht Tische und Stühle machen, auch nicht allgemein, aber er lehrt eine Technik von sehr allgemeiner Anwendung.

[24]A mode, or form, of reasoning is expounded which will later be used [this is in English.]. Man könnte sagen: Überlegungen nach diesem Schema werden später oft von Nutzen sein.

[25]Compare RFM V §24, quoted below at Footnote 29.

[26]In Putnam (2007), p. 243, the question is raised: "Is the assertion that there is a set of all real numbers simply a useless and misleading picture?", and Putnam seems to think that assertions of this sort are enunciated and endorsed by Wittgenstein. According to our interpretation, this is a misunderstanding. Wittgenstein does not reject the *notion* of 'all real numbers'. Accordingly, he can also easily accept the 'set' of all these numbers. What he wants to do instead is only to warn us about certain *pictures* we very naturally – perhaps inevitably – associate with our phraseology and which tend to mislead us. See Floyd (2020) for more an Putnam's misreading of Wittgenstein on this point

In his lecture "Philosophy for Mathematicians" in 1932/33, immediately after a long quotation from just this part of CPM, Wittgenstein had asked:

> [H]as the word "section" any meaning that passes beyond specific powers and specific numbers, e.g., $x^3 < 2$? Are Hardy's examples only for beginners, so that we could have the general calculus without these? Are the examples *essential*? (AWL, p. 215)

But the way in which Wittgenstein then went on to discuss these questions was – so far as the notes we have go – all too brief and rather lame. Ten years later, he makes a new attempt at answering the questions better. But the way he does this is not easy to understand. Sometimes he seems to be unsure of himself, unclear about the real substance of his own critique. As we have seen above, he warns of certain pictures like the geometrical picture of the straight line. Finally on p. 57 of MS 127 he explicitly asks, Why are such pictures *misleading*? That is, Why aren't they just innocuous heuristics?

He immediately replies: "Among other things, they conceal certain problems." – But this is not of much help. We have to work carefully to extract the reasons that might be given for such a response.

We can say this. The picture of the straight line is misleading when (for example) it lets us understand the notion of a "section" in a more straightforward way than is really appropriate. This is precisely what is hinted at in the passage of RFM V §34 (quoted in Sect. 5.3 above). As we have seen in Sect. 3.7, Wittgenstein warns us about Hardy's idea of geometrical *illustrations* because these may suggest altogether phantasmagoric *applications* of analysis to geometry. Immediately after this warning he writes:

> The idea of a 'cut' [i.e., 'section'] is one such dangerous illustration.[27]

What he refers to here is not the technical term "section" as defined on pp. 8–14 of CPM$_8$, but instead the geometrical *image* associated with it. His phrase "idea of a 'cut'" may resonate with Hardy's "idea of a 'section'" in fn. * on p. 28 of CPM$_8$. Here Hardy explains that this idea was

> first brought into prominence in Dedekind's famous pamphlet *Stetigkeit und irrationale Zahlen* [. . .]

and adds that it

> is one which must be grasped by every reader of this book [i.e., CPM].

Now Dedekind had in fact initially explained this idea purely geometrically, just where he was first explicating what he calls the "essence of continuity", something he enunciates in language that he himself puts into scare quotes:

> "If all points of the straight line fall into two classes such that every point of the first class lies to the left of every point of the second class, then there exists one and only one point which produces this division of all points into two classes, this severing of the straight line into two portions" (Dedekind (1872), p. 771; Dedekind's own quotation marks in the original).

[27] MS 126, p. 128; in RFM V §29. Compare MS 162a, p. 30 from 1938.

Unfortunately we do not know whether Wittgenstein ever read this important Dedekindean text, though he does mention Dedekind explicitly in remarks we will canvas below (RFM V §40). We may only assume his familiarity with Hardy's book. Nevertheless, we can use our knowledge of Dedekind to explain what may have been going on in his annotations to Hardy. The idea of a cut, i.e., a section, can be seen as a "dangerous illustration" for the following two reasons.

First, the original examples that led to sections of the rational numbers are *modes of division*. This is Hardy's own term: see, e.g., pp. 11, 14 and 28, based on certain techniques of constructing decimal expansions of singular numbers like $\sqrt{2}$, π, etc. But then one quickly detaches oneself from such techniques and talks about "sections" *in general*, about *all* sections of the rational numbers – as if the step to them were an altogether innocuous one. For the familiar singular numbers like $\sqrt{2}$, π, etc. – and accordingly for the sections corresponding to them – what we possess are identity criteria which are rather down-to-earth. But then we factor these out when we subsume the numbers under the general notion of a *section*. This is the step to the *extensionalist* standpoint: a cut is just a kind of set. We create a totally different, remodeled conceptual situation. There is here a deep conceptual gap to be bridged. And the picture of the straight line may simply mislead us in our attempt to negotiate it.

The second reason for the dangerousness of the geometrical idea of a section turns up when we apply it not only to the rational, but also to the real numbers. Wittgenstein alludes to this in his annotation on the bottom of p. 28, which we will discuss in Sect. 5.6.

Hardy's geometrical imagery lets the extensionalist standpoint appear all too natural. And this is certainly a defect of his exposition. A more serious defect of Hardy's exposition is the all too unreflective step into extensionalism, a step opened up by geometrical imagery. We should try to clarify, therefore, that which Wittgenstein deems as problematic in extensionalism as such, irrespective of all geometrical allusions and underpinnings.

Extensionalism suggests that the real numbers in their entirety, as an "aggregate" – as Hardy calls it – are simply *there*, waiting to be dealt with. In reality, however, within our mathematical practice of presenting this "aggregate", we speak and think about it in markedly schematic terms, similar to the way in which we speak when using the schematic language of formal logic in presenting logicist reductions of mathematics.

Wittgenstein hints at this analogy between logicism and extensionalism in the following passage of RFM V §40, a passage that originated in MS 127, pp. 57–60 and which spells out a bit more what may be "dangerous":

> Dedekind gives a general pattern of expression; so to speak a logical form of reasoning.
>
> A general formulation of a procedure. The effect is similar to that of introducing the word "correlation" with a view to the general definition of functions. A general way of talking is introduced, which is very useful for the characterization of a mathematical procedure (as in Aristotelian logic). But the danger is that one will think one is in possession of the complete explanation of the individual cases when one has this general way of talking (the same danger as in logic).

> We determine the concept of *the rule* for the construction of a non-terminating decimal further and further.
>
> But the content of the concept?! – Well, can we not complete the construction of the concept as a receptacle for whatever application may turn up? May I not complete the construction of the *form* (the form for which some content has supplied me with the *stimulus*) and as it were prepare a form of language for possible employment? For, so long as it remains empty, the form will contribute to determining the form of mathematics.
>
> For isn't the subject-predicate form open in this way; and waiting for the most various new applications?
>
> That is to say: is it true that the whole difficulty about the generality of the concept of a mathematical function is already to be found in Aristotelian logic, since we can no more survey the generality of propositions and of predicates than that of mathematical functions?

Here Wittgenstein sees danger in the idea that grammatical form *per se* can determine a certain sort of content, and danger also in the idea that a certain content determines altogether the form that is relevant to individuating it. He also worries about presuming that we are in final possession of all the forms we will ever need. He is wondering – here in connection with Dedekind – about the *grammatical* (in his broad sense, *logical*) significance of the fact that a mathematical procedure has been set up and works smoothly.

The parallel he draws, and worries about, is this. Aristotelian logic's "subject-predicate" structure works smoothly with its procedures. But what about the further idea that its syllogistic forms are empty "forms", unprejudicial and available for extension into every possible context? This is truly a dangerous assumption – one which Kant infamously made when he wrongly declared, in light of syllogistic logic, that general logic is a "completed science".[28] Frege's extension of the mathematical notion of function across all of language refuted Kant's claim, or at least made the efforts to hold onto this view of the priority of syllogistic logic problematic. But one could also equally well wonder whether his notion of "concept" – or indeed Dedekind's notions of "system" and "cut" – will serve foundationally, so to speak, all possible extensions into the future. Frege's unbridled notion of *concept* got him into trouble, after all, with his Axiom V: there *are* dangers of contradictions. More generally, there is something portentous and perhaps ill-omened about the intrusion of logic into mathematics: the specific differences and similarities among the variety of mathematical techniques may be obscured or mis-categorized or glossed over.[29]

---

[28] Kant (1998), Preface to B edition, at viii.

[29] This is the point of the notorious remark Wittgenstein made at RFM V §24:

> "The disastrous invasion" of mathematics by logic.
>
> In a field that has been prepared in this way *this* is a proof of existence.
>
> The harmful thing about logical technique is that it makes us forget the special mathematical technique. Whereas logical technique is only an auxiliary technique in mathematics. For example it sets up certain connections between different techniques.
>
> It is almost as if one tried to say that cabinet-making consisted in gluing.

It is worth recalling here just how central to Wittgenstein's early philosophy these questions were. A useful example is the controversy surrounding Frege's effort to draw a fundamental distinction between the general notions of *concept* and *object*. Kerry objected to Frege that statements such as the following are problematic, given Frege's sharp and general distinction between concepts and objects:

> The concept *horse* is a concept easily attained.

Here "the concept *horse*" is *used* as an object expression. Yet what it refers to is, presumably, a concept. – Kerry argued that this defeats the universality and fundamentality of Frege's own dichotomy.

In response, Frege admitted that his language constantly falls into figurative use as he attempts to draw this fundamental distinction; in spite of his best efforts, he asks his reader for a "pinch of salt" in using his characterizations to grasp what he tries to convey.[30] One lesson Wittgenstein drew from Frege was this: when it comes to the basic, fundamental logical distinctions, one must always apply to one's own language consideration of the standpoint from where a dichotomy is drawn. Frege himself spoke of "elucidations" [*Erläuterungen*]: metaphors, hints, winks and so on that will help an audience catch on to the fundamental notions.

Russell faced this difficulty of fundamentality differently. He drew no dichotomy between "concepts" and "objects", insisting that any logical form could figure as the subject of a sentence, something that could be directly and immediately grasped. Yet Russell had explicitly worried about how many different "forms" post-Aristotelian logic might require, i.e., how many different kinds of relations logic might have to be able to express, and how on earth we could be in a position to anticipate answers to these worries.

In the *Tractatus* Wittgenstein had attempted to forge a path between the Fregean and the Russellian perspectives. He allowed that there *is* a general form of proposition, and this form embodies no dichotomy between concepts and objects. And yet he insisted that logic cannot and must not anticipate: it must not take any stand on the number or kinds of "forms" that logic might have to confront (5.554ff.) The latter is to be figured within empirical and conceptual needs and logical developments as they arise. Our grasp of "forms" must come through reflection on our uses of words and sentences in inferences: it is not a matter of immediate "acquaintance" in Russell's sense, the kind that would take us into all future contexts.

In the above-quoted remarks on Dedekind, Wittgenstein has of course surrendered many Tractarian ideas (see Chap. 1). One point he is insisting on, however – just as he did in the *Tractatus* – is that individual cases and step-by-step procedures must be accounted for at the foundation. "The theory of classes", he had written, is "completely superflous" for the foundations of mathematics (6.031). And when a totalizing perspective sets itself up, one had better be careful about assuming that one is in possession of one, overarching way of thinking that is now *necessarily*

---

[30]Frege (1892a), p. 205.

fundamental. One must take into account the uses of one's own words when enunciating a logical distinction and make sure the perspective is thoroughgoing.

Now in the case of the extensionalist viewpoint we can say this much. If we imagine not being able to appeal to the dimension of the point of what is being done in a particular context – and thus a human context – we may delude ourselves into thinking that all that is necessary for understanding and meaning may be packed into a form of words alone. Under the sway of such a delusion, we may come to imagine that we may "prepare a form of language" – here the concept of a *rule for the construction of a non-terminating decimal* – as if it may serve as "a receptacle for whatever application may turn up".

Yet as we have seen, it is *this* that Wittgenstein is questioning: all the way through his annotations to CPM and the surrounding manuscripts, he rejects this whole idea of a "form of language" as something relevant to the foundations of logic and mathematics. The very notion of a "form of language" is far more plastic, in the dimensions that interest him, than we might suppose.

There are countless examples where Wittgenstein expounds this idea. An especially interesting and deep place to where we see it developed and applied occurs in a tantalizing and insightful remark from 1947. In this remark Wittgenstein revisits the diagonal argument in Turing's paper "On computable numbers, with an application to the *Entscheidungsproblem*" (1936/1937) – the *specific* one Turing used to show the unsolvability of the *Entscheidungsproblem*. This argument did not invoke the law of excluded middle, but instead works *directly*, without appeal to any specific logical forms or structures (such as negation).[31]

For Wittgenstein what was most significant here is that

> Turing's "machines": these are *humans* who calculate.[32]

The key to Turing's analysis of the notion of a "logical step" was his *comparison* between his "machines" (which are mathematical objects) and the step-by-step behavior of a human calculator. Adapting Turing's own diagonal argument – as if he knew it from memory – to apply to "the form of the law for the development of decimal fractions", Wittgenstein argues that when we imagine all the rules for such expansion placed on a list, we have created the wherewithal for a rule to be defined that we cannot follow in the same way we follow the rules on our list. Thus we really should admit that, though definable, this rule is utterly different in character from any we have so far framed.

For Wittgenstein this kind of limitative argument has the following significance:

> That we *calculate* with some concepts and with others do not, merely shows how different in kind conceptual tools are (how little reason we ever have to assume uniformity here).[33]

---

[31] See Floyd (2012c) and Sect. 8.5 below.

[32] See RPP I §§1095–7, from MS 135, p. 118.

[33] RPP I §1095; see Z §347, drawn from MS 135, pp. 117–118.

The particulars of this later remark are discussed by one of us elsewhere.[34] But the general, "non-extensionalist" train of thought appears to be in evidence already in the annotations to CPM, i.e., four to five years earlier. For on pp. 131f. of MS 126, Wittgenstein explicitly says – as he does in the annotations we are discussing – that he *accepts* the Dedekindean "phraseology" because of its potential usefulness. Yet he also adds interesting remarks about the relation of this "phraseology" to *individual* real numbers.

What he has in mind are our paradigmatic examples of irrational numbers like $\sqrt{2}$ and π. And these throw more light on the "danger" he warns against here. As he writes,

> How would it be if I said: the general theory of real numbers makes ready a phraseology which then in the special case can be of great use. – But by making this phraseology ready, it [the phraseology] is either a self-contained piece of mathematics or it can treat the real numbers in *vague* generality by examples. The exactness would lose nothing thereby, for the *application* of these general pointers to each special case would always establish complete determinateness (see MS 126, pp. 131f.).

This passage deserves careful explication.

What follows after the introductory "How would it be if I said" is a general concession to the extensionalist phraseology mentioned: that it may be of great use in special cases. Wittgenstein does not say what this use might be, but he does say something, albeit different. He admits that the "phraseology" might be a self-contained piece of mathematics. But at the same time, it awaits our application of it to specific real numbers.

Wittgenstein has the tendency to think – as further passages quoted below will show – that this sort of application is also required to give the "phraseology" real substance. Such would be Wittgenstein's comment upon Hardy's talk, at the beginning of Hardy's Section 16, about "two different views" from which the real numbers may be regarded: the view of an *aggregate* of all the real numbers, called the "arithmetical continuum", or the view of singular numbers *individually*. These two views or aspects of the continuum may be taken up.[35]

But there is a principled asymmetry between them. According to the aggregate view (which represents extensionalism) the real numbers possess only *schematic* properties; according to the individual view, the properties assigned to the numbers are very *specific*: they appear as solutions of certain equations, as having characteristic geometrical meanings, or what have you. This latter is a *non*-extensionalist perspective on the numbers, because it refers to the specific ways in which the numbers are given.

Furthermore, the characteristic *precision* of the extensionalist phraseology is not relevant to its application to the special numbers, and so its generality with regard to individual numbers is only a "*vague*" one, as Wittgenstein says. When

---

[34]Floyd (2012c).

[35]For a general discussion of the theme of aspects see Floyd (2010, 2012a, 2018a) and Sect. 8.3 below.

applying the phraseology to individual numbers, the precision – in the sense of completely determining the numbers – stems from the different ways the numbers are given, depending on the specific case at hand. Accordingly, the gap between the extensionalist's phraseology and the individual numbers appears deep, and Hardy's exposition, though mentioning the difference, doesn't really face up to it.

On p. 10 of MS 127, Wittgenstein writes:

> The extensional explications of the functions, the real numbers, etc. pass over everything intensional – although they presuppose it.[36]

Hardy would certainly agree, if this passage is appropriately understood (and reformulated accordingly). It is the official extensional *definitions* of the functions and the real numbers which pass over everything intensional. And yet the *explications* of these definitions presuppose intensional approaches to individual functions and numbers. Precisely this is the usual approach, and it is also present in Hardy's book. Wittgenstein is attending to what Hardy *does* in his book, and seeing how it measures up against what he *says*.

Yet Wittgenstein tends to go farther than this. Isn't the extensionalist phraseology of numbers and functions – like the general technique of planing, which aims at the production of individual tables, chairs, etc. – ultimately aimed at individual mathematical entities: at specific numbers and functions? And here the latter would of course all be considered non-extensionally, i.e., as given to us in specific ways. Isn't this is the actual *raison d'être* of the phraseology?

As far as we know, Wittgenstein never goes so far as to explicitly state this view, but he comes near it. At least we can say that he sees a philosophical problem at precisely this point. On p. 144 of MS 126, he writes:

> The role of the examples, that is our problem. And this shows of which kind the philosophical problems of mathematics are.

And on p. 23 of MS 127:

> The irrational numbers are – one might say/so to speak – individual cases.[37]

Although these passages are only shots in the dark, they at least bespeak the sort of disquietude Wittgenstein feels in the present context.[38]

---

[36]The German original reads:

> Die extensionalen Erklärungen der Funktionen, der reellen Zahlen, etc. übergehen alles Intensionale – obwohl sie es voraussetzen.

This passage appears in RFM V §35, but with a mistranslation: instead of "extensional explications" – which would correspond to Wittgenstein's "extensionalen Erklärungen" – Anscombe chose "extensional definitions". But the extensional *definitions* of the functions and the numbers do not presuppose intensional approaches to them; this is only so when one *explicates* these definitions.

[37]This is in RFM V §37. Again, we have modified Anscombe's translation.

[38]The concentration on individual numbers and functions presumably belongs to Wittgenstein's fight against the direction of mathematics as it is represented in CPM; see pp. 129f. of MS 126, as

Incidentally, one should note that immediately after the passage about his fight against a certain direction in mathematics, Wittgenstein writes:

> However strange it sounds: The further expansion of an irrational number is a further expansion of mathematics.[39]

This passage is on p. 129f. of MS 126. Astonishingly, it also turns up in RFM V §9, a section which is otherwise constituted by notes that occur on pp. 55–60 of MS 126. The context of these pages suggests an intimate connection with Wittgenstein's rule-following considerations. Wittgenstein's assertion that the further expansion of an irrational number is a further expansion of mathematics has in fact received impressively extensive interpretations in terms of these considerations (see Wright (1980), pp. 40f. and 145). In reality, however, the context in which this assertion actually appears is far from that of "rule-following". It is the context of Wittgenstein's struggle against the dominance of the extensionalist phraseology and on behalf of the appreciation of individual numbers. This context suggests a much more down-to-earth interpretation: *if* one is interested in individual numbers (for example in an individual irrational number), *then* the further expansion of such a number would be considered as a further expansion of mathematics.

All this throws further light on Wittgenstein's very first annotation to Hardy's CPM, the one where the applications of analysis are called "essential". This can now be better understood as follows. If we look at analysis as it is really *done*, including the usual explications of its terms and our subsequent actual understanding of them, its general phraseology – "sections", and nowadays a whole apparatus of set theory – has to be understood together with its application to individual numbers and functions; only then does the phraseology get real traction and substance within analysis. When Wittgenstein refers to *geometrical* applications in particular (in the passage on p. 127 of MS 126, which we have quoted in Sect. 3.1 and which turns up in RFM V §29), he certainly has in mind numbers like $\sqrt{2}$ and π or the trigonometric functions, with their obvious geometrical meaning.

Yet in his first annotation he is, or appears to be, more open. There he only mentions applications in general, whether geometrical or not, and this may be the appropriate wording. Perhaps geometry can be totally avoided in the practice of analysis. This may seem perverse – though it actually has been propagated, e.g. in the Bourbaki school. But examples of individual numbers and functions *are* necessary for an adequate understanding of *all* "fields of variation". We can view these examples as *applications* of the general phraseology, and then Wittgenstein

---

quoted in a footnote above. That Wittgenstein may not be alone in this fight is documented, e.g., in the following assertion by the mathematician Carl Ludwig Siegel, as reported on p. 53 of Weil (1992) (with thoroughgoing emphasis in the original!): "*Ein Bourgois, wer noch Algebra treibt! Es lebe die unbeschränkte Individualität der transzendenten Zahlen*" [*It's a bourgeois, who still does algebra! Long live the unrestricted individuality of transcendental numbers*].

[39]In the German original: "So seltsam es klingt: Die Weiterentwicklung einer irrationalen Zahl ist eine Weiterentwicklung der Mathematik".

is certainly right in saying that they are "essential": they are essential to our mathematical practice as it is, which without doubt goes far beyond the phraseology.

Wittgenstein has the tendency *not* to consider said phraseology as essential to analysis: it is a piece of mathematics which he accepts, but not an essential basis of analysis. If we understand him correctly, he reproaches Hardy for suggesting the opposite view: that the phraseology is essential and that the reference to individual numbers might be regarded as only illustrations of what in principle the phraseology already contains. According to Wittgenstein, this point of view turns the real situation upside down. For him the touchstones of individual cases are essential for determining our words about what is going on, while the phraseology presents an additional mathematical development. He admits that this development can be useful, also useful with respect to the individual cases – as planing is useful for making tables and chairs – but it is not "essential".[40]

We are not sure that Wittgenstein's reproach of Hardy is justified. Hardy seems to be rather anxious to emphasize the two aspects at issue – the approach *via* individual numbers and the general Dedekindean approach *via* sections of the rational numbers. And in any case it seems rather plausible to take *both* of these to be essential to analysis. They represent a certain sort of conflict, but it seems to be a conflict characteristic of analysis: it exists between (a) the consideration of real numbers as *individual* entities, each with its specific properties, and (b) the view of the whole aggregate of real numbers as a *homogeneous* one as exemplified by the geometrical image of the continuum.

Paul Bernays, in his (1957), p. 4, has given a beautiful description of this conflict, as already quoted above, near the end of Chap. 2. It seems to us that Wittgenstein would have been well advised to simply accept it without wishing to play the aspect of the individual numbers off against the aspect of homogeneousness (in fact his notion of "aspect" expresses a plurality of perspectives having a certain unity as an object of perception all the same). As the concept of the *arithmetical* continuum shows, one can take account of this homogeneousness without falling prey to geometrical images.

Bernays explicitly recognizes the role of the individual numbers as measures of magnitudes, i.e., as involving the application of analysis, and this would presumably make his point of view especially attractive to Wittgenstein. As we have seen (compare Sect. 5.3 above) Dedekind too insists on *not* identifying our 'familiar' real numbers like $\sqrt{2}$ and π with his cuts of rational numbers. With the cuts he captures the continuum in a non-geometrical way – that is, to use Hardy's terminology, he conceives of it as the arithmetical continuum. And yet he doesn't seem to bother much about this sort of distinction. And in fact: doesn't this double-aspect view simply give a realistic account of a characteristic tension or plurality of perspectives in our actual practice of arithmetic? According to his own philosophical

[40] This attitude appears to be similar to what Wittgenstein conveys in RFM III about logicist (or set theoretic) reductions of mathematics: he doesn't dispute that the logicist phraseology may have its own value, but *not* as an essential basis of mathematics; see the exposition in Mühlhölzer (2010).

perspective, Wittgenstein need not attempt to reconcile the conflicting tendencies – as Bernays actually demands – at least not in a mathematical way. It is clear from what we have seen that Wittgenstein is more concerned about the extensionalist perspective subsuming and occluding the non-extensionalist one than he is about the non-extensionalist perspective giving us a truncated perspective. And in the end a sensitive description of our actual mathematical practice, paying careful attention to differing remodelings of the real numbers, just might be enough.

5.6 Picturing “Gaps” and “Cuts”

After his definition of sections of the rational numbers, having identified real numbers (= numbers) with sections the lower classes of which do not have a greatest element, in §17 (“Sections of the real numbers”) Hardy introduces sections of these real numbers analogously. The best characterization of these occurs in Hardy’s formulation of *Dedekind’s theorem* on p. 30.[41] According to it, such a section is a division of the real numbers into two classes L and R in such a way that (i) every number belongs to one or other of the two classes, (ii) each class contains at least one number, (iii) any member of L is less than any member of R.

Of course, with this same formulation, by merely replacing “real numbers” by “rational numbers”, the sections of the rational numbers are also defined. In their case, the introduction had proceeded as follows – and we now revert to the (“dangerous”) image of “gaps” between numbers, for this was done by Dedekind and many others, and it is very suggestive, although Hardy avoids it, at least explicitly. Examples of certain individual numbers showed that we have to acknowledge important gaps in the aggregate of rational numbers, called “irrational” numbers; a gap of this sort can be identified with what is left to it, a class L, and what is right to it, a class R – and this is a really beautiful idea! Such a gap is nothing but a section of the rational numbers where L doesn’t have a greatest and R doesn’t have a least element. In a subsequent step one identifies the rational numbers themselves with sections, now consisting of classes L and R where R has a least element (= the rational number with which this section is identified). And with this move we have immediately gotten at the totality of *all* real numbers, of irrational and rational ones: namely, as *all* sections of the rationals – in the original sense of this term – whose lower classes do not have a greatest element.

When considering sections of the real numbers themselves, we may revisit the idea of “gap” again. The introduction of real numbers as just described had the aim of “closing the gaps” in the aggregate of rational numbers, but now the question arises whether such gaps may still be present in the aggregate of real numbers. It was

[41] Hardy’s own explications of the central notion of a “section” are surprisingly diverse (compare pp. 11, 14, 28 and 30), but they are always meant to define the same sort of division of an aggregate, be it the aggregate of the rational or that of the real numbers.

Dedekind's idea to understand the so-called *continuum* as precisely *not* containing such gaps anymore.

Initially, Dedekind presented this idea geometrically, with respect to the straight line. But then he transferred it into the purely arithmetized context of the rational and real numbers, as we find it now also in CPM. The aim was to show that in this way the aggregate of the real numbers turns out to be a continuum. As Dedekind before him, so also Hardy shows that this aim is achieved; the proof is on pp. 28–30. He states this central result on p. 30 and calls it *Dedekind's theorem*.

In the formulation of this theorem, Hardy does not say that the lower and upper classes L and R of the sections considered are defined by any "properties" of the reals; but on p. 28 he expresses himself precisely in such a way. He writes:

> Let us suppose that P and Q are two properties which are mutually exclusive, and one of which is possessed by every real number. Further let us suppose that any number which possesses P is less than any which possesses Q. We call the numbers which possess P the *lower* or *left-hand class* L, and those which possess Q the *upper* or *right-hand class* R.

In smaller letters Hardy explains this with the example of the real number $\sqrt{2}$:

> Thus P might be $x \leq \sqrt{2}$ and Q be $x > \sqrt{2}$. It is important to observe that a pair of properties which sufficed to define a section of the rational numbers may not suffice to define one of the real numbers. This is so, for example, with the pair '$x < \sqrt{2}$' and '$x > \sqrt{2}$' or (if we confine ourselves to positive numbers) with '$x^2 < 2$' and '$x^2 > 2$' . Every rational number possesses one or other of the properties, but not every real number, since in either case $\sqrt{2}$ escapes classification.

In response to all this Wittgenstein now remarks on the lower margin of p. 28:

> One can divide the rat. numbers into greater and smaller in such a way that the division functions like a cut [i.e., a Hardyan *section*], but one cannot divide the real numbers in the same way. One can cut them, but the cut doesn't bring about such a division.[42]

We interpret this remark[43] as follows: With the term "division" – a term used by Hardy himself when he explains sections as 'modes of division' (on pp. 11 and 28, for example) – Wittgenstein means procedures and step-by-step techniques such as the purely calculative technique to decide for any given positive rational number p/q whether $(p/q)^2$ is below or above 2. This corresponds to respective properties P and Q of the rational numbers. In this case, P is true of p/q iff

$$p/q \leq 0$$

or

[42]Man kann die rat. Zahlen so in größer und kleiner einteilen, daß die Einteilung wirkt wie ein Schnitt, man kann aber nicht die reellen Zahlen ebenso einteilen. Man kann sie schneiden, aber der Schnitt bewirkt keine solche Einteilung.

[43]It is in a way anticipated in formulations of the Lecture "Philosophy for Mathematicians" from 1932/1933 (see AWL, pp. 219f.). These remarks, however, are not worth presenting here because they possess much less precision than Wittgenstein's subsequent annotations and manuscripts.

$$(p/q)^2 < 2$$

for

$$p/q > 0;$$

Q is true of p/q iff $p/q > 0$ and $(p/q)^2 > 2$; and these properties give rise to the respective sections

$$L_{\sqrt{2}} = \{p/q \mid p/q \leq 0 \text{ or } (p/q)^2 < 2 \text{ for } p/q > 0\}$$

and

$$R_{\sqrt{2}} = \{p/q \mid p/q > 0 \text{ and } (p/q)^2 > 2\}.$$

This sort of reasoning is understood by Wittgenstein in the rather down-to-earth sense that there exists a mathematically respectable procedure – in the case of $\sqrt{2}$ consisting of simple calculations – according to which it is decided whether p/q belongs to $L_{\sqrt{2}}$ or to $R_{\sqrt{2}}$.

Hardy would of course say that not every section of the rational numbers is defined in such a way, but important ones are, and they in fact belong to Hardy's paradigmatic examples of sections of the rational numbers. In the case of the real numbers, however, there is *never* such a way! There is simply no mathematically respectable procedure to decide for every real number whether it belongs to the lower or the upper class of a section of the real numbers.

Cantor would say that there are "too many" real numbers to allow this. But Wittgenstein, instead, would say that the expression "every real number" is not apt for procedures. To think otherwise would involve phantasmagoric ideas of mathematics, as well as phantasmagoric ideas of procedures. It would look from this perspective as if infinite decimal expansions of real numbers were quite similar to finite expansions. It would also look as if our inability to produce such expansions were only a sort of weakness on our part.

In fact mathematicians often express themselves in such a way. In Chap. 2 we already quoted the set theorists Hrbacek and Jech's statement that

> it is possible to write down decimal expansions $0.a_1a_2a_3 \ldots$ where "$\langle a_i \rangle_{i=1\ldots\infty}$" is an arbitrary sequence of integers between 0 and 9 (Hrbacek and Jech 1999, p. 86).

Now of course in one sense, they clearly know better. Yet, seemingly, they do not care about such peanuts. This is to a certain extent also true of Hardy. And Wittgenstein is bothered by this sort of carelessness. For example, there are his remarks in MS 126, pp. 133f.:

> Nowhere is it pointed out by Hardy that the irrational number does not possess a numeral in the sense in which the rational number does possess it. The fiction seems to be that is has an infinitely long one. If anything, of course, the sign for the rule of expansion might count as the numeral. – But this absence of the numeral means an infinitely fundamental

> difference. And in a certain sense Cantor's diagonal proof very well shows that it *cannot* have a numeral.

To mathematicians, this remark will not say anything new (at best it might lead them to say merely that they should express themselves a bit more carefully). But for the interpretation of Wittgenstein's philosophy of mathematics it is interesting, and the last sentence in particular. Wittgenstein refers to Cantor's diagonal proof as a matter of course, and he makes sensible use of it. So obviously he doesn't have any qualms about it, as is sometimes claimed in the literature.[44]

Cantor himself would respond that Wittgenstein here implicitly accepts the uncountability of real numbers and therefore the result that there are in fact "too many" real numbers in the case at hand, just as he expressed the situation to himself at the outset of his work on the higher infinite. At this point Wittgenstein might even agree with Cantor. But, philosophically speaking, *only* if Cantor didn't mean more with this way of expressing his result than what is contained in his proof, and only if he would acknowledge that his remarks about "too many" themselves contain a potentially misleading use, as well as an important conceptual remodeling of our original notion of cardinality.[45]

As we understand it, the last sentence in Wittgenstein's annotation on p. 28, i.e.,

> One can cut them, but the cut doesn't bring about such a division,

refers to the geometrical picture of the numbers as points on the straight line. About this Wittgenstein says in RFM V §34 (as already quoted in Sect. 5.3) that it lets us see the divisions

> through the model of a cut made somewhere along the straight line, *hence extensionally*. For if I *cut*, I can of course choose where I want to cut.

As just explained, this picture is inadequate if one associates with those "divisions" the *modes of division* that we meet when considering paradigmatic examples like $\sqrt{2}$ and π. Freed from this picture and from these examples, the "sections" talked about in the general Dedekindean phraseology turn out to be deeply different in the two cases of the rational and the real numbers.

It is the almost paradoxical character of the extensionalist viewpoint that, on the one hand, it all too easily brings us to see the transition from the rational to the real numbers as a simple extension of the domain of numbers, but that, on the other hand, if taken seriously – that is, freed from geometrical pictures and individual examples – this transition appears to bridge a deep chasm. As we understand it, this is precisely the point of Wittgenstein's annotation on the lower margin of p. 28.

---

[44] See Hodges (1998).

[45] A Wittgensteinian reaction of this sort may be gathered from his attitude shown in RFM II. See the discussion in Mühlhölzer's Chap. 7 below.

5.7 The Prudishness of Hardy's Proof

On pp. 28f. Hardy gives a proof of *Dedekind's theorem*, which is stated on p. 30 as follows (and presented entirely in italics):

> **Dedekind's Theorem** *If the real numbers are divided into two classes L and R, in such a way that*
>
> *(i) every number belongs to one or other of the two classes,*
> *(ii) each class contains at least one number,*
> *(iii) any member of L is less than any member of R,*
>
> *then there is a number* α*, which has the property that all the numbers less than it belong to L and all the numbers greater than it to R. The number* α *itself may belong to either class.*

Hardy's proof is rather clumsy.[46] Fortunately we need not go into it though, because Wittgenstein makes a short remark on its style and its style alone, without really considering its inner mechanism.[47] Beyond this he merely remarks on its result, i.e., on Dedekind's theorem as it is presented and discussed by Hardy.

Nowadays, when as usual set theoretic language is employed, the proof of the theorem appears more or less trivial. The only structure on the real numbers needed for the proof is the relation $\leq$, simply defined as $(L_1, R_1) \leq (L_2, R_2)$ iff $L_1 \subseteq L_2$, for any real numbers (L_1, R_1) and (L_2, R_2) (i.e. sections of the rational numbers such that L_1 and L_2 do not have greatest elements). One then observes that the set-theoretic union of the lower classes of all the sections of rational numbers that make up the lower class L of our given section of the real numbers doesn't have a greatest element, and it obviously is the lower class of a section of the rational numbers. Therefore this section is a real number; call it α. One may almost immediately see

---

[46]The reason is that it contains superfluous and confusing elements which conceal what is really essential to it. For example it makes use of algebraic structure (e.g. by constructing the arithmetic mean of numbers) which is not necessary; it makes shifts between considering the rational numbers as originally presupposed and then as embedded in the set of real numbers which is constructed out of these presupposed rational numbers, but without calling the reader's attention to these shifts. Above all, it obscures the simple fact that the sought for number α can simply be defined via the union of the lower sets of the $\mathbb{Q}$-sections that make up the given L. – That Hardy presents such a clumsy proof, and this from the second edition onwards through all further editions of CPM, is astonishing because he was perfectly aware of the much more adequate sort of proof which we shall present in a moment. On this see his perceptive commentary in Hardy (1929), pp. 8 and 22, where Hardy discusses in particular the impredicativity involved in the proof. This is something that he totally ignores in CPM.

[47]In his letter to Schlick of July 31, 1935 (quoted in Floyd 2012a, p. 252), Wittgenstein writes that Schlick can only understand what a mathematical theorem says until "you go through the proof from A to Z". Here, Wittgenstein refers to Gödel's incompleteness theorem, but his advice is meant quite generally. Unfortunately, Wittgenstein himself all too often doesn't adhere to his own advice, as in our present case, the proof of Dedekind's theorem. Wittgenstein's sensitive reflections on the concepts involved – "number", "section", etc. – may suffice for his critique of Hardy's comments on the theorem, but what the theorem actually says, when stripped of these comments, remains poorly explicated by Wittgenstein.

then that according to the way "$\leq$" is defined for the real numbers, α must be either the greatest element of L or the least element of R. – This is the proof.[48]

Now immediately after his proof – even before stating the theorem itself – Hardy explains its significance thus (pp. 29f.):

> Any section of the real numbers therefore 'corresponds' to a real number in the sense in which a section of the rational numbers sometimes, but not always, corresponds to a rational number. This conclusion is of very great importance; for it shows that the consideration of sections of all the real numbers does not lead to any further generalization of our idea of number. [. . .] The discussion which precedes shows that [. . .] the aggregate of real numbers, or the continuum has a kind of completeness which the aggregate of the rational numbers lacked, a completeness which is expressed in technical language by saying that the continuum is closed.

On p. 29 we now find the following three annotations by Wittgenstein. On the upper margin:

[A] The proof seems to show that we now are finished and that on the line there are no more points; for if we divide them we <u>must</u> cut through a real point.[49]

On the right margin:

[B] The prudishness in this line of proof![50]

On the lower margin:

[C] Here it is supposed that the section cannot hit a real number and at the same time a "number" of quite another kind.[51]

With **[A]** Wittgenstein surely means this: What the proof *seems* to show is something geometrical, namely, that the gaps on the line left by the rational points are now closed by adjoining the irrational ones. Then there are no more points left. We see this when we divide the line, because we will now *always* cut through a real point, i.e., a rational *or* irrational one.

---

[48] See, e.g., the lucid exposition in Hrbacek and Jech (1999), pp. 86–89. – This straightforward train of thought can in a sense also be detected in Hardy's proof in CPM, but, as already said, only under a superfluous and needlessly complicated sheathing which actually conceals it.

[49] Der Beweis scheint zu zeigen, daß wir jetzt am Ende sind und es auf der Linie keine anderen Punkte mehr gibt; denn wenn wir sie teilen <u>müssen</u> wir durch einen reellen Punkt schneiden.

[50] Das Prüde an dieser Beweisführung!

[51] Hier ist angenommen, daß der Schnitt nicht eine reelle Zahl und <u>zugleich</u> eine "Zahl" einer ganz anderen Art treffen kann.

Wittgenstein's annotation remarks that we "*must* cut through a real point", as Hardy presents things. Such may in fact be a necessity contained in the picture associated by many with the proof: the extensionalist one. Yet this picture models something very far from what the proof actually shows. Wittgenstein puts his finger on this.

However, Hardy wouldn't disagree, and no reflective mathematician would. Hardy in fact succinctly explains what is actually shown by the proof, and he does so quite independently of any geometrical imagery. The "adjoining" of the real numbers to the rational ones happens *via* the detour of considering all the sections of the rational numbers whose lower classes do not have a greatest element, and by next identifying the original rational numbers with just those sections whose upper classes have a least element (this is the element with which the section is identified).

In this way the rational numbers become a part of the real numbers such that the domain of the real numbers can in fact be considered as an *extension* of the domain of the rational numbers. But the important point is that this extension gives rise to a *new structure* with new properties: any Cauchy sequence of numbers now has a limit, any set of numbers with upper bound now has a supremum, and so on (and Cantor, of course, would furthermore emphasize the greater cardinality of the new set).

An *analogous* procedure may then be entertained with the set of real numbers. Consider sections of them whose lower classes do not contain a greatest element (where the latter condition is merely imposed for reasons of uniqueness) and then identify a real number α with the section whose upper class has this number α as its least element. Dedekind's theorem then says that in this way *every* such section of the real numbers is captured, and we do not get a new structure analogous to the new structure produced by the transition from the rational to the real numbers. In precisely this sense, as said by Hardy on pp. 29ff.,

> the aggregate of real numbers, or the continuum has a kind of completeness which the aggregate of the rational numbers lacked, a completeness which is expressed in technical language by saying that the continuum is closed.

Wittgenstein doesn't seem to appreciate this beautiful train of thought. On the contrary, in his remark **[B]** he calls the underlying proof "prudish". MS 126 indicates what he means by this and that he means it to be pejorative, or at least indicative of philosophical unclarity about our attitudes toward how people present their own scrupulousness.[52] There, on p. 130, he writes:

> A line of proof is prudish, if one anxiously avoids the least logical ambiguity, but tolerates crude nonsense.

We are at a loss, though, to see what the nonsense, let alone a crude one, in Hardy's line of proof might be. This proof, as already said, is a bit clumsy, but Wittgenstein's remark certainly doesn't refer to this.

[52]Khan (2014) argues that the charge of prudishness in Wittgenstein's remarks has to do with his wider views about contradiction and especially mathematicians' remarks about infinity.

Instead, we take it, Wittgenstein means the general style of the proof, which is the same in Hardy's proof and in the better proof sketched above. But Wittgenstein has not shown that there is 'nonsense' in or around it in any obvious sense. Wittgenstein may have in mind here misleading geometrical pictures associated with the proof, but Hardy is careful enough to explain everything in a way that avoids such pictures. So, Wittgenstein's deprecation doesn't seem to be justified on that ground.

Nevertheless, the word "prudish" appears quite appropriate if it is understood from Wittgenstein's own, non-extensionalist point of view. And in this sense it is a better word for him to have used about Hardy's proof than what is said in his own gloss on the term itself (in MS 126, p. 30, just quoted). The proof of Dedekind's theorem, in whatever form, has without doubt the characteristic abstractness expressed in Dedekind's phraseology which, as we have seen, is uncoupled from the familiar invigorated and enriching life of individual numbers like $\sqrt{2}$ and π in a variety of mathematical areas. In this sense it can be called "prudish", if one conceives of it as lending rigor and precision to an otherwise flabby or unrigorous mode of procedure.

Incidentally, one *can* find something nonsensical in the standard proof of Dedekind's theorem, as for example Hermann Weyl did in his (1918). What bothered Weyl was the determination of the number α – see the proof sketched above. α is a section of the rational numbers whose lower class, call it l, is defined as the set-theoretic *union* of the lower classes of the sections making up the given section L of the real numbers. When one expresses this in logical notation, the definition involves a quantification over a set of classes of rational numbers to which the class l belongs.[53] Thus, the definition is *impredicative*; Weyl even saw a *vicious circle* in it, and accordingly he rejected the standard proof of Dedekind's theorem. Weyl uses very strong language to express his criticism, and it is remarkably similar to the way in which also Wittgenstein tends to express himself. Thus he talks about the "meaninglessness" [*Sinnlosigkeit*] of the Dedekindean approach, even of its being "clearly meaningless" [*evident sinnlos*].[54]

As we have already seen above in Chap. 1, Wittgenstein had already rejected Frege and Russell's analysis of the ancestral by accusing *them* of being entangled in a *circulus vitiosus*; see TLP at 4.1273. In his earliest work he was inclined to think of logical processes as formulatable in terms of step-by-step, iterative formal procedures with signs, procedures that are always well-founded, hence, logically speaking, predicative. And he was not inclined to accept full second order logic, with its impredicativities. (Nor of course were Russell and Whitehead in *Principia Mathematica* (1910), concerned as they were to present the theory of types.)

Weyl presented his own ideas about real numbers in, e.g., Weyl (1918, 1919, 1925, 1927). Wittgenstein was certainly influenced by him: we know that he read

[53]See the expositions in Mancosu (1998), p. 70, and Feferman (1998), pp. 54–58 and 254–258. As already noted, Hardy was also totally aware of this; see Hardy (1929), pp. 8 and 22.

[54]See Weyl (1925), p. 524 (p. 139 of the English translation), and Weyl (1919), p. 45 (p. 111 of the English translation).

at least Weyl (1925) and parts of Weyl (1927).[55] As we have already said in Chap. 2, however, one must not presume that Wittgenstein's sort of criticism in his annotations to Hardy and related manuscripts is the same as Weyl's. In contrast to Weyl, Wittgenstein here does not *reject* the Dedekindean phraseology and the standard proof of Dedekind's theorem: as we have seen, he only criticizes the "nonsense" that so often seems to be accompanied with it. Above all, he never attempts to develop an own elaborate brand of analysis as a seriously meant alternative to classical analysis, whereas this was done by Weyl – and this was in fact Weyl's most important achievement in the foundations of mathematics.[56]

Remark **[C]** sounds mysterious and we aren't really sure how to understand it. It seems certain to us, though, that in it Wittgenstein refers to an assertion by Hardy on p. 29, drawing beside it a squiggly line on the right margin. The assertion of Hardy's is this:

> [T]he idea of a section of the rational numbers led us to a new conception of a number [...].

This, as we have argued above, is Wittgenstein's own point of view. Furthermore, there are passages in MS 127, pp. 65–69, which look like elaborations of **[C]** and which suggest the following interpretation.

Wittgenstein looks at the real numbers in the same way one previously looked at the rational numbers. He considers pairs of classes L and R of real numbers where L has no greatest and R no least element but which otherwise are like sections of the real numbers. Let us call them "quasi-sections of the real numbers". (This is not Wittgenstein's term, but it seems convenient to us.) Then the proof of Dedekind's theorem shows that – in contrast to the case of the rational numbers – for any quasi-section L and R of the real numbers there is precisely one real number α which is between L and R. At this point, then, Wittgenstein seems to raise the question of whether there might not turn up another kind of "number" between the classes L and R. After all, this had analogously been the case with the rational numbers,

[55]See the footnotes in WVC, pp. 37, 81 and 103. Of course Wittgenstein knew about Weyl's thoughts also through Ramsey (1926), and presumably through his discussions with Ramsey as well. See also Schappacher (2008) for much textual evidence of this. – With Weyl's "own ideas about real numbers" we of course mean only the ideas presented in Weyl (1918) and in the 1920s, i.e., in what Schappacher calls the second and third phase in Weyl's thinking about the foundations of mathematics (2008, pp. 213–222). Through the influence of Brouwer, these ideas were in flux (see Schappacher (2008), and the useful exposition in Mancosu (1998), pp. 65–85), but this is not important for our assessment of Wittgenstein, who in any event picked up what he found valuable, sorted out the rest and transformed everything according to his own thinking. Marion (1998), pp. 85ff, argues that Wittgenstein followed Weyl in thinking of mathematical induction as a template, or schema, drawing on Weyl's notion of a "judgment extract" for claims about the existence of certain kinds of numbers; he in turn draws on Majer (1988) and Sahlin (1995) in contrasting Wittgenstein's own position with Brouwer's.

[56]Weyl's was an achievement that is taken seriously even today; see, e.g., the Preface to Feferman (1988, 1998). In Mancosu (1998), pp. 65–85, may be found a good description of the actual situation immediately after the appearance of Weyl (1918) and then also in the 1920s, as well as a clarification of the supposed 'vicious circle' stated by Weyl.

where the existing gaps were closed by constructing a new kind of number, called "real number", namely: the sections of the rational numbers whose lower classes do not have greatest elements.

Couldn't one imagine that there could be, in a similar fashion, such a new kind of "number" also in the case of quasi-sections of the real numbers? This is a kind of question about categorical distinctions that turns up throughout Wittgenstein's annotations and remarks, especially those concerned with the universal applicability of logical "forms". In RFM V §40 (an *analogy* to the situation with Dedekind cuts) we see a remark, as we have already discussed (Sect. 5.4), that originated in MS 127, pp. 57–60. And indeed: the question just posed *is* suggested in MS 127, pp. 65–68. On p. 68 of MS 127 Wittgenstein comments on it as follows:

> The question may perhaps be: "is there still a number lying between these classes [constituting a quasi-section] or not?". But one can't ask further: "and isn't it perhaps of quite another kind?" – because a quite new kind may in fact lie between them, after all, but this wouldn't be investigated at all.

Yet this is a strange assertion. Hardy precisely alludes to such a "new kind" when on p. 29 he writes that

> the considerations of all the real numbers does not lead to any further generalization of our idea of number.

The "new kind" would be precisely one that would incorporate the sections of real numbers. Yet, as we have said above, in one clear sense this would not lead to something *really* new because the structure consisting of this "new" kind of "number" would be structurally isomorphic to the original one and in that sense would not have any new properties that are mathematically relevant.

At the beginning of p. 69 of MS 127, Wittgenstein immediately adds to the passage just quoted:

> On the other hand the real numbers really close a chapter – but is this proved by the proof [of Dedekind's theorem]?

We are puzzled by this question. For Hardy very clearly explains to the reader of CPM in just what sense the "closing of a chapter" *is* in fact proved by the proof: namely in the sense, just stated in our quotation from p. 29, that the aggregate of the sections of the real numbers (with lower classes which do not possess greatest elements) doesn't bring about anything new. This is in fact precisely the point of Hardy's explanation of the closedness of the real numbers.

On p. 67 of MS 127, Wittgenstein writes:

> The idea that [Dedekind's theorem] shows the 'closedness of the continuum' appears crude [*abgeschmackt*] to me.

His annotations on p. 29, however, cannot give this feeling much support, so we must probe further.

5.8 Hardy's "Crudeness"?

On p. 30 Hardy's explanation of the significance of Dedekind's theorem (quoted and discussed in Sect. 5.6 above) culminates with the assertion that

> the continuum has a kind of completeness which the aggregate of rational numbers lacked, a completeness which is expressed in technical language by saying that the continuum is closed.

This is continued by Wittgenstein, on the same line on which it ends and then on the right and upper margin of p. 30, with the following remark:

> `This seems to be an enormous result. For it seems as if we had found out something crucial about the ultimate structure of the continuum, where we imagine the continuum as space, say.`[57]

This again is a puzzling remark: how can a result be "enormous" when the proof is so straightforward (as we have constructed it above in Sect. 5.7) – and so "prudish", as Wittgenstein himself says? And in fact on p. 132 of MS 126, Wittgenstein writes what appears to take up the *opposite* point of view:

> One might ask: what might a child of 10 years *not* understand of the proof of Dedekind's theorem? – Isn't this proof much easier than all the calculations which the child has to master?

This statement, however, is puzzling as well. Does Wittgenstein really want to claim that the proof of Dedekind's theorem in its extreme abstractness and prudishness could be easily understood by a 10-year-old child?

Both remarks attain a rather good sense, however, becoming coherent when we interpret them as follows: The *result* appears "enormous" if it is understood as showing the *underlying essence* of the continuum, where "the continuum" is something known through the whole history of mankind in the form of geometric space. For centuries, there have been mysteries surrounding the concept of the continuum. And now (this idea would run) they are dispelled through Dedekind's analysis. At the same time the *proof* of Dedekind's theorem, apparently leading to this result, is of a rather schematic nature and altogether straightforward, consisting of steps which are quite easily understood – maybe not by a 10-year-old child (this is a Wittgensteinian exaggeration[58]), but in any case by someone with only a minimal familiarity with mathematical thinking. And now we seem to have a sort

---

[57]`Dies scheint ein ungeheuerliches Resultat. Denn es scheint als hätten wir etwas Entscheidendes über die letzte Struktur des Continuums herausgefunden, wobei wir uns das Continuum etwa als den Raum vorstellen.`

[58]Continuing this remark about the 10-year-old child, Wittgenstein writes: "–And if now somebody would say: the deeper content of the sentence it [= the child] cannot understand – then I ask: how does this law/this sentence attain a deep content?"–This may be a legitimate question, but as Wittgenstein rightly says, it hasn't anything to do with the child's understanding of the individual steps of the proof. These are in fact simple – though certainly not for a 10-year-old child.

of paradoxical situation in analysis again: something extraordinarily deep gets a proof that is straightforward and easily understood.

But of course this paradox is only a seeming one. For Dedekind's theorem and its proof proceed on a purely arithmetical level, deliberately uncoupled from geometric space as historically understood. Wittgenstein's allegation, however, if we understand him correctly, is that this uncoupling is obscured in the usual expositions. For here the tendency prevails of presenting Dedekind's result as an extraordinarily deep one. And this tendency is typical of the mathematician's and all other scientists' aspirations to produce awe and excitement in an audience during the exposition of fundamental results.

Wittgenstein despises this tendency to overflourish and market mathematical results as if from on high. This may be the reason why he writes, as quoted above, that

> The idea that [Dedekind's theorem] shows the "closedness of the continuum" appears crude to me (MS 127, p. 67).

We, however, do not see such crudeness in Hardy's exposition. Hardy clearly and explicitly distinguishes between the *arithmetical continuum* and the *linear continuum*, and when on p. 30 he talks about the completeness and closedness of the continuum, he has only the arithmetical continuum in mind. Furthermore on p. 28 he writes:

> It is plainly possible to apply the same idea [i.e., the idea of considering sections of the rational numbers] to the aggregate of all real numbers [which is nothing but the "arithmetical continuum"], and the process is, as the reader will find out in later chapters, of very great importance.

But here Hardy does not allude to the historical importance according to which now, through Dedekind, the concept of the continuum has been clarified (as if there were only one concept of this sort). Rather, he thinks of the *technical* importance with respect to proofs of countless theorems of analysis, proofs that rely on the sort of completeness stated in Dedekind's theorem and involve the arithmetical continuum alone.

It is typical of Wittgenstein's philosophizing about mathematics that it very often places itself all too far from the actual statements of real mathematicians. His annotations to Hardy's book are an exception: here he responds in a direct way to what a mathematician actually says. And yet, as we see, even now his own picture of what mathematicians do and say has a tendency to develop along lines, and with corresponding demands, that lie at a great distance from what is actually done by a real mathematician. The value of his questions are, instead, more purely philosophical. But he does not bring this out himself in his annotations: we have had to do it for him.

Chapter 6
Functions and Limits: The Annotations on pp. 40–47 and 117–121, with Commentary

> Must *a function have a behavior for* $x \to \infty$*?*
> *The* extensional *view makes it appear so. (Annotation on p. 117 of CPM)*

Abstract An analysis of Wittgenstein's annotations to Hardy's *Course of Pure Mathematics* pp. 40–47 and 117–121, with commentary on each remark. Notions discussed include the idea of a function (from a non-extensionalist and extensionalist perspective), the notion of a limit, the importance of different methods of handling limits, and the very definition of a limit.

6.1 The Idea of a Function

Hardy begins his §20 ("The idea of a function"), the first section of his Chapter II ("Functions of a Real Variable"), by explaining what it means to say that y is a function of x, where x and y are "continuous real variables" (as he writes). His explanation is that the value of x *determines* that of y, period. Simple examples of such functions are given, like $y = x$, $y = 2x$, $y = \frac{1}{2}x$, $y = x^2 + 1$, and Hardy states on p. 40:

> This notion of functional dependence of one variable upon another is perhaps the most important in the whole range of higher mathematics.

Furthermore, he emphasizes that this notion is much more general than the examples given might suggest. As he writes on pp. 40f.:

> [W]e must point out that the simple examples of functions mentioned above possess three characteristics which are by no means involved in the general idea of function, viz.:
>
> (1) y is determined *for every value of* x;
> (2) to each value of x for which y is given corresponds *one and only one value of* y;
> (3) the relation between x and y is expressed by means of *an analytic formula*, from which the value of y corresponding to a given value of x can be calculated by direct substitution of the latter.

J. Floyd, F. Mühlhölzer, *Wittgenstein's Annotations to Hardy's Course of Pure Mathematics*, Nordic Wittgenstein Studies 7,
https://doi.org/10.1007/978-3-030-48481-1_6

> [...] All that is essential is that there should be some relation between x and y such that to some values of x at any rate correspond values of y.

From our contemporary point of view this is an astonishing explication. For nowadays it is precisely condition (2) which mathematicians consider to be the essential characteristic of a *function* in contrast to other relations. But Hardy sees it otherwise, from the first to the last edition of CPM, and we shall stick to Hardy's notion. Wittgenstein's remarks about Hardy's conception are not specifically directed to conditions (1), (2) and (3) anyway.

On the margins of pp. 40f. there are three annotations in Wittgenstein's hand. On the left margin of p. 40, with continuation on the upper margin:

[A] `H.'s procedure is here quite different from the case of the numbers. There he began with the rational numbers and then by and by introduced new kinds of numbers. Here, he introduces first a vague general notion of a function and only later kinds of functions!`[1]

On the lower margin of p. 40:

[B] `The important thing here is that there is a mathematical and a non-mathematical correspondence!`[2]

And on the right margin of p. 41, with an arrow from his word "these" [diese] to Hardy's sentence "All that is essential is that there should be some relation between x and y such that to some values of x at any rate correspond values of y":

[C] `By these explanations the idea of a function is not defined but merely illustrated.`[3]

[A] remarks on an obvious difference between the way Hardy introduced numbers (= real numbers) in contrast to functions in his earlier chapters. Before, he began with the rational numbers as already understood, and then introduced different kinds of numbers. He gave relevant geometric constructions, for example, or said that they are solutions of certain equations, stressing that the constructions and the equations themselves can be of different kinds. Hardy could also have mentioned –

---

[1] H.'s Prozedur ist hier ganz anders als bei den Zahlen. Dort fing er mit den rationalen Zahlen an und führte nach und nach neue Zahlenarten ein. Hier führt er zuerst einen vagen allgemeinen Begriff der Funktion ein und erst später Arten von Funktionen!

[2] Das Wichtige ist, daß es hier eine mathematische und eine nicht-mathematische Zuordnung gibt!

[3] Durch diese Erklärungen wird die Idee der Funktion nicht definiert, sondern bloß illustriert.

though he did not – numbers presented *via* Cantor's diagonal method and by other "higher-order" methods.

All this culminates, as we have seen, in the rigorous definition of numbers *via* sections (= cuts), formulated with Dedekindian phraseology. Now this latter Dedekindian definition also could be taken to be an introduction of new kinds of numbers, just as we have argued in Chap. 5. Nevertheless, in **[A]** it seems to us that Wittgenstein is only mentioning the initial ways of introducing numbers, the ones that do not involve Dedekindean phraseology.

In the case of functions of real variables, one might have expected a procedure of introduction that would be similar or parallel to what came before in Hardy's Chapter I. Yet here, at the outset of his Chapter II, Hardy proceeds quite differently. He commences with certain hints as to what functions are, giving a few initial examples, and then comments on his hints by emphasizing the great generality of his notion of a function. It is this way of proceeding that Wittgenstein responds to in his annotations.

In **[A]** Wittgenstein is certainly right to note that everything Hardy says here is rather vague – at least in comparison to what happened in the case of numbers in Chapter I. For there a precise general definition was given in the end. Yet one should not think that such a goal is striven for by Hardy in the case of functions of real variables. On the contrary, all that he produces are different kinds of functions: polynomials, rational functions, algebraical functions, transcendental functions, and so on. A precise general definition is never given.

Wittgenstein's comment on this in **[C]** is certainly to the point and quite unproblematic. Now with regard to **[A]** it is important to note that when Wittgenstein calls Hardy's comments on generality "vague' this is not necessarily meant as a critique! Let us look to the manuscripts to see why. In MS 127, pp. 5f., Wittgenstein asks:

> If we now, as I perhaps want to say, have an unclear notion of the 'principle of correspondence' (e.g.), – can/should one replace it by a clear one? Or is it precisely the vagueness which renders a good service?

And on pp. 13f. of MS 127 (RFM V §38) he simply states:

> The generality of functions is so to speak an *unordered* generality. And our mathematics is built upon such an unordered generality.[4]

These remarks show that Wittgenstein's annotations in remarks **[A]** and **[C]** may be read as simply highlighting what Hardy does, and involve no criticisms at all.

But why did Hardy choose such different procedures in the case of numbers in contrast to the case of functions? The reason may lie in his decision to avoid the phraseology of set theory, a decision that presumably reflects didactic motives. The introduction of Dedekind cuts (sections) seemed unavoidable because this was an

[4]The context in MS 127 (less than the context in RFM V) shows that what Wittgenstein has in mind here is not the generality of every single function (a y is assigned to *any* x ...) but the generality of our *notion* of a function.

established part of analysis. The case of functions may have appeared to Hardy to be quite different. As we have already seen, however, this avoidance of set theory leads Hardy to present a rather clumsy proof of Dedekind's theorem. And it now prevents him from rigorously defining functions in the standard set theoretic manner: as specific sets of ordered pairs. From Hardy's standpoint this would actually have been advantageous, but he avoids it – and does so from the first to the last edition of CPM.

Of course, Wittgenstein might have cut in here once again. He could have pointed to the contrast between the set theoretic sort of general phraseology and the individual examples. Yet, as we have said, Wittgenstein does not reject this phraseology. He only problematizes it when it is used in a manner that underplays the conceptual moves intrinsically belonging to the extensionalist perspective.

Now whereas Wittgenstein's remarks **[A]** and **[C]** are quite straightforward, we are not sure what he precisely has in mind with **[B]**. Why does he write *this* down here? With "non-mathematical correspondence" Wittgenstein might mean correspondences that are non-mathematical from the outset, as the physical correspondences mentioned by Hardy on pp. 41f. representing Boyle's law or van der Waal's law. Or he might mean mathematical cases like $y = x$, $y = 2x$, $y = \frac{1}{2}x$, and $y = x^2 + 1$. These are mentioned at the beginning of Section 20 but could at this point be regarded in a non-mathematical way. For example, one might see them in terms of the idea that a person actually uses these formulae in order to assign specific values of y to specific values of x. The respective correspondence would then be an empirical or anthropological one, depending on what the person actually does. – We are not sure, however, which of these interpretations, or some other, is the one Wittgenstein actually intended here.

Why does this sort of remark occur on p. 40? Maybe it is the following thought that underlies it, and with respect to this it may not be very important which of these meanings of "non-mathematical correspondence" is chosen: such non-mathematical correspondences as we have mentioned seem to be simply there, independently of any specific rules or techniques that we use to represent them; in other words, they may fuel an *extensionalist* view of functions. And this is just what Wittgenstein warns against. It may be that Wittgenstein had this warning in mind when he wrote down **[B]**.

6.2 Graphical Representations of Functions

On p. 46, in his Section 23 ("Further examples of functions and their graphical representation"), Hardy raises the question as "to what is meant by a continuous curve, a curve without breaks or jumps". As an example he presents a picture of the graph of the function $y = x^2$, to which he adds the following comment (pp. 46f.):

> There is, however, one fundamental question which we cannot answer adequately at present. The reader has no doubt some notion as to what is meant by a *continuous* curve, a curve without breaks or jumps; such a curve, in fact, as is roughly represented in [our figure

> of a parabola]. The question is whether the graph of the function $y = x^2$ is in fact such a curve. This cannot be *proved* by merely constructing any number of isolated points on the curve, although the more such points were construct the more probable it will appear.
>
> This question cannot be discussed properly until Ch. V [the chapter entitled "Limits of Functions of a Continuous Variable. Continuous and Discontinuous Functions"]. In that chapter we shall consider in detail what our common-sense idea of continuity really means, and how we can prove that such graphs as the one now considered [the graph of the function $y = x^2$] and others which we shall consider later in this chapter, are really continuous curves. For the present the reader may be content to draw his curves as common sense dictates.

Here Wittgenstein put a squiggly line under the expression "such graphs" (which occurs on p. 47), and that is all. More, however, is not necessary, for the point of his squiggle is obvious: what Hardy really presents later in his Chapter V is not at all an explanation of what our "common-sense idea of continuity" means with respect to such graphs, but rather the classical ϵ–δ definition of the continuity of functions invented by Bolzano and Cauchy at the beginning of the nineteenth century and subsequently elaborated by Weierstraß. The whole question of in what sense we *have* a "common sense" notion of the continuum is not explored or specified. Referring to a graph as "really" continuous thus hangs in the air here.

The preliminary of this definition occurring at the beginning of Section 23 of CPM is the typical *prose* to be seen in many mathematical textbooks, with all the dangers of misleading us about what to say about the content of the mathematics that is actually developed. The ϵ–δ definition of continuity is an ingenious innovation in mathematics, not at all easily understood by beginners, and required by mathematicians themselves, for their work. Hardy's prose not only doesn't say this. It actually obscures it.

6.3 The Notion of a Limit: *Must* a Function have a Behavior for $x \to \infty$?

The next image of Wittgenstein's annotations that we have is on p. 117. We are now in Hardy's Chapter IV ("Limits of Functions of a Positive Integral Variable"). Here, after having discussed many different kinds of functions in Chapter II and introduced complex numbers in Chapter III, Hardy finally gets to the notion of a *limit*. This notion may be taken to be, in fact, the actual center of analysis as a branch of mathematics. The "functions of a positive integral variable" mentioned in the title of Chapter IV are what are normally called "sequences".

Hardy represents these by functions ϕ, expressed by formulae like $y = \phi(n)$, defined for all values of the positive integral variable n and with real values y. He speaks of properties P "which $\phi(n)$ may or may not have" (p. 113) and on p. 115 he says that "$\phi(n)$ *has the property* P (which will as a rule be a property expressed by some relation of inequality) *for large values of* n" if and only if we can determine some definite number, n_0 say, such that $\phi(n)$ has the property for all values of n greater than or equal to n_0.

Now after showing how this precisification can be tethered down by further specifications given the examples he uses, Hardy writes (p. 115f.):

> Thus we may say that 'all large primes are odd', or that '$\frac{1}{n}$ is less than .001 for large values of n'. And the reader must make himself familiar with the use of the word *large* in statements of this kind. *Large* is in fact a word which, standing by itself, has no more absolute meaning in mathematics than in the language of common life. It is a truism that in common life a number which is large in one connection is small in another; 6 goals is a large score in a football match, but 6 runs is not a large score in a cricket match.

In Section 55, "The phrase 'n tends to infinity'", he explains said phrase as simply meaning "that n assumes successively the values 1, 2, 3, ...", where the ellipsis dots "..." indicate the "unending growth of n" (p. 116). Instead of saying "n tends to infinity" he also uses the common notation "$n \to \infty$".

At the end of this section, Hardy brings up the question as to which properties a given sequence $\phi(n)$ has for large values of n, which he puts this way:

> 'how does $\phi(n)$ behave as n tends to ∞?' (p. 117).

As an example, he mentions in Section 56 ("The behaviour of a function of n as n tends to infinity") the property of being smaller than a certain real number $\delta > 0$, which the sequence $1/n$ has for large values of n, irrespective of the chosen $\delta > 0$, however small it may be (which shows that $1/n$ converges to 0). In response to all this, on the lower margin of p. 117, Wittgenstein writes:

> Must a function have a behaviour for $x \to \infty$? The extensional view makes it appear so.[5]

This is a perceptive remark. Since, as noted on p. 40, the generality of the notion of a function is a vague one, it is not at all settled a priori for any stated property P whether or not it makes sense to ask whether a given $\phi(n)$ has it for large values of n or not. If it doesn't, then the question about the respective behavior of $\phi(n)$ is senseless as well.

The extensional view, though, suggests that functions are simply *there* and that our approach to them – the rules and techniques that we associate with them – are only of a secondary nature. Then in every case it may seem decided beforehand whether a stated property P applies to the values of a function for large values of n or not. Hardy himself actually doesn't say this explicitly, but his text suggests it. And in any case, Hardy doesn't point to the problems that turn up at precisely this point.

It would have been helpful if Wittgenstein had given an appropriate example for problems of this sort, but as far as we see, he did not do so. In MS 127, pp. 11f., he ponders the story of people who are brought to assign $\phi(n)$'s to given n's, giving

[5] Muß denn eine Funktion ein Verhalten für $x \to \infty$ haben? Die extensionale Vorstellung läßt es so erscheinen.

"$\phi(n)$" its sense in an anthropological or empirical way, but then "$x \to \infty$" isn't at issue from the outset.[6] So his remark on p. 117 has merely the character of a shot in the dark.

However, the question of whether a given $\phi(n)$ has a property P for large values of n or not has a typically intuitionistic flair, and intuitionism abounds with cases where it has to be left open. So Wittgenstein's question "*Must* a function have a behaviour for $x \to \infty$?" is one that also intuitionists might raise. Again, from a philosophical point of view Wittgenstein may be said justly to criticize Hardy for remaining silent about this whole issue.

With this we do not mean to insinuate that Wittgenstein endorsed intuitionist ideas himself here. More generally, we see no tendency within Wittgenstein's non-extensionalist viewpoint to approach the intuitionistic point of view. As we have described above in our Chap. 4, Wittgenstein presents an extensive discussion of the law of excluded middle on many pages of MS 126 (which underlie RFM V §§9–23). Yet he nowhere rejects this law, as intuitionists do. He wants to know, instead, what this law means. As shown throughout his work, Wittgenstein is hostile to many elements of the intuitionist position. This is so even if there remain critical questions about the extensionalist viewpoint that he shares with intuitionism. We have come across such a question here in his annotation on p. 117 of CPM.

6.4 Teaching One's Opponent

On p. 118 of CPM Hardy explains that the statement "$1/n$ tends to 0 as n tends to ∞" is equivalent to the more explicit statement: "if δ is any positive number, however small, then we can find a number n_0 such that $1/n < \delta$ for all values of n greater than or equal to n_0". On p. 119 he then comments on this idea as follows:

> The reader should imagine himself confronted by an opponent who questions the truth of the statement. He would name a series of numbers growing smaller and smaller. He might begin with .001. The reader would reply that $1/n < .001$ as soon as $n > 1000$. The opponent would be bound to admit this, but would try again with some smaller number, such as .0000001. The reader would reply that $1/n < .0000001$ as soon as $n > 10000000$, and so on. In this simple case it is evident that the reader would always have the better of the argument.

To this Wittgenstein writes on the right margin of p. 119, with an arrow from his own word "reader" ["Leser"] to the last occurrence of the word "reader" in Hardy's passage:

[6] Wittgenstein's actual example and his actual notation in MS 127, pp. 11f., are a bit different, but this is irrelevant to the point at hand.

> `And the reader would prove nothing by this, unless he taught his opponent the technique of producing a number that is always smaller.`[7]

In fact, it is remarkable that in his – *prima facie* – quite neat description Hardy actually left out the decisive aspect of a successful interaction with the imagined opponent: to communicate the *technique* underlying the reader's ways of producing the numbers. When Hardy writes, "it is evident that the reader would always have the better of the argument", he must have such a technique in mind. But the *extensionalist* point of view, which permeates his presentation, makes him repress the non-extensional aspects of the reader's mathematical practice, i.e., the techniques, rules or methods of proof which let him act as he does. This is astonishing. For it is precisely the relevant *methods of proof* that the novice needs to learn when he or she is trained to get the idea of a limit.

It is not the mathematician Hardy who says this, however, but instead the philosopher Wittgenstein, who writes in MS 127, p. 17 (published as end of RFM V §36):

> The concept of the limit and of continuity, as they are introduced nowadays, depend, without its being said, on the concept of *proof*. For we say
>
> $lim_{x \to \infty} f(x) = \ell$ when it can be *proved* that ...
>
> This means that we rely in reality on concepts which are far more complicated to communicate and to grasp than are those that we use in making the concept of limit explicit.[8]

One might speculate at this point that it was Hardy's conscious intent in CPM to go against the previous habit of British mathematicians, who overemphasized the training of students in proof techniques at the expense of the underlying general mathematical ideas. Nevertheless, Wittgenstein is certainly right in emphasizing that this countermove would now have its own deficits.

6.5 The Importance of Methods

On p. 119, at the beginning of Section 57 (a continuation of Section 56), "The behaviour of a function of n as n tends to infinity", Hardy introduces statements of the form "n^2 *is large when n is large*". He writes that such statements are "equivalent to the more formal statements":

---

[7] `Und der Leser würde dadurch nichts beweisen, es sei denn, daß er seinen Gegner die Technik lehrt eine immer kleinere Zahl zu erzeugen.`

[8] As usual, the presentation in RFM has to be altered: the function is $f(x)$, not $F(x)$ as in RFM, the limit meant by Wittgenstein is not 1 but ℓ, and the word "proved" is underlined in his manuscript.

'if Δ is any positive number, however large, then $n^2 > \Delta$ for sufficiently large values of n',

and

'we can find a number $n_0(\Delta)$ such that $n^2 > \Delta$ for all values of n greater than or equal to $n_0(\Delta)$'.

On the lower margin of p. 119 Wittgenstein remarks:

I can exceed any Δ by an n^2 and I must still exceed any n by some Δ.[9]

This remark is, as it stands, very puzzling. The first part of it (before the "and") simply seems to repeat what Hardy says. And the sense or point of the second part is, as it stands, incomprehensible to us. In MS 127, pp. 45f., however, Wittgenstein restates the first part, leaves out the second, and then explains what he aims at with the first:

"I can exceed any Δ by an n^2". – What sort of sentence is that?

One can easily imagine here the infinite extension of the squares, or the method of finding for any number another one whose square exceeds it.

Here we see that Wittgenstein's criticism is the same as in the previous annotation: Hardy has a tendency to skip the *non-extensional* aspects of our mathematical practice which involve the *methods* we use in order prove the statements at hand (for example the statement 'n^2 is large when n is large').

6.6 Definition of a Limit

On p. 120, at the beginning of the important Section 58 ("Definition of a limit"), Hardy defines the central notion of analysis, the notion of a *limit*, as follows (his sentence is written in italics throughout):

The function $\phi(n)$ is said to tend to the limit l as n tends to ∞, if, however small be the positive number δ, $\phi(n)$ differs from l by less than δ for sufficiently large values of n; that is to say if, however small be the positive number δ, we can determine a number $n_0(\delta)$ corresponding to δ, such that $\phi(n)$ differs from l by less than δ for all values of n greater than or equal to $n_0(\delta)$.

Wittgenstein corrects this by crossing out the two occurrences of "however small be the" and writing in each case "for any" above, and he draws squiggly lines under "we can determine" and "corresponding to δ".

This can again be understood as a criticism of Hardy's prose. For the real mathematics that lies behind these formulations does not involve talking about "small" δ's and about "determining" a number $n_0(\delta)$ "corresponding" to δ. Instead,

[9] Ich kann jedes Δ durch ein n^2 übersteigen, and jedes n muß ich für irgendeine Δ noch übersteigen.

it says that for any $\delta > 0$ there is a number $n_0(\delta)$ such that for all n greater than or equal to $n_0(\delta)$ the distance between $\phi(n)$ and l is smaller than δ.

There are similar Wittgensteinian corrections to Hardy's shortening of his definition on p. 121:

> [T]he definition may be stated more shortly as follows: '*if, given any positive number,* δ, *however small, we can find* $n_0(\delta)$ *so that* $|\phi(n) - l| < \delta$ when $n \geq n_0(\delta)$' [...].

Here Wittgenstein crosses out Hardy's "however small" and draws a squiggly line under "find", apparently criticizing this sort of prose.

Part III
More on Wittgenstein and the Real Numbers

Chapter 7
Wittgenstein on Cantor's Diagonal Method (Felix Mühlhölzer)

Abstract In CPM Hardy completely dispenses with set-theoretic language and cardinality questions do not turn up at all. Wittgenstein shows the same abstinence in his annotations, but apart from that he repeatedly discusses cardinality and in this connection Cantor's diagonal method. This can be seen, above all, in Part II of his *Remarks on the Foundations of Mathematics*, and the present Chapter is mainly concerned with this Part. RFM II is dominated by the following reasoning: Cantor's way to construct a new decimal expansion that sequentially dodges all the expansions in a given list can appear rather cheap, as a straightforward trimming of the given expansions in order to get another one. At the same time this is hailed by mathematicians as leading to particularly substantial and deep results. There is an obvious disparity between the simplicity of Cantor's idea and the supposedly amazing mathematical theorems following from it. How can that be? This sort of puzzlement is the central issue in RFM II, and its §10 represents the core of Wittgenstein's considerations, with its provocative beginning: "It means nothing to say: 'Therefore the X numbers are not countable'". Here Wittgenstein talks about someone who has understood Cantor's diagonal method, seeing that to any list of decimal expansions of real numbers a further expansion of a real number exists, the "diagonal number", that is not in the list. But when then saying: "Therefore the real numbers are not countable", one simply repeats what was already clear: that to any list of decimal expansions of real numbers a further expansion of a real number exists, viz. the diagonal number, that is not in the list. Consequently, the "Therefore" means nothing. Obviously, Wittgenstein here accepts the notion of uncountability, and he in particular accepts the uncountability of the real numbers. However, he doesn't understand this notion simply in the sense of the negation of the well-known notion of countability, because this alone would only be a meagre formal scaffold that lets us down completely if we look for real mathematical substance. So, the concept of uncountability should be understood with explicit reference to the diagonal method as defined in §10. But then the essence of this method does not consist in letting us discover new mathematical facts but merely in the invention of a conceptual novelty.

J. Floyd, F. Mühlhölzer, *Wittgenstein's Annotations to Hardy's Course of Pure Mathematics*, Nordic Wittgenstein Studies 7,
https://doi.org/10.1007/978-3-030-48481-1_7

> Asked: "Can the real numbers be ordered in a series?", the conscientious answer might be: "For the time being I can't form any precise idea of that". (RFM II §16)

> Here it is very useful to imagine the diagonal procedure for the production of a real number as having been known long before the invention of set theory, and familiar even to school-children, as indeed might very well have been the case. (RFM II §17)

7.1 Introduction

In his annotations on pp. 40f. of his copy of Hardy's *A Course of Pure Mathematics* (CPM) Wittgenstein remarks that the notion of a function, introduced by Hardy in his Section 20 ("The Idea of a Function"), where it is called "perhaps the most important [notion] in the whole range of higher mathematics", is not really defined, but instead merely illustrated in CPM – and that this stands in striking contrast to Hardy's meticulous procedure with respect to the real numbers which are explicitly defined as Dedekind cuts of the rationals (termed by Hardy "sections" of the rationals). Wittgenstein does not raise the question of why Hardy's approach to functions is so different, and Wittgenstein may in fact have been quite content with that because it avoids the sort of general phraseology that he struggles with in the case of Dedekind cuts. Hardy himself does not explain why he proceeds in such different ways, but one of the reasons certainly lies in the fact that in CPM he consistently avoids making use of set theoretic parlance. In the language of set theory it would be straightforward to define explicitly what a function is, and this in fact happens in almost all textbooks about analysis that are more 'modern' than CPM, such as Dieudonné's *Foundations of Modern Analysis* (1960), or Hewitt/Stromberg's *Real and Abstract Analysis: A modern treatment of the theory of functions of a real variable* (1965).

According to these books, a *function* f from set X to set Y is nothing but a subset of the cartesian product $X \times Y$, with the property that for every $x \in X$ there is one and only one $y \in Y$ such that $(x, y) \in f$. To be in line with tradition, one then writes $f(x) = y$ instead of $(x, y) \in f$. It is true that in more recent courses on analysis there has been a tendency to avoid this strict set theoretic definition by now saying that a function f is an object that, given any input $x \in X$, *assigns* an output $f(x) \in Y$, and this object is no longer considered a set.[1] But the language of set theory is still used in such a definition. This can be so even in the case of historically oriented approaches to analysis, as in Hairer/Wanner's *Analysis by Its History* (1996), where one reads on p. 202:

> **(3.1) Definition** (Dirichlet 1837) *A function $f\colon A \to B$ consists of two sets, the domain A and the range B, and of a rule that assigns to each $x \in A$ a unique element $y \in B$. This correspondence is denoted by $y = f(x)$ or $x \mapsto f(x)$.*

[1] See for example Tao (2006), pp. 55–58. The corresponding subset of $X \times Y$ is then called the *graph* of f.

Here, the language of set theory is projected into the year 1837, which is more than thirty years before Cantor actually started his expedition into sets.

We find nothing of that in CPM, and so cardinality questions, for example, which are raised as a matter of course in every modern treatment of analysis, do not turn up in it at all. The same is true of Wittgenstein's annotations to CPM. In the expansions of these annotations, that is, in MSS 126 and 127 (which are the basis of RFM V), Wittgenstein adds some allusions to set theory, but they occur rather seldom[2] and there is only one with a direct connection to CPM. It is on pp. 133f. of MS 126 and has already been quoted and discussed briefly in Sect. 5.6 of this book. It is worth quoting again in the present chapter on Cantor's diagonal method:

> Nowhere is it pointed out by Hardy that the irrational number does not possess a numeral in the sense in which the rational number does possess it. The fiction seems to be that it has an infinitely long one. If anything, of course, the sign for the rule of expansion might count as the numeral. – But this absence of the numeral means an infinitely fundamental difference. And in a certain sense Cantor's diagonal proof shows very well that it *cannot* have a numeral.[3]

I read the last statement as a sort of *ad hominem* argument towards Hardy: although Hardy does not show set theoretic thinking in CPM, he is nevertheless someone who accepts the usual set theoretic results, and in particular the uncountability of the set of real numbers established by Cantor's diagonal proof, with the consequence that not every real number can have a numeral because the set of numerals is countable (where "numeral" can be understood in a very general way, as arbitrary symbolic expression for a real number, say, like "$\sum_{n=0}^{\infty}\frac{(-1)^n}{2n+1}$" for $\frac{\pi}{4}$, and so on).[4]

This argument should not be taken at face value, however, because, as we will see, Wittgenstein's actual attitude towards Cantor's diagonal proof is a scrutinizing one that does not allow us to make use of it in the straightforward way just presented. The passage quoted above is from 17 December 1942. Nearly four years earlier, in his manuscript MS 162a written in January 1939, Wittgenstein mentions Cantor's diagonal argument in connection with CPM in another way:

> Why doesn't Hardy, for example, use the C.ian demonstration when he wants to show that besides the rational ones there are still other numbers? – Because the argument at this level is too difficult? Not rather for the reason that what thereby is emerging doesn't actually look like a further *number*? (He also could have mentioned the number 0.01001000100001..., which obviously isn't a periodic decimal number.) He wanted to present something with which one *measures*. (MS 162a, pp. 29f.)

When, at the beginning of CPM, Hardy shows that there exist irrational numbers, he at first discusses a geometric example involving the number $\sqrt{2}$ (see pp. 3–7 of CPM), but he *might* have referred also to a Cantorian demonstration: all the rational numbers can be arranged in an infinite list, each number represented by its decimal

[2]See MS 126, pp. 11, 35, 51 and 134, MS 127, p. 53, and RFM V §§5–7.

[3]The translations of Wittgenstein's German *Nachlass* passages are always my own.

[4]As in many similar cases, I have replaced Wittgenstein's "irrational number" by "real number" because this makes it easier to present his thoughts.

expansion, and Cantor's diagonal construction then exhibits a number that cannot occur in this list and consequently must be irrational. Why didn't Hardy proceed in *this* way?

Wittgenstein's answer is that Hardy wanted to present something with which one *measures*, and this was in fact the predominant idea of real numbers that prevailed before Dedekind and Cantor offered their more abstract approaches (see Sects. 3.1 and 5.3 of this book). This is not, however, Hardy's actual procedure at the beginning of CPM. When developing his example of $\sqrt{2}$, Hardy doesn't mention measurement at all but makes use of geometry as it is practised in the Euclidean tradition, with its own way of referring to numbers. This way can be presupposed in CPM as a starting point for the Dedekindian approach because it presents real numbers as entities that are familiar to the readers. It presents, as it were, our familiar numbers. In contrast, a number constructed by Cantor's diagonal method may not appear familiar at all because its construction presupposes a whole infinite list of already given numbers and therefore resides at some sort of 'higher level' than the familiar ones. To say, as Wittgenstein did in the passage quoted above, that what is emerging from Cantor's demonstration doesn't even look like a *number*, is a much coarser claim, but we will see that Wittgenstein's actual position was more nuanced.

When thinking about the notion of real numbers, we should of course discuss Cantor's diagonal argument quite independently of Hardy's neglect of it in CPM. This is done repeatedly by Wittgenstein, beginning with remarks in his MS 106 of 1929 until the year 1947, when he developed his own version of a diagonal argument, presumably inspired by passages in Alan Turing's "On computable numbers, with an application to the *Entscheidungsproblem*" (Turing (1936); see Floyd (2012c)).[5] There are at least the following two reasons why Cantor's diagonal method is interesting for Wittgenstein. First, this method is usually presented as a way of showing that the set of real numbers is much "greater" than the set of rational or even algebraic numbers. Of course, what "greater" should mean here can be explained via the "cardinality" of these sets, but Wittgenstein is still bothered because of the extensionalist point of view thereby presented, which he deems inadequate or at least too far removed from our actual practice in mathematics, as has been argued above in Chap. 2 of this book. This anti-extensionalist motive is already visible in 1929, when Wittgenstein wrote:

> If one says, "the set of all transcendental numbers is greater than the set of the algebraic ones", then this is nonsense, it is of another nature. It is not "no longer" countable but simply not countable!
> [...]
> The diagonal method shows the *kind of infinite possibility*.
> [...]
> I cannot arrive at an extensionalist view of the infinite sets. (MS 106, pp. 58–72)

[5]However, and rather astonishingly, in the *Big Typescript*, where real numbers are extensively discussed and where Wittgenstein wrote a chapter specifically about set theory, Chapter 137, entitled "On Set Theory", Cantor's diagonal argument is not granted a single mention.

Wittgenstein's second important reason for scrutinizing Cantor's diagonal method, intimately connected with the first one, has to do with the common view that Cantor's method is a method of *discovery*: that it lets us discover not only new numbers but also new mathematical facts, like the fact that the set of real numbers is of greater cardinality than the set of algebraic numbers. This, however, contradicts Wittgenstein's notorious slogan: "The mathematician is an inventor, not a discoverer" (RFM I §168; see also §2 of Appendix II to RFM I). Again, this is too coarse a statement in view of Wittgenstein's actual approach, and there are instances where Wittgenstein expresses himself more cautiously, for example in MS 126, p. 130: "The principal lack of clarity in mathematics concerns what is *discovered* and what is determined." A yet more refined way of putting this may be to say that the principal lack of clarity in mathematics concerns what aspects of mathematical results resemble discoveries and what aspects resemble inventions; or, alternatively, to say that there are mathematical results that more closely resemble discoveries than inventions and results that more closely resemble inventions than discoveries. To my mind, on the side of discoveries is the age-old result that there are only five regular polyhedra (the five Platonic solids), but I do not know whether Wittgenstein would agree. Most of the cases he discusses are those where he sees mathematical results on the side of inventions,[6] and this is especially the case with results proved by Cantor's diagonal method. On p. 38 of MS 162a the method itself is explicitly called "the Cantorian invention" ("die Cantorsche Erfindung").

There is another way to read Wittgenstein's slogan that the mathematician is an inventor and not a discoverer, a more *therapeutic* one. This is suggested by a remark like the following one on p. 19 of MS 162a:

> That is, I wish you to think of mathematics as complete in each phase and each new calculus not as a discovery but as a new invention.

According to this formulation, one may admit that there are aspects of discovery in mathematics, but that it may be advisable to concentrate on what appears as invented in order to correct misrepresentations of mathematics and to solve or dissolve philosophical problems arising from them. One should then, from case to case, diagnose such problems and try to treat them with a therapy of this sort.

In the following I will be mainly concerned with Wittgenstein's reflections on Cantor's diagonal method published as Part II of RFM, which is based on passages in MSS 117 and 121 written in 1938. To be more precise: the first 22 sections of RFM II are the complete reproduction (and translation) of a connected part of MS

---

[6]I say "most of the cases" since Wittgenstein's decidedly non-dogmatic way of philosophizing allows for exceptions. A beautiful example of this can be seen in RFM VII §6, where he writes: "There is no doubt at all that *in certain language-games* mathematical propositions play the part of rules of descriptions, as opposed to descriptive propositions". Here, already the word "certain" is remarkable because Wittgenstein's philosophy of mathematics is often described as being based on the general view that mathematical propositions do play the part of rules of descriptions, full stop. What is still more remarkable is Wittgenstein's immediate continuation: "But that is not to say that this contrast does not shade off in all directions". I comment on this passage in Mühlhölzer (2012), p. 104.

117, ranging from p. 97 to p. 110, which Wittgenstein himself headed "Ansätze", a word that may be translated as "beginnings"; and the remaining sections of RFM II, the §§23–62, are a selection of remarks on pp. 27r-92v of MS 121, of which I will only discuss those that are explicitly concerned with the diagonal method. There are more than a few remarks on the diagonal method also in the wonderful MS 162a, written in January 1939, which may be seen as continuations of those in RFM II, and I will bring them into play when they appear important to me. In the same way I will treat other relevant Wittgensteinian remarks scattered in his texts. I will refer to sections of RFM II simply in the form "§section-number"; sections in other parts of RFM will be tagged by "part-number §section-number". Before going into RFM II, it will be advisable, though, to say something about Cantor's diagonal method itself.

7.2 Cantor's Diagonal Method

This method can appear in quite different guises, but the paradigmatic one may be the following: Suppose there is given an infinite list of infinite sequences of natural numbers that belong to a specified finite domain, as the domain, say, of the numbers from 0 to 9. Instead of numbers one can also consider the numerals denoting them. This makes no difference in the present context, as long as by "numeral" one means type and not token, and for reasons of simplicity I will only talk of 'numbers'. Let us call the first sequence in this list $a_{11}, a_{12}, a_{13}, \ldots$, the second $a_{21}, a_{22}, a_{23}, \ldots$, and so on.[7] One can think of these sequences as listed one below the other so that a sort of 'square' arises which is infinitely extended rightwards and downwards:

$$
\begin{array}{c}
a_{11}a_{12}a_{13}\ldots \\
a_{21}a_{22}a_{23}\ldots \\
a_{31}a_{32}a_{33}\ldots \\
a_{41}a_{42}a_{43}\ldots \\
a_{51}a_{52}a_{53}\ldots \\
\cdot \\
\cdot \\
\cdot
\end{array}
$$

[7]This notation is sloppy because ambiguous when indices are greater than 9 (what is a_{123}? Is the first index 1 or 12?), but it is an ambiguity that can be easily dispelled (in a later case I will use brackets, for example). Furthermore, in the list of sequences shown below I dispense with the commas of the sequences. I'm sure that both of these abbreviations, done for reasons of lucidity alone, will not lead to difficulties.

In this sort of quadratic arrangement the sequence $a_{11}, a_{22}, a_{33}, \ldots$ forms an infinite "diagonal" of which one can now imagine that it will be altered at any place; that is, for every i the number a_{ii} can be replaced by a b_i with $b_i \neq a_{ii}$. A sequence $b_1, b_2, b_3, \ldots$ that is constructed in this way is called the *diagonal sequence*, or simply the *diagonal*, of the given list of sequences (although the more appropriate name would be "altered diagonal", but this isn't the way one is talking here). Obviously, this diagonal sequence is different from any sequence in the list, since if you suppose that it occurs at list-place k, then this would mean: $b_i = a_{ki}$, for all i, and in particular: $b_k = a_{kk}$, but precisely this is ruled out by the construction of the b_i.

This method of building a sequence different from all the sequences of a given infinite list is constructive through and through if each of the listed sequences is produced by a rule and if the listing itself is produced by a rule, that is, by a rule listing the rules that represent the listed sequences. Because then also the diagonal can be produced by a rule by simply standardizing the alteration of the a_{ii}, for example by defining the diagonal sequence $b_1, b_2, b_3, \ldots$ as: $b_i = 5$, if $a_{ii} = 4$, and $b_i = 4$, if $a_{ii} \neq 4$. In this case, to each i the corresponding b_i can be effectively constructed by first going to list-place i, then developing the sequence $a_{i1}, a_{i2}, a_{i3}, \ldots$ up to a_{ii} and finally checking if a_{ii} is equal to 4 or not: if it is, then b_i should be 5, if not, b_i should be 4.[8]

Such a constructivist view of the diagonal method is the appropriate one from Wittgenstein's non-extensionalist standpoint, and we will see how it is elaborated in RFM II. From this standpoint to talk of a "list", as done so far, is actually infelicitous because what is always meant in the context of Cantor's diagonal method is an *infinite* list, which can never be completed like the lists we normally talk about, and therefore it is too easily understood in an extensionalist manner, without giving much thought to how the lists are produced, as if corresponding rules or methods of production need not exist. Wittgenstein avoids the word "list" and instead uses the word "series" (see, for example, the first motto of this chapter). In the following I will adopt Wittgenstein's way of speaking. In usual mathematical parlance, a "series" is actually an object symbolized as $\sum_{k=1}^{\infty} a_k$ and interpreted as the sequence of the finite sums $\sum_{k=1}^{n} a_k$ with $n = 1, 2, \ldots$, but in what follows this sense of "series" will play no role and can be ignored.

The following is an important example of an application of the diagonal method. Consider equations of the form

$$a_n x^n + a_{n-1} x^{n-1} + \ldots + a_1 x + a_0 = 0,$$

[8]The constructive potential of Cantor's method is expressly emphasized in Fraenkel (1935), and like Fraenkel I use words like "constructive", "rule" and "effectively" in a non-technical sense. Within the frameworks of elaborated versions of constructive mathematics they can, of course, be made precise in different ways, and the word "effective" has a technical sense in the theory of recursion. But in the present context it is not necessary to adopt any of these ways. Further specifications concerning constructivity, the affinity with rules and effectivity may be given when particular examples are considered, but great formal precision cannot be our aim anyway, because Wittgenstein's thoughts do not demand it.

with rational coefficients a_i. A real number x that is a solution of such an equation is called an *algebraic* number, otherwise a *transcendental* number.[9] Now, as is easily shown, all the algebraic numbers can be ordered in a series, and there is actually a variety of simple rules that produce such an ordering.[10] Consider any ordering of this sort. Each algebraic number has a decimal expansion, and the number at place i of the ordered series can then be represented, say, by $\ldots .a_{i1}a_{i2}a_{i3}\ldots$. The three points at the left side of this representation are to be understood as the finitely many symbols in front of $.a_{i1}a_{i2}a_{i3}\ldots$ that are necessary for the decimal representation, but they need not interest us in the following because they do not play any role in the process of diagonalization.[11] In this way, as in our paradigmatic case of diagonalization, one arrives at a 'square' extended infinitely rightwards and downwards:

$$\begin{array}{l} \ldots .a_{11}a_{12}a_{13}a_{14}a_{15}\ldots \\ \ldots .a_{21}a_{22}a_{23}a_{24}a_{25}\ldots \\ \ldots .a_{31}a_{32}a_{33}a_{34}a_{35}\ldots \\ \ldots .a_{41}a_{42}a_{43}a_{44}a_{45}\ldots \\ \ldots .a_{51}a_{52}a_{53}a_{54}a_{55}\ldots \\ \qquad\qquad \cdot \\ \qquad\qquad \cdot \\ \qquad\qquad \cdot \end{array}$$

If one then constructs the diagonal as before – for example with $b_i = 5$ if $a_{ii} = 4$ and $b_i = 4$ if $a_{ii} \neq 4$ – then the number $0.b_1b_2b_3\ldots$ say (or any number of the form $\ldots .b_1b_2b_3\ldots$), which is now called a *diagonal number* of the series, cannot occur in the series and must therefore be a transcendental number.

In such an easy way the existence of transcendental numbers can be proved and this, of course, was well-known to Wittgenstein. Already in 1929, in the passage of MS 106 quoted above, he alludes to it, and it is in fact one of Wittgenstein's

[9]In the present context a *real number* may be understood in the way common in school: as a number with arbitrary decimal expansion, like $0 = 0.000\ldots$, $3 = 3.000\ldots$, $1/7 = 0.142857142857\ldots$, $\sqrt{2} = 1.414213\ldots$, $\pi = 3.141592\ldots$, and so on (the corresponding negative numbers included, of course). This representation is not unambiguous because, for example, 1.2 and 1.1999… (with the 9's going on *ad infinitum*) are the same number, but one can repair this by choosing always the form 1.1999… instead of 1.2, and so on. This has the nice, because unifying, consequence that now every real number is represented by an infinite decimal expansion. In what follows I will make no further mention of this, because it is always easy to take it into consideration.

[10]See Gray (1994).

[11]Of course, the fact that I simply write three points to the left of $.a_{i1}a_{i2}a_{i3}\ldots$ does not indicate that always the same finite sequence of characters is meant, but only that we are not interested in this sequence. And at the left-most place of this sequence a minus sign may occur – which need not interest us either.

standard examples of an application of Cantor's diagonal method.[12] In the literature it is sometimes claimed that the Cantorian proof of the existence of transcendental numbers is an essentially non-constructive one, starting with the assumption that all real numbers are algebraic and then deducing a contradiction from it. But this is an error, because the ordering of all algebraic numbers, and also the decimal expansion of every algebraic number, can be given by a rule such that one also obtains a rule for the diagonal number, which is transcendental, as described above. This proof is without doubt a constructive one and, as expounded in Gray (1994), its construction of the transcendental number can easily be realized by computer programs, even ones that are not only of theoretical interest but turn out to be really feasible.

The proof that is essentially non-constructive and that most people may associate with the diagonal method is Cantor's second proof of the uncountability of the real numbers.[13] It starts with the assumption that the set of all real numbers is countable, considers the elements of this set as ordered in a series (to put it in a Wittgenstein-like manner), albeit not necessarily in a series defined by a rule, and then concludes that the diagonal number of this series cannot occur in it, in contradiction to the assumption; consequently, the real numbers cannot be thought of as ordered in such a series. In other words: they are not countable.

7.3 The Simplicity of Cantor's Idea

This is the very beginning of RFM II:

> 1. In what way does the diagonal method prove that there is a number which – let's say – is not a square root? – It is of course extremely easy to show that there are numbers that aren't square roots – but how does *this* method show it?

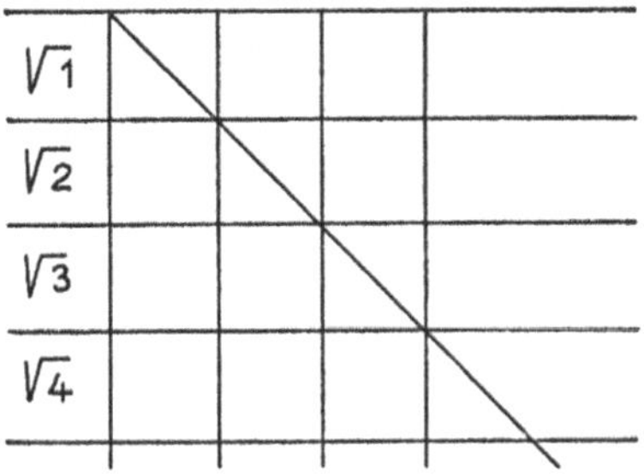

[12]See MS 117, pp. 106f. (the relevant passage published as §16 of RFM II); MS 121, pp. 29v-30r and 38v-39r; MS 162a, pp. 23f. and 54f.; TS 208, p. 33; and TS 209, p. 91.

[13]See Cantor (1891). Cantor's first proof of the uncountability of the reals in Cantor (1874) is a topological one and not as generalizable as the second one, though already in this first proof the scaffolding framework, as it were, of the second one can be discerned; see Kanamori (2012a), pp. 401–409.

> So do we have a general concept of what it means to show that there is a number that is not included in this infinite set?[14]

Wittgenstein's example with the series of square roots (by which he means square roots of natural numbers), which is mathematically ridiculous, of course, functions as the particularly simple basis for several questions that direct his subsequent investigations. These questions are the following: Given an infinite series of real numbers, there may be different methods to show that a number exists that is not included in this series, and Cantor's method is one of them. Wittgenstein wants to know what is *specific* to this method, and there are several aspects of it that aroused his suspicion. One of them is hinted at in the last statement of the passage just quoted: Cantor's method is a totally new one; nobody before him had come up with this idea. Is it nevertheless sensible to say, then, that this method is an answer to an already existing concept of what it means to show that there is a number not included in such an infinite series? That perhaps such a concept was already there at the time when Cantor presented his special method? Do we have a general concept of what that means?

This is Wittgenstein's *first question* concerning Cantor's diagonal method, and his entire subsequent considerations amount to answering it negatively. The *second question* emerges in §2. It is not explicitly formulated there, but this is unnecessary because it suggests itself in view of the beginning of this section:

> 2. "Name a number that agrees with $\sqrt{2}$ at every second decimal place!" What does this task demand? – The question is: is it satisfied by the answer: It is the number got by the rule: develop $\sqrt{2}$ and add 1 or -1 at every second decimal place?

As it stands, Wittgenstein's formulation of this task does not work because $\sqrt{2}$ itself agrees with $\sqrt{2}$ at every second decimal place, and it also does not fit the specific performance of the task mentioned immediately afterwards. What Wittgenstein meant is certainly the following task: Name a number that does not agree with $\sqrt{2}$ precisely at every second decimal place. However, there is a trivial way to perform this task by simply giving the answer of §2, that is, by naming the number produced by the rule: develop $\sqrt{2}$ and add 1 or -1 at every second decimal place (and I understand the word "or" in such a way that, say, one has to add 1 if the place is unequal to 9 and -1 if the place is 9). Wittgenstein asks whether the task stated is satisfied by the given answer, and of course it is not. Obviously, the given answer is simply trimmed in the cheapest way to fulfil the task at hand. If seriously meant, this task must be aiming at quite different answers, answers in which numbers are named in ways that do not have obvious connections to their decimal expansions, as with "$\sqrt[m]{n}$" (for specific numbers m and n), "π", "$\sum_{n=0}^{\infty} \frac{(-1)^n}{2n+1}$", and so on. In this case, performing the task can be extremely difficult, if indeed it is possible at all.

[14]As always I have relied on the Anscombe translation of Wittgenstein's German in BGM, but made changes when it seems appropriate or actually necessary. So already Anscombe's "How far" as translation of Wittgenstein's first two words, "In wiefern" ("In wiefern beweist die Diagonalmethode [. . .]"), is a mistake, clearly shown not only at the end of Wittgenstein's second sentence: "how [!] does *this* method show it?", but also by the whole context.

Now there is an obvious connection of this with naming a number which is not a square root by using Cantor's diagonal method, as imagined in §1: Isn't this method also an all-too-cheap way to perform the task at hand, again by simply trimming the answer to satisfy it? Certainly, the task stated at the beginning of §1 was not meant in such a way that anybody would have thought of the diagonal method at all. Strictly speaking, the actual formulation of this task might even be considered incomplete because its fulfillment could be all too trivial; but even if one accepts this formulation, the solution *via* the diagonal method appears like a lame joke.

That precisely this should be regarded as the point of §2 is made explicit by Wittgenstein himself in §32, which stems from p. 37r of MS 121, supplemented by the unpublished passages subsequent to it up to p. 41v.[15] Wittgenstein does not believe that we possess, as he says in §1, "a general concept of what it means to show that there is a number that is not included in this infinite list", i.e., the infinite list of square roots, and in §31 he introduces the following specific concept adapted to Cantor's diagonal method: the decimal expansion of a real number (and also the real number itself) with decimal places b_i ($i \in \mathbb{N}$) is called *diagonally different* from the decimal expansions of the numbers of a given series of real numbers (and also of the numbers with these expansions itself) with decimal places a_{ik} ($i, k \in \mathbb{N}$) iff $b_i \neq a_{ii}$ for every i (see the explanation in Sect. 7.1).[16] In §32 Wittgenstein then formulates the following task:

> Find a number whose expansion is diagonally different from those in this system.

By "this system" Wittgenstein means an infinite series of decimal expansions of real numbers, and he introduces his formulation with the words: "Thus, there is a *task*", with emphasis on the word "task". The German original is: "Es gibt also eine *Aufgabe*". Amazingly, Anscombe's translation of this is: "Thus it can be *set* as a question", by which she totally gives away Wittgenstein's obvious reference to what he wrote in §§1–2, where, with a similar intent, he uses the word "task" ("Aufgabe") as well. Evidently, Anscombe has not understood what Wittgenstein is doing here, and this is shown also by the fact that the subsequent passages in MS 121, which are necessary for an adequate understanding of §32, are omitted in RFM. There, among other things, Wittgenstein says the following:

> Now there is a big difference / But here is a difference /, whether the rule, starting with the *expansions*, produces the new expansion by their alteration, or whether it has another

[15] I presume that these passages are written *later* than §2, without being totally sure of this, however. According to Wittgenstein's own dating in his manuscripts, it is known that §2, coming from MS 117, was written between 11 September 1937 and 27 June 1938, and the relevant passages from MS 121 are from 12 to 17 June 1938. These dates would in principle allow the converse succession of Wittgenstein's writing of §2 and §32, but in view of their actual content I do not think this very probable and therefore agree with the order that the RFM-editors have chosen.

[16] To be precise, in §31 Wittgenstein doesn't define "diagonally different" for "expansions" but for "extensions"; see my discussion of §31 below. In the present context, however, this difference is of no importance.

starting point but a proof exists to show that the expansion differs step by step from those of the system. (MS 121, p. 37v)

One can operate with an extension in two ways: By operating with the *number* or directly with the places of the extension. "Add 0.1 to π" is the first case, "Interchange the second place of π, n, with $n + 1$" is the second. (MS 121, p. 40v)

There are rules that deal with the pattern of the expansion, and rules that deal with the arithmetic source of the expansion. (MS 121, p. 40v)

"tamper with the extension"[17] (MS 121, p. 40v)

Obviously, to tamper with the extension as in the case of Cantor's diagonal method, analogously to the case described in §2 with regard to the decimal expansion of $\sqrt{2}$, is something trivial, and the more so if compared with the task to find a proof showing that the expansion of a number "with another starting point" (which is sufficiently far away from this expansion) is diagonally different from those of a given system.

Thus, Cantor's diagonal method can only be considered to be a sensible method with regard to specific, appropriately chosen tasks, and it would be useful to get a clear view of them. One sensible kind of such a task may be seen in the case where the construction of a series of the infinitely many given numbers is not an all-too-trivial affair. It is then the combination of this construction with the diagonal method that should interest us. This is so in the case of the Cantorian construction of a transcendental number, as described in Sect. 7.1 of this Chapter. It actually seems to be the easiest method to show the existence of transcendental numbers and even to construct one,[18] and so Cantor's method is certainly valuable here. Another sensible kind of task emerges when one asks whether certain numbers can be ordered in series at all, possibly irrespective of the question of whether the ordering can be constructed or not, as in the case of Cantor's proof of the uncountability of the set of all real numbers. It starts with the assumption that this set *has* an ordering in a series (which is simply what "countability" means) and then gets a contradiction via the diagonal number, which is a real number but nevertheless cannot occur in the series.

So Cantor's method seems to be important in order to obtain substantial mathematical results. Wittgenstein, however, is suspicious and asks whether the usual understanding of these results is adequate. The first two sections of RFM II can be read as a sort of entrance to Wittgenstein's articulation of his suspicion. What he sees is an obvious disparity between the simplicity of Cantor's idea, which may even be taught to school-children (see the second motto to this Chapter), and the supposedly deep and amazing mathematical results gained by it. How can that be? As I see it, this is the central issue of the philosophical investigations concerning Cantor's diagonal method in RFM II, clearly an issue of the form "I don't know my way about", which according to Wittgenstein is characteristic of philosophical

---

[17]This phrase is written in English by Wittgenstein himself.

[18]In the case of the so-called Liouville numbers the construction of the numbers may be considered to be yet more straighforward, but the proof that the constructed numbers are in fact transcendental ones is much more difficult than in the Cantorian case; see the Wikipedia entry "Liouville number" and the beautiful presentation in Kac and Ulam (1968), pp. 9f.

problems in general (see PI §123). Wittgenstein does not explicitly *say* that this is his central issue here, but it can be sensed in many of his formulations, and to my mind it presents the most fruitful perspective on his investigations.[19]

His method to find "his way about" then consists in giving surveyable representations of the use of our words and signs (see PI §122). In the *Nachlass* passage immediately after the text of §32, which up to now I have quoted only in part, Wittgenstein tries to approach such a surveyable representation by recognizing the difference between *essential* and *inessential* real numbers, a difference that shows itself in the way we actually use the relevant words and signs.[20] Interestingly, this attempt starts with an (implicit) reference to CPM:

> If in a textbook like Hardy's someone wants to present examples of irrational numbers, he presents π, e, $\sqrt{2}$, but not $\sqrt{2}$ and what becomes of it if you replace every 5 in its expansion by 3. (MS 121, pp. 37v-38r)

With regard to the last-mentioned number, Wittgenstein then remarks:

> But, we would like to say, we do not know whether it is an *essential* number. It appears to us as if it were a game of writing more and more new places, but as if there were not one number to which they all belonged. As if there were no *essential* operation that would produce all these pieces of an expansion. (MS 121, p. 38r)

What could be meant by "essential" here? In the case of $\sqrt{2}$ one might mention its geometric meaning, as the length of the diagonal of the unit square, and its role as the paradigm of an irrational number: the simplest one and the one possessing the simplest irrationality proof.[21] As for π, this number has a geometric meaning as well, as the ratio of a circle's circumference to its diameter, say, and above all it plays an almost inexhaustible systematic role in mathematics. Such a role can also

---

[19]Precisely the same sort of issue turns up again when, four years later, Wittgenstein discusses Dedekind's theorem; see the exposition in Sects. 5.6, 5.7, and 5.8 in Part I. There is, on the one hand, the presumably deep result concerning the completeness of the continuum, a seemingly "enormous result", as Wittgenstein writes, and, on the other hand, the fact that the proof leading to this result might be understood by a 10-year-old child, as Wittgenstein also says (with typical exaggeration). This tension, however, can be resolved by observing that the presumed notion of "the continuum" is just trimmed in such a way that the result, Dedekind's theorem, immediately follows.

[20]The following quotation again only reproduces parts of the passage after §32. The rest of this passage appears rather dark to me and seems to be of an only tentative character. I refrain from discussing it here because this would require too much commentary.

[21]There are actually almost countless proofs of the irrationality of $\sqrt{2}$; see Sects. 3.4 and 3.5 in Part I of this book and the collection of proofs in <http://www.cut-the-knot.org/proofs/sq_root.shtml>. It can easily be shown, by the way, that for any natural number n the square root $\sqrt{n}$ is irrational if $\sqrt{n}$ is no natural number: Let $\sqrt{n}$ be no natural number and assume that it is rational. Then there exist positive natural numbers a and b such that $\sqrt{n} = a/b$ and that, furthermore, b is minimal in its role as the denominator of a fractional representation of $\sqrt{n}$. We then have: $a/b = \sqrt{n}\,(\sqrt{n}/\sqrt{n}) = n/\sqrt{n} = nb/a$; and *via* division with remainder: $a/b = k + b'/b$ with $b' < b$ and $nb/a = l + a'/a$ with $a' < a$. Because of $a/b = nb/a$, the remainders b'/b and a'/a have to be equal, that is: $b'/b = a'/a$, and this is equivalent to $a/b = a'/b'$. However, because of $b' < b$ this contradicts the presupposed minimality of b.

be recognized in the case of e, and there is the magical-looking equation $0 = 1+e^{i\pi}$ in which π and e are connected, together with the fundamental numbers 0, 1 and i and the elementary operations of addition, multiplication and exponentiation.[22] In comparison to this, numbers defined by tampering with extensions – like: what becomes of $\sqrt{2}$ if you replace every 5 in its expansion by 3 – can certainly be called "inessential".

Now, in its own specific way also Cantor's diagonal method consists in a "tampering with extensions", and Wittgenstein seems to suggest that the diagonal numbers produced by this method may be 'inessential' as well. In the passage from MS 162a, pp. 29f., which is the second quote at the beginning of this chapter connecting Cantor's method with CPM, he even asks whether the diagonal number constructed out of the series of rational numbers looks like a *number* at all. As it happens, Wittgenstein *struggles* with the question of how the diagonal numbers should be assessed in comparison to other numbers, especially to our paradigms of irrational numbers like $\sqrt{2}$ or π, and how the difference is best described. In MS 162a, he calls the diagonal number of a system S, i.e., of a series S of real numbers in decimal expansion, "$D'S$" and on pp. 33–36 his struggle looks as follows:

> Whence the peculiar feeling that indeed $\sqrt{2}$ is a number because – as I am tempted to say – it foresees its whole infinite expansion, whereas $D'S$ is not actually a number because there are always only the parts that I just construct?
>
> But wherein does it lie, then, that $\sqrt{2}$ is – as it were – already *complete*, already completely there, even if its expansion is still under way?
>
> Or: How does it happen that we are not tempted to say that we do not precisely know the number π because we can't write down its numeral entirely. Why does the expansion here appear to us quasi as a byproduct of the number, whereas in the case of $D'S$ it appears to be everything to the number?
>
> Because, obviously, in the case of π the construction of the expansion is in fact only a small part of the calculus with π, because thus the sign π is embedded in an extensive calculus (which gives it its meaning), whereas $D'S$ is nothing but a technique to construct decimal places.[23]

Seen in this way, it may in fact seem adequate to call diagonal numbers "inessential". Even the transcendental number constructed *via* Cantor's method is interesting primarily because of its particularly simple construction and its generalizability, but not with regard to other roles it may play.

However, there is also another way to understand Wittgenstein when he raises the question of whether diagonal numbers are *numbers* at all, be they "essential" or not. A central point of RFM II concerns the fact that Cantor's method of constructing the diagonal numbers turned up as a totally *new* method, in no way anticipated by our preceding practice of talking about numbers, and therefore it may not be *a matter*

[22] With regard to π and e see the beautiful Chapter 5, "What is π?", in Ebbinghaus (1996).

[23] The actual passage is a bit more complicated because it contains formulation-variants that Wittgenstein has not crossed out, but I think that the version I quote here nevertheless presents Wittgenstein's train of thought adequately.

of course to consider this method from the outset as delivering constructions of "numbers". We should now turn to this argumentative strategy.

7.4 The Novelty of Cantor's Idea

In §3 Wittgenstein imagines someone who, at least initially, doesn't conceive the diagonal method as producing in each case only *one* diagonal number, the number with infinite decimal expansion $\ldots .b_1b_2b_3\ldots$, as described above, but, step-by-step, the numbers with finite decimal expansions $\ldots .b_1, \ldots .b_1b_2, \ldots .b_1b_2b_3, \ldots .b_1b_2b_3b_4$, and so on. This is the first paragraph of §3:

> If someone says: "Show me a number different from all these", and is given the rule of the diagonal for answer, why should he not say: "But I didn't mean it like that!"? What you have given me is a rule for the step-by-step construction of numbers that are successively different from each of these.

At first this idea may seem somewhat artificial because Wittgenstein does not himself object to infinite decimal expansions, at least not if they are produced by rules. Accordingly in the second paragraph of §3 he immediately raises the obvious question: "But why aren't you willing to call this too a method of calculating a number?" – that is, calculating *one* single number, namely, the number with infinite expansion $\ldots .b_1b_2b_3\ldots$. Wittgenstein calls it "the number D". So up to this point everything may sound quite innocuous, but Wittgenstein in a way distances himself from the question just stated by setting it in quotation marks and by finishing §3 with the following comment:

> But what is the method of calculating, and what the result, here? You will say that they are *one*, for it makes sense now to say: the number D is bigger than ... and smaller than ...; it can be squared etc.
>
> Is the question not really: What can this number be *used* for? True, that sounds strange. – But what it means is: in which mathematical surroundings is it placed?

According to the extensionalist standpoint that dominates our usual mathematical thinking, people would *not*, of course, say: "they are *one*", whereas Wittgenstein himself regularly proposes doing precisely this, that is, identifying method and result. It is thus puzzling when he writes: "You [!] will say that they are *one*"; but it may be that in this context with the word "you" he is referring to himself. In that case, the justification given in §3 fits well with his own point of view: if the number D is conceived in the form of a method or rule of production, it can nevertheless be used in a way that is characteristic of numbers – to rate it as bigger or smaller than other numbers, to calculate with it, and so on –, so that there doesn't seem to be a sensible reason not to identify this method with the number itself. If I understand Wittgenstein correctly, this is precisely what he means when pointing to the number's "mathematical surroundings": these are the surroundings of talking about "bigger" and "smaller", of calculation, and so on. In §6 he repeats this point with the admonition: "Take a *wider* look round!"

In §4 Wittgenstein further explicates his viewpoint by saying that he is comparing not numbers that emerge as results of certain methods of calculation, but the methods themselves:

> So I am comparing methods of calculating – but there are certainly very different ways of making comparisons. However, I am supposed in some sense to be comparing the *results* of the methods with one another. But this is enough to make everything unclear, for in *one* sense they don't each have a *single* result, or it is not clear in advance what is to be regarded here as *the* result in each case. I want to say that here we are afforded every opportunity of twisting and turning the meanings. –

The examples Wittgenstein has in mind here are those given previously in RFM II: in §1 he mentions methods for obtaining numbers that are not square roots of natural numbers, as, for example, to extract cube roots, like $\sqrt[3]{2}$, or to tamper with the extension of $\sqrt{2}$ by subtracting 1 from the first decimal and leaving the rest unchanged; and there are countless other methods as well, such as forming the arithmetic mean of $\sqrt{2}$ and $\sqrt{3}$, and so on; or, as is the prime subject in §1, by means of Cantor's diagonal method applied to the series of all square roots of natural numbers, producing the diagonal number D. In §3, as just described, Wittgenstein then also mentions the possibility of regarding Cantor's method not as a rule for constructing *one* number, but for the step-by-step construction of numbers that are successively different from each of the given series of square roots.

So Wittgenstein is interested in the comparison of these methods. When adopting the extensionalist point of view, however, one is more interested in the *results* of our methods. Wittgenstein now transforms this into the questions: what should comparisons of methods with precisely this aim look like, and what can be said about the equality or inequality of the results and about other ways of comparing the results? According to Wittgenstein's approach the results are not given in advance, in some sort of realm of abstract entities, say, to be grasped now by appropriate methods, but it is the methods themselves that are considered as primary, and their results are understood by way of the question of when different methods have *the same result*. Wittgenstein does not object to our speaking of "the results", considered as genuine mathematical entities we are talking about. It is true that in his middle phase there are passages that sound as if this were the case, as when in BT he writes that "mathematics is a calculus and therefore is really about nothing",[24] but the later Wittgenstein's view is more differentiated. In the important §10 of PI, for example, he explicitly says that number-words "signify numbers", and the philosophical task then is to better understand this sort of signification.[25] Accordingly, we should now try to gain a fuller understanding of the ways in which methods of calculation are compared with regard to their results.

[24] BT, p. 532. And Wittgenstein then concludes that therefore "there isn't any metamathematics"! In Mühlhölzer (2012) I try to show that this *prima facie* incomprehensible statement, if appropriately understood, can very well be defended, even in the light of Wittgenstein's later philosophy.

[25] On this, see Mühlhölzer (2014a).

Wittgenstein diagnoses some lack of clarity and possible indeterminacy here because, as he now says in §4, "in *one* sense [the methods of calculation] don't each have *a single* result, or it is not clear in advance what is to be regarded here as *the* result in each case". What he has in mind is the reaction described at the beginning of §3, that the diagonal method might be considered not as producing a single number – *the* diagonal number $\ldots.b_1b_2b_3\ldots$ – but successively an infinite sequence of finite decimal expansions $\ldots.b_1, \ldots.b_1b_2, \ldots.b_1b_2b_3\ldots$, and so on. In this case one might say that, on the one hand, a view of this sort presents a sequence of different results; on the other hand, the view suggests itself that it produced a single result only, for aren't the infinite expansion $\ldots.b_1b_2b_3\ldots$ and the sequence of finite expansions $\ldots.b_1, \ldots.b_1b_2, \ldots.b_1b_2b_3, \ldots$ representations of *the same number*? And this, of course, is the view of our normal professional mathematics, according to which the number with expansion $\ldots.b_1b_2b_3\ldots$ is simply identical with the limit of the sequence of numbers with expansions $\ldots.b_1, \ldots.b_1b_2, \ldots.b_1b_2b_3, \ldots$.

Seeing different methods as leading to the same results is in fact on the agenda of normal mathematics. The number with expansion $\ldots.b_1b_2b_3\ldots$ *is* the limit of the sequence of numbers with expansions $\ldots.b_1, \ldots.b_1b_2, \ldots.b_1b_2b_3, \ldots$, even if different methods of calculation may be involved. A real number is considered to *be* – let's say – a certain equivalence class of Cauchy sequences of rational numbers,[26] and the Cauchy sequences of such a class may be produced by quite different methods; and so on. When, at the end of §4, Wittgenstein says: "here we are afforded every opportunity of twisting and turning the meanings", he shouldn't be understood as proposing new ways of practicing mathematics with quite different meanings than the usual ones. Rather, he wants to point out that what we consider to be "the usual meanings" relies on meaning-determinations based on deliberately chosen methods, and not on entities given in advance which we approach and grasp by means of our methods.

Wittgenstein's orientation towards methods of calculation is a central trait of his philosophy of mathematics, consisting of the admonition: look more at what's really *done* in mathematics and less at what we then *say* about it and especially about its results. In §7 he articulates this attitude as follows:

> The result of a calculation expressed in verbal language ["Wortsprache" in Wittgenstein's German] is to be regarded with suspicion. The *calculation* illuminates the meaning of the verbal expression. It is the *finer* instrument for determining the meaning. If you want to know what the verbal expression means, look at the calculation; not the other way round. The verbal expression casts only a dim general glow over the calculation: but the calculation a brilliant light on the verbal expression. (As if you wanted to compare the heights of two mountains not by the technique of measurement of heights, but by their apparent relation when looked at from below.)

[26] See the beautiful exposition in George and Velleman (2002), pp. 73–76. There are other ways to define real numbers, such as the Dedekindean one discussed in Part I of this book, and Wittgenstein's approach can be applied to all of them.

Wittgenstein shows this suspicion towards using verbal language in mathematics already at the end of 1929 and then over and over again. Often he calls the verbal language "prose"[27] and sees it as opposed to mathematics proper, conceived as a sort of "calculus", with the abacus as Wittgenstein's paradigm. From his conversations with members of the Vienna Circle, for example, the following statement is reported, uttered on 19 June 1930:

> Mathematics is always a machine, a calculus. The calculus does not describe anything. [...] The calculus is an abacus, a calculator [*ein Rechenbrett*], a calculating machine; it works by means of strokes, digits, etc. (WVC, p. 106)

Wittgenstein goes as far as to call "prose", and treating with suspicion, even sentences that *prima facie* may appear totally harmless, like "There are infinitely many prime numbers".[28]

The reason for Wittgenstein's suspicion is very simple: a person who doesn't know the proof of a mathematical proposition which proceeds within a certain calculus, may all too easily get the wrong impression of what the proposition's meaning actually is. In the case of the infinitely many prime numbers, for example, the impression may be that the corresponding proposition is a sort of confirmed hypothesis, similar to hypotheses in the natural sciences, which have content already before their confirmation.[29] In reality, however, only the proof of a proposition may convey its proper mathematical sense, in the case of the prime numbers, for example, by seeing that when finitely many prime numbers $p_1, p_2, \ldots, p_n$ are given, the number $p_1 \cdot p_2 \cdot \ldots \cdot p_n + 1$ must be divisible by a prime number that cannot belong to the $p_1, p_2, \ldots, p_n$; therefore this must be a further prime number, and consequently there cannot exist only finitely many. As Wittgenstein says in §7, only by some proof with such a genuinely mathematical character[30] will the proposition

[27] For the first time, if I see it correctly, on p. 14 of MS 108, written on 16 December 1929: "I want to say that a mathematical proposition is not the prose but the exact expression." But the word "prose" ("Prosa") also occurs much later, for example when in VII §41, written on 16 March 1944 on p. 139 of MS 124, Wittgenstein is speaking of "the curse of prose, and particularly of Russell's prose, in mathematics". The subject "prose" is an important and in part controversial one with regard to Wittgenstein's philosophy of mathematics, discussed, for example, in Floyd (2001b) and Mühlhölzer (2012); see also Sect. 3.8 in Part I above.

[28] See, for example, WVC, p. 137, and BT, p. 636. Unfortunately, the English translation in WVC is not correct. Wittgenstein's German statement: "Ebensowenig, wie man sagen kann: Es gibt unendlich viele Primzahlen. Das ist Prosa", is translated as: "Any more than you can say, 'There are infinitely many numbers.' That is everyday prose". Of course, it must be "prime numbers" instead of "numbers", and, above all, Wittgenstein did not speak of "everyday prose" but only of "prose". It is important to understand that with "prose" he means *not* only everyday formulations but also, for example, the sort of quasi-verbalization by Russell's language in *Principia Mathematica*. The opposition Wittgenstein has in mind is not between what is everyday and what is not, but between verbal language and the calculi of mathematics.

[29] Precisely this danger is mentioned in BT, p. 636.

[30] The proof just given, by the way, need not be understood as a proof by contradiction, dealing with the "totality of prime numbers" and involving some notion of actual infinity. It is enough to

about the infinity of prime numbers be adequately illuminated; only then has the finer instrument that determines its mathematical meaning been displayed.

In §8, referring back to §3, Wittgenstein considers the prose sentence "The diagonal method produces an expansion that is different from all the expansions in the given series". This is what he says:

> 'I want to teach you a method as to how with an expansion you can sequentially *dodge* all these expansions ["allen diesen Entwicklungen nach der Reihe *ausweichen*"].' The diagonal procedure is such a method. – "Therefore it produces a series [Wittgenstein actually means an expansion!] that is different from all these." Is that right? – Yes; if, that is to say, you want to apply these words to the case described above.[31]

The sentence "I want to teach you a method as to how with an expansion you can sequentially *dodge* all these expansions" is a beautiful pictorial description of the idea underlying Cantor's method. Wittgenstein puts it in single quotation marks, which I do not understand as presenting this sentence as a prose sentence to be treated with suspicion, but as highlighting the procedural character of Cantor's method. The following sentence in verbal language, however, is put in double quotation marks, and it is this sentence that is the object of Wittgenstein's suspicion: "Therefore it produces [an expansion] that is different from all these." That's the way we talk, and Wittgenstein, of course, doesn't object to it; but he adds: "if [...] you want to apply these words to the case described above".

What is the point of this qualification? In order to understand it we should go back to §1. There, Wittgenstein suggests a comparison of Cantor's method with other methods to exhibit numbers that are not square roots of natural numbers, and the important point is not only, as set out in §2 and then in §32 and the *Nachlass* passages following it, that this method is just trimmed to produce the desired result, but also that it is a totally new one in comparison to methods like extracting cube roots or forming the arithmetic means of consecutive square roots. These methods belonged to our mathematical practice already before Cantor's introduction of the diagonal method, while the latter one was an innovation that extended this practice. So the word "therefore" in "Therefore it produces [an expansion] that is different from all these" *cannot* mean something like "according to our preexisting, established practice"; what it means is rather that we "want to apply these words to the case described above". As I interpret §8, this is the point of Wittgenstein's qualification at the end.

---

show, in the way described, that to any finite set of prime numbers there is another prime number not belonging to them.

[31] I have considerably altered the Anscombe translation of this section, and in particular the word "So" at the beginning of her translation of the sentence immediately after the first dash. In Wittgenstein's German the sentence reads as follows: "Also erzeugt sie eine Reihe, die von allen diesen verschieden ist." The word "Also" is again the start of the prose sentence at the end of §9, again with "So" as Anscombe's translation, and then the start of the important prose sentence referred to at the beginning of §10. There, in §10, Anscombe has translated the "Also" as "Therefore", and I now do precisely that also in the other two cases in order to make clear the thematic connection between these three sections.

Of course, we do not just "want" this on a whim. Perhaps it is even the case that we 'cannot help but' saying: "it produces an expansion that is different from all these", but in doing so we cannot draw on something that was previously regulated. On pp. 12–14 of MS 162a Wittgenstein says it as follows:

> With forming a word for a concept [...] and having also a use for this word that is familiar to us, we are not yet capable of saying how, given deviant circumstances, we will be tempted to apply it.
>
> That is, if I come to you with a new situation, you will have to make a new decision as to its use, or as well recognize that you 'cannot help but' use it like this, follow *this* route. (To say, for example, "there is no construction of a triangulation of an angle with ruler and compass".)
>
> And in this way also Cantor's proof brings a new situation to us and the question is: "What should we say now?"

So in §8 Wittgenstein asks how we want to apply the expression "production of an expansion that is different from all those of series so-and-so" when we are in the new situation opened up by Cantor. And in the succeeding sections, Wittgenstein scrutinizes the following expressions in a similar way: "the real numbers cannot be ordered in a series" (§§16–19); "the expansion given by the diagonal procedures is a real number and it is different from all the members of the series" (§§20–29); "extension which is different from all the extensions of a system" (§§30f.).

As we shall see, this sort of philosophical line of argumentation runs through the entire discussion of Cantor's diagonal method in RFM II, and on pp. 32f. of MS 162a Wittgenstein explicitly counters the objection that it is rather the *truth* of the respective statement that Cantor reveals and not mere conceptual determination:

> Now, how about the following technique of expansion: 'Add 1 to each stage of the diagonal of the system.' After all, this *is* a technique of expansion. And isn't it true that its results do not diagonally agree with the expansions of the system? – The circumstances are totally different ones here (I have introduced a completely new sort of rule), therefore I *need not* (*would not need to*) say the very same. But I *can* say the very same and therefore call the $D'S$ a number different from all the numbers of the system.

I *can* say the very same, and I in fact *want* to say it, but there is nothing fixed in advance that I can appeal to. Therefore, it is misleading to talk about "truths" that are revealed.

At this point it is important to see that the really innovative aspect of Cantor's method is not so much the replacement of the unaltered diagonal places a_{ii} by the b_i with $b_i \neq a_{ii}$, but rather the construction of the decimal expansion $\ldots .a_{11}a_{22}a_{33}\ldots$ of the unaltered diagonal itself. It is this that Wittgenstein is actually concerned with in the context of diagonal arguments. In 1938, when he wrote the text making up RFM II, this was not yet sufficiently clear to him, but on July 30, 1947, he explicitly asserted it: "I have [...] always had the feeling that the Kantorian proof did *two* things, while *appearing* to do only one" (MS 135, p. 60v).[32] What he means by the "first thing" is the expansion of the unaltered diagonal

[32] Wittgenstein rather often writes "Kantor".

of the given a_{ij}, that is, the expansion $\ldots .a_{11}a_{22}a_{33}\ldots$, and by the "second" the alteration to the number $\ldots .b_1b_2b_3\ldots$ with $b_i \neq a_{ii}$ for every i, that is, to what is normally called the "diagonal number". In a sense this may appear to be a trivial remark, but Wittgenstein is in fact making an interesting observation with respect to the "first thing", for if Cantor's method is seen non-extensionally, it turns out that already the unaltered diagonal expansion $\ldots .a_{11}a_{22}a_{33}\ldots$ cannot occur in the given series of expansions $\ldots .a_{k1}a_{k2}a_{k3}\ldots$ $(k = 1, 2, \ldots)$![33] What we get thereby is a Wittgensteinian variant of Cantor's diagonal argument that Juliet Floyd called "Wittgenstein's Diagonal Argument", to be discussed below in Sect. 8.8 at length.[34] Wittgenstein considers the given series as given by rules – or, as he also says, "laws" – for the expansions $\ldots .a_{k1}a_{k2}a_{k3}\ldots$, which he calls $\phi(k, \ldots)$. According to this notation, $\phi(k, n)$ is the nth decimal place determined by the kth rule. Wittgenstein then argues as follows:

> A variant of Kantor's diagonal proof:
> Let $v = \phi(k, n)$ be the form of the laws for the expansion of decimal fractions. $\underline{v}$ is the nth decimal place of the $\underline{k}$th expansion. The law of the diagonal then is: $v = \phi(n, n) =^{def.} \phi'(n)$.
>
> It is to be proven that $\phi'(n)$ cannot be one of the rules $\phi(k, n)$. Assume it is the 100th. Then we have the formation rule
> of $\phi'(1)$: $\phi(1, 1)$
> of $\phi'(2)$: $\phi(2, 2)$
> etc.,
> but the rule for the formation of the 100th place of $\phi'(n)$ is/becomes $\phi(100, 100)$, that is, it tells us only that the 100th place is supposed to be equal to itself, and so for $n = 100$ is *not* a rule. (MS 135, pp. 60r-60v)[35]

In order to understand this sort of proof, one should read the law $\phi'(n)$ as an instruction. For $n = 1$ it says: calculate the first decimal place provided by the law $\phi(1, \ldots)$; for $n = 2$: calculate the second decimal place provided by the law $\phi(2, \ldots)$; $\ldots$; for $n = 100$: calculate the 100th decimal place provided by the law $\phi(100, \ldots)$. But we just presupposed that the law $\phi(100, \ldots)$ is the *same* as $\phi'(n)$! Therefore, this instruction, namely "Calculate $\phi'(100)$ by calculating $\phi(100, 100)$", is identical with the instruction: "Calculate $\phi(100, 100)$

[33] I use the index k now in order to accord with Wittgenstein's notation as it will be presented shortly.

[34] See Floyd's meticulous investigation in her (2012c), which includes much of the historical environment of Wittgenstein's variant. It is probable that the source of it were either discussions leading up to the writing of Turing (1936) with Alister Watson and Turing, or Turing's text itself, but in any case Wittgenstein gave them a form that is more down-to-earth than what the mathematician Turing actually presented.

[35] With minor notational alterations this passage is repeated in TS 229, and I think that the explanation for these alterations (aside from the change of "Kantor" to "Cantor") is merely the fact that the typewriter used did not possess the sign "ϕ". In TS 229, however, for reasons unknown to me, Wittgenstein has deleted the statement about his "feeling" concerning the "*two* things", quoted above, which in the manuscript occurs immediately after Wittgenstein's proof. The typescript version is then published in Z §694 and RPP I §1097.

by calculating $\phi(100, 100)$", which is totally empty; that is: it is "*not* a rule", as Wittgenstein says.

This argument does not work, of course, if one considers the decimal expansions *extensionally*, that is, if one severs the results of the expansion rules from the rules themselves. Then all the expansions lie, so to speak, simply spread out before of us and nothing seems to prevent the unaltered diagonal $\phi'(n), n = 1, 2, \ldots$, of the given series from occurring in the series itself. As soon as one thinks of the rules, however, to act as instructions, the situation changes radically, as shown by Wittgenstein's argument. And this argument shows the richness of Wittgenstein's non-extensionalism in a really beautiful way, because from the extensionalist point of view we are simply blind to what Wittgenstein lets us see.

Nine years earlier, when, at the end of §8, Wittgenstein is saying: "if [...] you want to apply these words to the case described above", he hasn't yet seen this, but his qualms are already directed to our way of seeing the construction of a diagonal as such, be it altered or not. This involves a certain handling of infinity, while the alteration itself – the simple condition $b_i \neq a_{ii}$ for every i – isn't the essential aspect. So when in §3, he imagines someone who does not want to consider the diagonal construction as producing *one* infinite decimal expansion but a sequence of infinitely many finite expansions instead, the alteration of the diagonal isn't actually the important point. The same happens in §9. There, Wittgenstein indulges in certain ways of playing around with variants of the diagonal method, such as the following[36]: Imagine the given series in such a way that the decimal expansion in each row is only developed to the diagonal figure a_{ii}. We then get the following "infinite triangle":

$$\begin{array}{l} \ldots .a_{11} \\ \ldots .a_{21}a_{22} \\ \ldots .a_{31}a_{32}a_{33} \\ \ldots .a_{41}a_{42}a_{43}a_{44} \\ \ldots .a_{51}a_{52}a_{53}a_{54}a_{55} \\ \quad\quad . \\ \quad\quad . \\ \quad\quad . \end{array}$$

Imagine further, that the altered diagonal, that is, the number $\ldots .b_1b_2b_3\ldots$ with $b_i \neq a_{ii}$ for every i, is constructed by respectively setting $b_i = a_{ii} + 1$ or $b_i = a_{ii} - 1$, *depending* on what ensues at a later stage of the

[36]I render what Wittgenstein says somewhat more precisely, but without altering the point he wants to make. This §9 is not important enough to be quoted, with the exception of its end (see below), and I only describe what is really relevant in it.

development of the series, for example, when we go ten steps forward, from $\ldots .a_{i1}a_{i2}\ldots a_{ii}$ to $\ldots .a_{(i+10)1}a_{(i+10)2}\ldots a_{(i+10)(i+10)}$. So, depending on what $\ldots .a_{(i+10)1}a_{(i+10)2}\ldots a_{(i+10)(i+10)}$ looks like, b_i will be $a_{ii} + 1$ or $a_{ii} - 1$, respectively.[37] Of course, such a construction of a diagonal number can also be expressed by a rule, similar to the usual one.

In this case, however, someone might say that *even though* the diagonal number constructed as just described *does* sequentially dodge all the expansions of the given series, the construction is *not* one "that produces an expansion that is different from all these", as the statement in §8 that is set in double quotation marks says. For, as Wittgenstein now writes in §9, one might put forward the following claims: "the development of the diagonal series [that is, the expansion of the diagonal number] never catches up with the development of the other series [that is, the expansions of the given series]"; "there is *always* one of the series [that is, of the expansions of the given series] for which it is not determined whether or not it is different from the diagonal series [once again, Wittgenstein means the expansion of the diagonal number]"; "they run after one another to infinity, but the original series is always ahead [that is, the expansions of the original series are always ahead]".

Someone who talks that way obviously hasn't understood that the diagonal method deals with an infinite series of decimal expansions and that the decimal number is one with infinite expansion as well. And precisely this objection occurs at the end of §9:

> But your rule already reaches to infinity, therefore you already know definitely that the diagonal series [that is, the expansion of the diagonal number] will be different from any other! – – –[38]

But why is this obvious answer set in quotation marks, and what do the three – really three! – consecutive dashes at the end mean?

It is again a sign, I think, analogous to what happened in §8, of Wittgenstein's suspicion towards the prose formulation of this sentence, and the three dashes may be read as shorthand for what Wittgenstein already said at the end of §8: "Is that right? – Yes; if, that is to say, you want to apply these words to the case described above." We *want* this because there isn't a preexisting, established practice that we can refer to in order to justify the sentence,[39] and again, as in §8, the "therefore" in it cannot mean more than that we in fact *do* want this. It is true that for almost all of us this reaction seems to be a matter of course, but it is not a reaction based on a preexisting practice we can refer to and hence can well be considered as the expression of something we simply "want" in the situation at hand. Perhaps we would like to say that it shows that one has *understood* what in mathematics is meant by "reaches to infinity", but according to Wittgenstein it should be seen as

[37]What if a_{ii} equals 9 or 0? Then we should set $b_i = a_{ii} - 1$ if $a_{ii} = 9$ and $b_i = a_{ii} + 1$ if $a_{ii} = 0$, irrespective of the later development of the series.

[38]Again, I have changed the Anscombe translation considerably.

[39]Imagine the situation when Cantor's method was first introduced.

the expression of a new aspect of this understanding which is created for the first time.

One may sense a similarity of this situation with the beginning of Wittgenstein's rule-following considerations in PI §185, where a pupil learns to develop the sequence of even numbers and acts as expected until he reaches the number 1000, but then goes on with "1004", "1008", "1012". He obviously hasn't *understood* what "sequence of even numbers" means, and this may appear to be similar to the person imagined in §8 who hasn't understood what "reaches to infinity" means. However, there is a decisive difference between the two cases. The deviant pupil of PI §185 is already placed in a preexisting, established practice of developing the requested sequence, and the attempts at correcting his deviant behavior can very well refer to this practice. In the case of Cantor's diagonal method, on the other hand, we should imagine the situation when the practice of using it is a totally new one. Nothing previously done by mathematicians anticipated it, and consequently, in contrast to the situation in PI §185, the justification of acting in accordance with it cannot refer to such a practice. Precisely this is the reason why Wittgenstein replies to the "therefore" in the quoted sentences in §§8–9 with the words: "if you want to apply these words to the case described above."

Nevertheless, the rule-following considerations *are* important also with regard to this case, because they should prevent us from looking for things beyond this practice that might be alluded to when we say "therefore". According to Wittgenstein, there is nothing of this sort, and he not only expressed this with the famous words: "That's why 'following a rule' is a practice", with which PI §202 begins and thereby gives a sort of summing-up of the considerations before, but in many other passages as well, for example in VI §28, where we read the succinct statement: "Following according to the rule is on the BOTTOM of our language-game".[40] Consequently, if a *new* practice of following a rule is initiated, as with Cantor's diagonal method, we cannot justify the corresponding new ways of argumentation by referring to anything that was already established before. It is not a case of rule-*following* but of *introducing* a rule. And the question then becomes especially urgent as to how this new rule will actually be used.

Wittgenstein's considerations in §§8–9 are continued in §11, where further ways of looking at the 'reaching to infinity' relevant for Cantor's method are discussed. There, Wittgenstein again refers back to §1, this time to the drawing he had presented there, and he asks:

[40] "*Das Folgen nach der Regel ist am GRUNDE unseres Sprachspiels*", in Wittgenstein's German.

Since my drawing is after all only the *indication* [*Andeutung*] of infinity, why must it be like this:

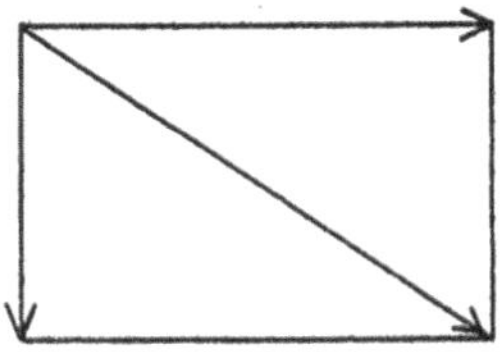

and not like this:

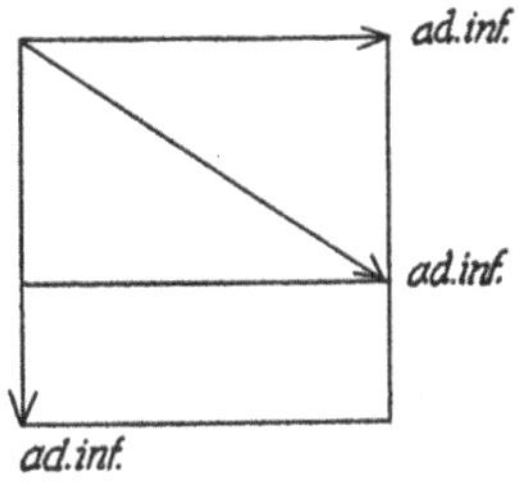

If I understand Wittgenstein correctly, what he has in mind is this: the drawing of §1 is to be interpreted as a 'square' infinitely extended rightwards and downwards, as shown in the following schema:

$$
\begin{array}{c}
\ldots .\mathbf{a_{11}}a_{12}a_{13}a_{14}a_{15}\ldots \\
\ldots .a_{21}\mathbf{a_{22}}a_{23}a_{24}a_{25}\ldots \\
\ldots .a_{31}a_{32}\mathbf{a_{33}}a_{34}a_{35}\ldots \\
\ldots .a_{41}a_{42}a_{43}\mathbf{a_{44}}a_{45}\ldots \\
\ldots .a_{51}a_{52}a_{53}a_{54}\mathbf{a_{55}}\ldots \\
\cdot \\
\cdot \\
\cdot
\end{array}
$$

Likewise, its unaltered diagonal – marked by bold letters – is slanting downwards without end, and it will now be altered at every place so that the "diagonal number" $\ldots .b_1b_2b_3\ldots$, with $b_i \neq a_{ii}$ for every i, emerges. This may be pictured in a drawing like the first one of §11. Infinity is indicated there by the three arrows, while in the a_{jk}-schema above it is indicated by dots, and in the drawing of §1 by lines that we automatically recognize as having no end.

Now, for many of us, maybe even for most of us, it is rather natural to understand the infinities involved as being actually there, as extensions, spread out before us like a particularly large a_{ik}-carpet. The mentioned *indications* – a word emphasized by Wittgenstein himself – are then automatically understood extensionally as well. But according to Wittgenstein's non-extensionalist point of view they must be conceived differently, as pointing not to one infinite square but to the successive production of squares that become bigger and bigger but always have finite extension. Hence, what Wittgenstein envisages is a sequence of finite schemata of the following sort (where, for reasons of simplicity, I now omit the "..." in front of the a_{ik}-rows):

$$\begin{array}{llllll}
\boldsymbol{a_{11}} & \boldsymbol{a_{11}}a_{12} & \boldsymbol{a_{11}}a_{12}a_{13} & \boldsymbol{a_{11}}a_{12}a_{13}a_{14} & \boldsymbol{a_{11}}a_{12}a_{13}a_{14}a_{15} & \ldots \\
 & a_{21}\boldsymbol{a_{22}} & a_{21}\boldsymbol{a_{22}}a_{23} & a_{21}\boldsymbol{a_{22}}a_{23}a_{24} & a_{21}\boldsymbol{a_{22}}a_{23}a_{24}a_{25} & \\
 & & a_{31}a_{32}\boldsymbol{a_{33}} & a_{31}a_{32}\boldsymbol{a_{33}}a_{34} & a_{31}a_{32}\boldsymbol{a_{33}}a_{34}a_{35} & \\
 & & & a_{41}a_{42}a_{43}\boldsymbol{a_{44}} & a_{41}a_{42}a_{43}\boldsymbol{a_{44}}a_{45} & \\
 & & & & a_{51}a_{52}a_{53}a_{54}\boldsymbol{a_{55}} &
\end{array}$$

Here, the three dots going horizontally to the right no longer indicate actual infinity but merely potential infinity. That is, we have to imagine a process by which, as indicated, the squares become bigger and bigger but always remain finite ones. Then, also the unaltered diagonal is conceived in such a way, as a sequence of finite diagonals of these finite squares, and correspondingly the diagonal number actually appears as a succession of decimal expansions that become longer and longer but always remain of finite length.

However, measured against the *result* we are aiming at, the process of enlarging squares appears to be arbitrary, because the a_{ik}-carpet can be knotted in quite different ways, for example in the way indicated by the second drawing in §11. As an a_{ik}-schema it looks as follows:

$$\begin{array}{llllll}
\boldsymbol{a_{11}} & \boldsymbol{a_{11}}a_{12} & \boldsymbol{a_{11}}a_{12}a_{13} & \boldsymbol{a_{11}}a_{12}a_{13}a_{14} & \boldsymbol{a_{11}}a_{12}a_{13}a_{14}a_{15} & \ldots \\
a_{21} & a_{21}\boldsymbol{a_{22}} & a_{21}\boldsymbol{a_{22}}a_{23} & a_{21}\boldsymbol{a_{22}}a_{23}a_{24} & a_{21}\boldsymbol{a_{22}}a_{23}a_{24}a_{25} & \\
a_{31} & a_{31}a_{32} & a_{31}a_{32}\boldsymbol{a_{33}} & a_{31}a_{32}\boldsymbol{a_{33}}a_{34} & a_{31}a_{32}\boldsymbol{a_{33}}a_{34}a_{35} & \\
 & a_{41}a_{42} & a_{41}a_{42}a_{43} & a_{41}a_{42}a_{43}\boldsymbol{a_{44}} & a_{41}a_{42}a_{43}\boldsymbol{a_{44}}a_{45} & \\
 & & a_{51}a_{52}a_{53} & a_{51}a_{52}a_{53}a_{54} & a_{51}a_{52}a_{53}a_{54}\boldsymbol{a_{55}} & \\
 & & & a_{61}a_{62}a_{63}a_{64} & a_{61}a_{62}a_{63}a_{64}a_{65} & \\
 & & & & a_{71}a_{72}a_{73}a_{74}a_{75} &
\end{array}$$

According to this process, our speaking of "diagonals" no longer makes sense, at least with respect to the rectangles that are now its stages, but it is still the elements of the form a_{ii} which are referred to when we construct the diagonal number, and they form an endless sequence, as in the process before. Of course, we can imagine countless other alternatives to this first process, with its successive augmentation of squares, but all lead to the same a_{ik}-carpet and the same diagonal number. However, the infinity of this carpet no longer needs to be understood as actual infinity, but instead as represented by these processes as such, which, in accordance with what

Wittgenstein said in §4, are now considered as *equivalent* with regard to their result. In other words, the supposed actual infinity of the a_{ik}-carpet is now replaced by potential infinity, understood via the processes mentioned, *together* with the idea of an equivalence of them with regard to their result. In this way actual infinity, a notion belonging to the extensionalist point of view, has obtained a non-extensional simulation, as one might say.

The sequence of a_{ik}-rectangles as indicated in the second drawing of §11 may seem pointless if from the outset one thinks of the diagonal number $\ldots .b_1b_2b_3\ldots$, which is constructed with reference to the a_{ii}. But if the determination of the b_i depends on rows that are below the row of the relevant a_{ii}, as imagined in §9, a sequence of rectangles may in fact appear more appropriate than the usual sequence of squares. Hence, we can indicate situations of this sort by means of different pictures accompanied by different ways of talking, as Wittgenstein says in §11. However, he then immediately adds:

> But does anything useful emerge if we have a dispute about the justification of *them*? What is important must reside somewhere else; even though these pictures fire our *imagination* most strongly.

If I understand him correctly, he is pointing here to the fact that these pictures are only *indications* and that the really important thing is the sort of *use* we make of them. This is what he certainly has in mind when referring to what "must reside somewhere else".

But isn't it a strong exaggeration to say then that these pictures "fire our *imagination* most strongly"? I myself do not find them particularly exciting and my imagination remains relatively cool. However, at this point one may think, of course, of the *uncountability* of the real numbers which is shown by means of Cantor's diagonal method.[41] Instead of discussing the ridiculous use of this method via the example of §1, one can see it now in a more general way as showing that for *any* series of real numbers, represented by their decimal expansions, the diagonal number does not occur in it and that therefore the set of real numbers is uncountable. Isn't this a particularly exciting discovery, featuring aspects that in fact may fire our imagination most strongly? For example via the following train of thought:

Consider the closed unit interval of all real numbers from 0 to 1, that is, the set [0,1] defined as $[0,1] := \{x \in \mathbb{R} \mid 0 \leq x \leq 1\}$. Cantor's diagonal argument then shows that this set consists of uncountably many real numbers, but at the same time it has a finite length – or a finite "measure", as one says in mathematics –, that is, length (= measure) 1.

Now consider first only the rational numbers in [0,1]. They have two important properties: first, every irrational number in [0,1] is the limit of a sequence of rational numbers in [0,1]; that is, to each irrational number in [0,1] there are rational ones that are coming arbitrarily close to it. If one imagines this situation pictorially and if one furthermore thinks of surrounding each rational number in [0,1] by a small

[41] Wittgenstein shortly – but with a provocative formulation! – mentions uncountability already in §10 and then in later sections. All this will be discussed below.

interval (a small "neighborhood", as one says), one may get the impression that the union of all such intervals should cover the whole of [0,1], or at least almost the whole of it. Second, the rational numbers in [0,1] are countable, that is, they can be lined up as follows: $r_1, r_2, r_3, r_4, r_5, \ldots$. Now, around each of these r_i an interval of length $\epsilon/2^i$ may be laid, with arbitrary $\epsilon > 0$. If we then add together all the lengths of all these intervals we get the number $\epsilon/2^1 + \epsilon/2^2 + \epsilon/2^3 + \epsilon/2^4 + \ldots$, and this number is equal to ϵ. But ϵ may be arbitrarily small, and this is then also true of the length of the union of all these intervals, which cannot be greater than the sum of all the lengths of the intervals themselves. This, however, means that the length of the complement of this union in [0,1] approaches 1 if ϵ approaches 0, that is, this length approaches the length of [0,1] itself. But it consists of irrational numbers only – and we now seem to be forced to say that the irrational numbers comprise practically the whole of [0,1], despite the fact that each of them can be arbitrarily approximated by rational ones. Isn't this crazy?[42] Furthermore, since the irrational numbers in [0,1] are uncountably many – otherwise the union of them with the rationals would be countable, which is not the case, as revealed by Cantor's diagonal method –, doesn't this show the truly excessive, almost incomprehensible magnitude of uncountable sets?

Confronted with considerations of this sort that lead, or seem to lead, to utterly surprising results, one might develop a view like the following, described by Wittgenstein in §1 of Appendix II of RFM I and called "dominant at the present day":

> a view [...] which values the surprising, the astonishing, because it shows the depth to which mathematical investigations penetrates – as we might measure the value of a telescope by its showing us things that we'd have had no *inkling* of without this instrument. The mathematician says, as it were: "Do you see, this is surely important, this you would never have known without me." As if, by means of these considerations, as by means of a kind of higher experiment, astonishing, nay the most astonishing facts were brought to light.

7.5 Wittgenstein on Uncountability

If we are prone to indulge in considerations like the one just presented, we get a cold shower when reading §10:

> It means nothing to say: "*Therefore* the X numbers are not countable." One might say something like this: I call the number-concept X uncountable if it has been stipulated that, whatever numbers falling under this concept you arrange in a series, the diagonal number of this series is also to fall under that concept.

[42] A variant of the argument just presented, developed in the context of the proof of the irrationality of $\sqrt{2}$ and formulated without notions like "length" or "measure", can be found in Kac and Ulam (1968), pp. 6f. They comment on it with the words: "it is so surprising as to appear nearly paradoxical".

The German original reads as follows:

> Es heißt nichts zu sagen: "*Also* sind die X-Zahlen nicht abzählbar". Man könnte etwa sagen: Den Zahlbegriff X nenne ich unabzählbar, wenn festgesetzt ist, daß, welche der unter ihn fallenden Zahlen du in eine Reihe bringst, die Diagonalzahl dieser Reihe auch unter ihn fallen solle.

Anscombe translates Wittgenstein's "abzählbar" as "denumerable" which is certainly a possible translation but, at least nowadays, not the most common one among mathematicians. The more usual word is "countable". Furthermore, there is the subtlety that Wittgenstein himself deviates from the mathematicians by *not* calling (say) the real numbers "überabzählbar", as it is usual in German, but "unabzählbar" (translated by Anscombe as "non-denumerable"). It may be that Wittgenstein simply wasn't familiar with the German mathematical terminology, but it seems more probable that he deliberately changed the word "überabzählbar" into "unabzählbar" because the former encourages associations with "exceeding what is countable" or "beyond the countable", which he didn't like (I will come to this below). "Unabzählbar" is the more neutral term and it happily corresponds with the usual English word, "uncountable", which we can simply accept, therefore, as the adequate translation. To change it, in order to accord with Wittgenstein's deviation from what is usual in German, might create the impression that Wittgenstein wanted to propose his own sort of mathematics, but this would be totally wrong. Wittgenstein doesn't want to change mathematics itself but the way we often talk and think about it.

In §10 Wittgenstein for the third time looks critically at a "therefore" – in German: "also" – that is supposedly expressing a consequence of the diagonal method. In §8 he had considered the sentence: "Therefore it produces [an expansion] that is different from all these", and he says, or at least insinuates, that the word "therefore" with which it starts cannot mean more than that we "want to apply these words to the case described above", as he writes at the end of §8. In §9 it is the sentence: "But your rule already reaches to infinity, therefore you already know definitely that the [expansion of the diagonal number] will be different from any other!", the "therefore" of which he again treats with suspicion and to which he tends to say (at least according to my interpretation) the same as in §8: "if you want to apply these words to the case described above". Now, in §10, it is the sentence: "*Therefore* the X numbers are not countable", to which he says, even more strongly, that it "means nothing". It is, once again, the "Therefore" – the "Also" – that attracts his attention and which now is even italicized.[43]

---

[43]In the original source manuscript the "Also" is underlined; see MS 117, p. 102. – Strangely enough, this emphasis on "therefore" is totally missed by Hilary Putnam who, in his paper "Wittgenstein and the Real Numbers", indignantly comments on §10 as follows: "He [= Wittgenstein] tells us, amazingly, that it *means nothing* to say of any class X of numbers that it is nondenumerable [i.e. uncountable]" (Putnam (2007), p. 236) – a reaction that is hardly comprehensible because Wittgenstein here explicitly and clearly tells us what this means, namely (to adopt Putnam's "class"-terminology, which actually isn't, of course, Wittgenstein's terminology): that, whatever numbers belonging to the class X you arrange in a series, the diagonal

By "X numbers" Wittgenstein means numbers that fall under the number-concept X, and what he has in mind is any concept X to which the notion of uncountability may be applied. The definition of "uncountable" he then gives is the following:

> I call the number-concept X uncountable if it has been stipulated that, whatever numbers falling under this concept you arrange in a series, the diagonal number of this series is also to fall under that concept.

Here, for the first time in RFM II, the contrast between "countable" and "uncountable" is mentioned! Before, Wittgenstein merely dealt with the diagonal method as producing to a given series of numbers a further number not belonging to this series, while 'cardinalities', as one says in set theory, didn't yet play any role. Now this aspect comes, or seems to come, to the fore. It is certainly verbally harsh to call *concepts* "uncountable" and not their *extensions*, but one can get accustomed to this. The talk about an 'extension' suggests something well-defined, well-delimited, and precisely this will be called into question by Wittgenstein in several cases. So his verbal harshness is justified. The important point here is not that it is *concepts* that Wittgenstein calls "uncountable", but that he *accepts* uncountability. That is, his non-extensionalist view is not, as one might think, confined to concepts that apply only to countably many entities. What he is turning against is not uncountability as such, but certain misleading pictures and verbalizations accompanied by it and the diluting of the notion of uncountability that he diagnoses in set theory. We will come to that.

Wittgenstein's notion of uncountability as just defined obviously applies to the real numbers and, equally obviously, not to the square roots as considered in §1 or to the algebraic numbers. At the same time, however, it makes the "therefore" in "*Therefore* the X numbers are not countable" totally empty because we now have to say that *if* it is the case that whatever numbers falling under the concept of X numbers you arrange in a series, the diagonal number of this series is also to fall under that concept, *then* it is the case that whatever numbers falling under the concept of X numbers you arrange in a series, the diagonal number of this series is also to fall under that concept. Wittgenstein has defined "uncountable" in precisely such a way that, trivially, the statement "*Therefore* the X numbers are not countable" in fact *means nothing*.

Of course, Wittgenstein's proposal to understand "uncountable" doesn't accord with the usual one among mathematicians, which simply is: "not countable", and his judgment *prima facie* sounds utterly provocative, to say the least. But his reasons are actually quite straightforward. The set of real numbers, for instance, *is* certainly "uncountable" *in Wittgenstein's sense*. The mathematicians would add that *therefore*

---

number of this series is also an element of that class. Obviously, this is not nothing. What in fact means nothing, according to Wittgenstein, is the "Therefore", and we now should clarify what he has in mind when saying that. Perhaps Putnam's misunderstanding may have been prevented if he had read the German original of RFM II with its recurrence of the word "also" in §§8–10 that is obfuscated by the Anscombe translation. But his misunderstanding is remarkable nevertheless in view of the emphasis on the word "Therefore" in §10, which Anscombe *has* respected.

it is *not countable*; but it was precisely the point of §§8–9 to argue that *this* sort of "therefore" doesn't mean more than that we *want* to apply the words "not countable" to the case of 'uncountability' in Wittgenstein's sense. Then, however, the real mathematical substance of "uncountable" is captured by precisely this Wittgensteinian sense, and Wittgenstein is right in reasoning that it *means nothing* to say that *because of* Cantor's diagonal method the real numbers are uncountable. Mathematicians, with set theory at the back of their minds, will not accept this, of course, and there is the following obvious objection they can present: it is set theory that gives us the more *general* concept of uncountability, which is very well understood quite independently of Cantor's method, and it is precisely this method by which the uncountability of the real numbers is then revealed. Consequently, it is clear what the "therefore" in §10 means: it means that Cantor's method *shows* us that the number-concept X falls under the *general* concept of uncountability, a concept in which this method is not already included.

It is precisely this reaction against which §10 is directed, and in order to fully understand Wittgenstein's point, we should consider the standard set theoretic argumentation in some detail. It starts with the following definitions of the countability and uncountability of a set M:

(C) M is *countable* $:\Longleftrightarrow \exists f(f$ is a surjection from $\mathbb{N}$ onto $M)$.[44]

Uncountability is then nothing but the negation of countability:

(U) M is *uncountable* $:\Longleftrightarrow \neg\exists f(f$ is a surjection from $\mathbb{N}$ onto $M)$.

In order to show the uncountability of, say, the set $\mathbb{R}$ of real numbers, one then considers an arbitrary function $f\colon \mathbb{N} \to \mathbb{R}$. Such an f can be thought of as putting elements of $\mathbb{R}$ into the following series, with every number in the series represented by its decimal expansion:

$$\begin{array}{c} f(0) \\ f(1) \\ f(2) \\ f(3) \\ f(4) \\ \cdot \\ \cdot \\ \cdot \end{array}$$

[44]A *surjection* of a set A onto a set B is a function $f\colon A \to B$ such that to each $b \in B$ there is an $a \in A$ with $f(a) = b$. – There are other standard definitions of "countable", but in our usual set theoretic framework they are all equivalent. (This is not the case, for example, in Brouwer's intuitionism, but in the present context we need not concern ourselves with this sort of complication; see the perspicuous presentation in van Dalen (2013), pp. 318f.)

If one now *assumes* that $\mathbb{R}$ is countable, there must be such an f that is surjective, which means that in the corresponding series *every* number of $\mathbb{R}$ occurs. By applying the diagonal method, however, we get a real number not appearing in the series. *Therefore*, the assumption must be false, and *therefore* $\mathbb{R}$ cannot be countable, that is, it must be uncountable. *QED*

What fault may be found with this seemingly simple proof? – There are in fact well-known objections to it and Wittgenstein was aware of them. They come in particular from constructivist angles, for example against the very general use of the negation and the existential quantifier in it, against the view of the real numbers as forming a completed set and that functions may be as irregular as one pleases as long as they are functions (i.e., that to every element of their domain precisely one element of their range is assigned). Objections of this sort can be heard in some of Wittgenstein's considerations, but his "It means nothing'" in §10 goes in another direction.[45] What he has in mind may be delineated as follows:

The notion of *countability* has mathematical substance far beyond its definition in the meager set theoretical language as presented in (C), because there are so many cases of countable sets in which the relevant functions f from $\mathbb{N}$ onto M can be given by easily usable algorithms, for example, if M is the set of the rational or of the algebraic numbers. In these cases there are numerous methods of enumeration that show a pleasant sort of mathematical concreteness. This, however, is totally different in the case of (U), *if* one doesn't already know Cantor's diagonal method. In this case, we merely have the set theoretic definition, which is only a meager formal scaffold that, considered as such, lets us down completely if we look for real mathematical substance, for example, when we aim at a proof of the uncountability of the set of real numbers.

In itself, (U) merely presents what Wittgenstein calls a "phraseology" whose mathematical usefulness remains to be seen.[46] The problem is that before Cantor's introduction of his diagonal method, one did not know what to do with the negation occurring in (U) ("there is *no* surjection of $\mathbb{N}$ onto M").[47] There were various methods to show that a set is countable – but how could one give the negation mathematical substance? No mathematically respectable ideas existed to

[45]Nevertheless, his notion of uncountability has a clear constructivist complexion. For example, a quite similar notion was entertained by Brouwer in his dissertation of 1907. Brouwer had given it a name, "denumerably unfinished set", but he ceased to use this notion afterwards (see van Dalen (2013), p. 112). I think it can be ruled out that Wittgenstein knew of this Brouwerian conceptual invention.

[46]The term "phraseology" is used, for example, with respect to the conception of real numbers as Dedekind cuts; see Chap. 5 of Part I of this book.

[47]This claim may seem to be wrong in view of Cantor's *first* proof of the uncountability of $\mathbb{R}$, which didn't make use of the diagonal method, at least not explicitly. However, it is possible to see already in this proof a sort of skeleton of the diagonal method. For this reason, in what follows I will ignore this historical complication (on this, see Kanamori (2012a), Section 1.3), and from now on I simply pretend that it was the diagonal method that made it possible for the first time to prove the uncountability of $\mathbb{R}$. If I see it correctly, also Wittgenstein makes this presupposition. I do not think that it weakens his arguments in the least.

address this issue. The situation was similar to the question in Euclidean geometry with respect to whether the construction of a trisection of arbitrary angles or the construction of the regular heptagon is possible, given merely the methods permitted. How could one show that such constructions are *not* possible? Only with the advent of Cartesian analytic geometry, which allowed the question to be tackled using arithmetical methods, could they be answered mathematically.[48] Analogously with the question of how to show that $\mathbb{R}$ is not countable: only the diagonal method, newly invented by Cantor, allowed a mathematical treatment. Only then did this question attain mathematical substance.

According to Wittgenstein, the mistake with the usual point of view lies in a wrong conception of the *generality* of the set-theoretic notion of uncountability as defined by (U). He explicitly refers to the "generality" of set theory on pp. 15–18 of MS 162a:

> What does the 'generality' of the calculus of sets ["des Mengenkalküls"] consist of? Not in the fact that it is a simple picture of a class of other ('special') calculi? Is it also general without a relation to them?
>
> What does the 'enlightened simple-mindedness' that Littlewood is talking about consist of? Doesn't it consist of believing that what the so-called *general* mathematical investigation brings to light are truths that are then carried out by the more special investigations only in more detail? Whereas the general calculus is 'general' only by referring to special calculi. Because in mathematics nothing is in the *word* but everything in the calculus. That is, because the sign brings no other meaning into pure mathematics than what the calculus itself bestows on it.
>
> Mathematics doesn't consist of considerations. And a part of mathematics can be called 'abstract' only insofar as its application, albeit hinted at, is nevertheless left vague.

Wittgenstein doesn't explain what he means by "calculus of sets" here, but his explicit reference to Littlewood makes clear that what he has in mind must be Littlewood's *The Elements of the Theory of Real Functions* of 1926, in which the expression "enlightened simple-mindedness" occurs at the beginning of the Preface. Despite its title, which actually suggests an analysis book, Littlewood's book is a rather elementary introduction to set theory,[49] formulated in such a way that "calculus of sets" can be seen as an appropriate characterization. Littlewood doesn't present set theory in axiomatic form, but he nevertheless gives it the appearance of a particularly abstract sort of calculus. However, Wittgenstein's "calculus of sets" may also be interpreted to include axiomatizations of set theory, especially if they are presented with a universal and foundationalist intent similar to the one already raised by Russell/Whitehead's *Principia Mathematica*. Axiomatizations of this sort have the character of "calculi" insofar as they proceed according to precise rules of building their formulae and of inferring formulae from the axioms and from each other.

Wittgenstein is extensively concerned with *Principia Mathematica* in the MSS 122 and 117, written in 1939 and 1940, which are the basis of RFM III, and the

[48] As for the case of the regular heptagon see Mühlhölzer (2001).

[49] Actually, Littlewood doesn't speak of "sets" but of "classes", but this is only a verbal matter.

investigation he presents there can be transferred also to "calculi of sets", at least if these calculi are regarded as foundations of mathematics. In RFM III Wittgenstein stresses the unsurveyability of most of the proof-patterns occurring in endeavors of this sort, and he develops two argumentative strategies against them: First, he shows that there are problems with the criteria of identity for the unsurveyable proof-patterns, and second, he points out that by making these patterns surveyable, we rely on concepts and procedures that go beyond the foundational frameworks. When we take these concepts and procedures seriously, mathematics no longer looks like a uniform system, as portrayed by these frameworks, but rather appears to be a mixture of different techniques.[50]

In the passage of MS 162a just quoted, the relationship between the calculus of sets and the special mathematical calculi (calculi that involve special techniques, for example techniques to calculate the derivative of specific functions or to order certain numbers into series, etc.) is seen in another way: that the former calculus is "a simple picture of" a class of the latter ones and that precisely this constitutes its "generality". In other words: that within the calculus of set theory certain abstract structures can be identified that we may prescind from the special calculi, without however going into technical details, especially not details concerning particular examples. Wittgenstein's reference to Littlewood's "enlightened simple-mindedness" indicates that he means precisely this. As already said, this expression occurs in the preface of Littlewood (1926), which begins as follows:

> THIS book contains the substance of lectures I gave annually for some years at Trinity College. These lectures are intended to introduce third year and the more advanced second year men to the modern theory of functions. The subject-matter is very abstract, but hardly any of it involves genuinely difficult technique. The aim of the lectures, indeed, is to inculcate the proper attitude of enlightened simple-mindedness by concentrating attention on matters which are abstract but not complicated.

Here Littlewood gives the impression of believing that the *general* mathematical investigation, which he calls *abstract*, in fact, as Wittgenstein says, "brings to light truths that are then carried out by the more special investigations only in more detail". Contrary to that, Wittgenstein sees the relationship between general and special calculi the other way round, namely: that "the general calculus is 'general' only by referring to the special calculi". Maybe Wittgenstein's formulation that the applications of these abstract and general calculi are "left vague" is not very felicitous,[51] but what he means appears rather obvious. This sort of "vagueness" is evident, for example, in the case of the notion of uncountability when merely explained by the abstract definition (U), because (U) completely leaves us at a loss as to possible proofs of uncountability. The perspective of such proofs is only opened when Cantor presents the 'special calculus' that involves his diagonal method.

[50] See Mühlhölzer (2006, 2010).

[51] The manuscript shows that he struggled with the adequate formulation: he first wrote "unclarified" ("im Unklaren"), but then crossed it out and wrote "vague" above it. I do not think, however, that he was really content with that.

It is important to see here the difference with *empirical* concepts. Our empirical judgments essentially rely on perception or experimentation, whereas nothing analogous can be said in the purely mathematical case. When we recognize that the real numbers are uncountable, the mechanism of our recognition is totally different, far away from perception or experimentation. What it involves instead is the invention of a new mathematical method. For Wittgenstein this difference has the consequence that the particular *generality* cannot be the same. In the case of empirical theories it means that the general concepts apply to a large class of objects, typically comprising many subclasses, and that our understanding of these concepts is independent of our access to these objects and subclasses. This access is established only afterwards by perception and experimentation. In mathematics, on the other hand, our proper understanding of the general concepts relies on our competence concerning special cases, and precisely this constitutes their 'generality'. This is what Wittgenstein wants to convey here.

To adequately understand the difference between what is characteristic of mathematics, on the one hand, and empirical endeavors, on the other, is actually the basic motive behind Wittgenstein's philosophy of mathematics in its entirety. He thereby admits to feeling a strong tendency towards *empiricism* insofar as an empiricist approach – an approach that, for example, regards a calculation as a sort of experiment – may appear to be the only *realistic* one, in a non-philosophical sense of "realistic", and that everything else must be moonshine or obscurantism (see III §76). The challenge for him, then, is to recognize what is wrong with this tendency without becoming unrealistic. He explicitly says this in VI §23: "Not empiricism and yet realism, that is the hardest thing."[52] For him, to become unrealistic means in particular to assimilate mathematical propositions to descriptions of facts, making a result like, say, that the set of real numbers is uncountable, "look like a fact of nature", as he writes in §19. And the way he avoids both, empiricism and such unrealistic points of view, then consists in recognizing that what may appear to be a fact-oriented description is in reality "the determination of a concept – *concept-formation*", as said in §19 as well. With particular emphasis this sort of way-out is expressed in the first sentence of IV §29: "The limit of the empirical – is *concept-formation*." Wittgenstein's discussion of Cantor's diagonal method in RFM II can be understood as an exemplary presentation of this insight with regard to the uncountability of the real numbers.

Note, by the way, that in the long passage in MS 162a, pp. 15–18, quoted above, when talking about "application" Wittgenstein must allow *intra*-mathematical applications, for example applications of general calculi concerning functions to more special ones, as envisaged by Littlewood. And Wittgenstein's usage of the word "meaning" – "the sign brings no other meaning into pure mathematics than what the calculus itself bestows on it" – does not involve extra-mathematical applications either. This stands in striking contrast to the notorious "mufti"-passage in V §2, where Wittgenstein says that it "is the use outside mathematics, and so

[52] On this issue, see Diamond (1991), Chap. 1, and Mühlhölzer (2010), Chap. 7 of Part II.

the *meaning* of the signs, that makes the sign-game into mathematics". That this passage should not to be taken at face value is already emphasized in Sect. 3.2 of Part I of this book, an assessment that is now confirmed by the passage quoted above. What I still sense as puzzling, however, is the chronology: the latter passage is from January 1939, whereas Wittgenstein wrote the "mufti"-passage much later, in October 1942, without any mention of his earlier thoughts! One must say, I think, that Wittgenstein doesn't seem to attain a stable view of the role of extra-mathematical applications, that he wavers between different tendencies. This sort of uncertainty is not uncharacteristic of his philosophy of mathematics as a whole.

The most important point of the passage in MS 162a, pp. 15–18, however, concerns the idea of "generality". According to Wittgenstein, *before* Cantor's invention of the diagonal method the concept of uncountability, in the plain sense of "not countable", was not a *general* concept but an *empty* one. Conceptual emptiness is not the same as conceptual generality. And even by Cantor's method this concept does not gain real "generality" – one would need *more special* methods than only this one – but it gains substance.

There is a further possible danger that Wittgenstein sees in the "calculus of sets", due to the fact that its presentation may be associated with "considerations", as he writes in the passage quoted above. The set theoretic language and its axioms then appear as a sort of prose which he treats with suspicion. It is strictly formalized prose, but prose nevertheless, similar to the prose in Whitehead/Russell's *Principia Mathematica* which is condemned in VII §41, where Wittgenstein talks about "the curse of prose, and particularly of Russell's prose, in mathematics". The danger lies in its neglect of the real mathematical practice, in particular of the way we really devise mathematical proofs. Set theory is at most a faint image of this practice, and in any case one that leaves out the important point: the way we use our mathematical symbolism within this practice. In Wittgenstein's own almost poetic words: it misses the *life* of our signs in mathematics.[53]

There is also another way to see this sort of Wittgensteinian argumentation, one that, rather than making use of the notion of prose, deals with our real mathematical *interest*. It may be compared with a beautiful argument by Frege against formalist positions in mathematics that can be adapted to our present situation. Frege's argument occurs in §90 of Volume II of his *Basic Laws of Arithmetic*, where he writes:

> That [in the case of purely formalistic solutions of mathematical tasks] it would take mental effort is evident[[54]], just as in the similar task in chess of arriving at a given end position from the initial set up according to the rules of the game, where there would be no talk of thoughts expressed by different positions, and no move could be interpreted as an inference [in contrast to the thinking and inferring in mathematics when it is non-formalistically conceived]. Thus although mental effort would be involved, the train of thought which accompanied the matter for us and which, in fact, bestowed interest upon it, would be completely missing.

[53] On this see Mühlhölzer (2014b).

[54] In Frege's German: "dass dabei geistige Arbeit geleistet werden müsse, versteht sich von selbst".

It seems to me that Wittgenstein can agree with this sort of argument if only the idea of an "accompanying" train of thought is replaced by his own view of thoughts that brings behavioral criteria into the foreground. But the central idea expressed by Frege – the idea that we should look at the *right* trains of thought, those that really deserve our *interest* – is highly relevant also in our present context. Even if backed up by axioms of set theory, (U) as such doesn't give the notion of uncountability sufficient mathematical substance because it doesn't contain ideas that lead to proofs of uncountability, and these are the ideas mathematicians are really interested in.

As I understand him, it is considerations like these from which Wittgenstein draws the consequence of connecting the notion of uncountability intimately with the diagonal method as expressed in §10: "call the number-concept X uncountable if it has been stipulated that, whatever numbers falling under this concept you arrange in a series, the diagonal number of this series is also to fall under that concept". Seen in this way, he has in fact good reasons to say that the "Therefore" in the statement "*Therefore* the X numbers are not countable" *means nothing*.

It should be conceded, however, that this cannot be Wittgenstein's last word on the matter. The uncountability of the irrational numbers, for example, cannot be shown, or at least not shown without further ado, by applying the diagonal method. For if we have a series of decimal expansions of irrational numbers, represented as usual by an a_{ik}-schema, we must take care that, say, the a_{ii} are not all the same, because the usual method of constructing the diagonal number then would *not* lead to an irrational number. So in this case the diagonal method is not enough to show uncountability. The normal proof looks as follows: we *assume* that the irrational numbers are countable, representable as a series of the form $i_1, i_2, i_3, i_4, i_5, \ldots$, and then, together with one of the well-known series $r_1, r_2, r_3, r_4, r_5, \ldots$ of rational numbers, we can produce a series of all the real numbers, for example the following one: $i_1, r_1, i_2, r_2, i_3, r_3, i_4, r_4, i_5, r_5, \ldots$. This, however, is not compatible with the uncountability of the real numbers – the term "uncountability" not only understood quite abstractly, as in the definition (U), but also in the Wittgensteinian sense explicated in §10. This uncountability is shown in fact with the help of the diagonal method; in the case of the real numbers Wittgenstein's definition of "uncountable" in §10 is applicable; but it fails in the case of the irrational numbers.

Nevertheless, Wittgenstein is certainly right in highlighting the diagonal method as *the* method to give the notion of uncountability its mathematical substance, and his pointed statement in §10 – "It means nothing to say" – should be understood accordingly. The core question behind all this is expressed in §12: "What can the concept 'uncountable' be used for?" It cannot be used if we only rely on the definition (U), and it can be used if we explain it as proposed in §10. Then we can show in a mathematically respectable way that, for example, it applies to the real numbers.

In §13, however, Wittgenstein gives a much more concrete answer to the question just raised:

> One might say after all – if anyone tried day-in day-out 'to put all real[55] numbers into a series' we would say: "Leave it alone; it means nothing; don't you see, if you had established a series, I should come along with the diagonal series [that is, the decimal expansion of the diagonal number]!" This might get him to abandon his undertaking. Well, that would be a benefit. And it strikes me as if this were the whole and proper purpose of this method. It makes use of the vague notion of this man who goes on, as it were idiotically, with his work, and it calms him down by means of a picture. (But one could get him to resume his undertaking by means of another picture.)

At first glance this is an embarrassing passage, and it doesn't come as a surprise that people ridiculed it.[56] Does Wittgenstein really want to tell us that to block the described attempt to put all real numbers into a series, performed by a sort of idiot, is "the whole and proper purpose of [the diagonal] method"? Isn't the diagonal method a paradigm of a mathematical method that has multifarious applications? Not only does it have a central place in set theory but, for example, also recursion theory is almost unthinkable without it. And Wittgenstein must have been aware of the essential role of this method in the proof of Gödel's First Incompleteness Theorem. How is this passage then to be understood?

In the contexts just mentioned, note that the diagonal method is normally used as a method of proof only, not as a method with a direct and concrete impact on actual attempts at putting numbers into a series and on corresponding activities, not to mention quasi "idiotic" ones. There is an ambiguity here in the term "method" when talking about the diagonal method: On the one hand, this method can be used in the down-to-earth sense as presented by Wittgenstein in §13. Meant in this way, it is not at all incomprehensible when he writes: "And it strikes me as if this were the whole and proper purpose of this method" – to wit: to prevent someone from actually trying to put all the real numbers, say, into a series. On the other hand, this method can be viewed – and this is what is usually done in mathematics – as something merely *thought of*, as something purely *theoretical*, that plays an essential role in proving theorems. One can see the point of §13 to emphasize precisely this ambiguity.

Of course, it is absurd to imagine a person *actually* trying to put all real numbers into a list. One should interpret §13 in such a way that this person aims at a *procedure* for listing the real numbers. Maybe, as a modern person, she thinks of developing a computer program that represents such a procedure – and then pointing to Cantor's diagonal method or, as Wittgenstein says in §13, to a picture that corresponds to this method, may really prevent the person from pursuing her project.

Why does Wittgenstein again say, as in §10, "it means nothing" – now that it means nothing "to put all real numbers into a series"? Instead, one might want to say at this point something like the following: One cannot put all real numbers into

[55] Wittgenstein writes "irrational" here but, as explained above, in the case of the irrational numbers it is not guaranteed that the diagonal method leads to the desired result. In this and similar cases I tacitly change Wittgenstein's "irrational" to "real".

[56] See, for example, Hodges (1998), p. 6.

a series (in the sense of: one cannot find a procedure to put all real numbers into a series), for any series of real numbers wouldn't contain the diagonal number of it, which is a real number itself. Why doesn't Wittgenstein say that? What he balks at here is the "cannot". In *Zettel* we find the following illuminating remark:

> Do not say "one cannot", but say instead: "it doesn't exist in this game". Not: "one can't castle in draughts" but – "there is no castling in draughts"; and instead of "I can't exhibit my sensation" – "in the use of the word 'sensation', there is no such thing as exhibiting what one has got"; instead of "one cannot enumerate all the natural numbers" – "there is no such thing here as enumerating all the members".[57]

We cannot enumerate all the natural numbers – in the sense of "actually numerate" and not merely "think of being enumerated" – because there are infinitely many. But isn't this, contrary to what Wittgenstein says, an important factual truth and not only something that "doesn't exist in this game", i.e., the game of mathematics? As already quoted in Chap. 2 of Part I of this book, Russell even called this impossibility a merely 'medical' one.[58] There are, in fact, thought experiments that present our counting of the natural numbers as proceeding in such a way that the counting steps become faster and faster: from 1 to 2 in one second, say, from 2 to 3 in half a second, from 3 to 4 in a quarter of a second, and so on. In this way we would be finished in two seconds! Isn't the reason why we cannot do this simply that such a fact is a medical – or let's say: physical – impossibility? People even speculate about physical models of the universe in which this impossibility does not hold.[59]

From Wittgenstein's perspective, however, considerations of this sort are beside the point because they are far away from our actual mathematical practice, i.e., from what Wittgenstein calls "the game" which, in the present context, means our actual "game of mathematics". When doing mathematics we acknowledge characteristic standards of what is right and what is wrong, and these standards determine what has mathematical sense and what does not. According to this practice we do not bestow sense on our assertions about "all the natural numbers" by fantasizing, like Russell, a run through their totality, but by devising specific methods in order to gain insight into this totality, such as mathematical induction and other methods, methods that can become arbitrarily complicated. According to this practice there is no "enumerating all the natural numbers" – as there is no castling in draughts. In other words: it means nothing to talk about enumerating all the natural numbers, as it means nothing to talk about castling in draughts.

Similarly, Wittgenstein might have written in the present case, the case of §13: Do not say "one cannot put all real numbers into a series" but say instead: "it doesn't

[57]Z §134 (instead of "natural numbers" Wittgenstein actually has written "cardinal numbers" by which, however, he means, as almost always, finite cardinal numbers, i.e., natural numbers). The first occurrence of this passage is in MS 116, pp. 178f., which is presumably written in 1938, that is, in the same year as the text of RFM II; further occurrences are in MS 158, p. 57 (also written in 1938), and TS 222 §137.

[58]See Russell (1936), p. 143. A similar point of view is expressed in Hardy (1929), p. 5, also quoted in Chap. 2 of Part I.

[59]See, e.g., Copeland and Shagrir (2007).

exist in this game a putting of all real numbers into a series", or "there is no such thing here as putting all real numbers into a series". This simply doesn't belong to our game of doing mathematics and there is no "cannot" in the sense of a factual impossibility – "factual" in whatever sense. And the connection of this with Cantor's diagonal method also belongs to our mathematical game and has no factual basis, again in whatever sense of "factual". Wittgenstein has expressed this already in §§8–9 when writing: "if you want to apply these words [namely: "Therefore it produces an expansion that is different from all these"] to the case described above [i.e., the case involving the diagonal method]." We in fact *want* this, we incorporate it into our game of doing mathematics, and also the person referred to in §13 wants this when abandoning her undertaking of putting all real numbers into a series. She abandons it because she accepts that it *means nothing*.

This interpretation is confirmed by §§19–20.[60] In §19 Wittgenstein says the following:

> The dangerous, deceptive thing about the formulation: "One cannot order the real numbers into a series", let alone "The set ... is not countable", is that it makes what is the determination of a concept, the formation of a concept, look like a fact of nature.

My manner of speaking of above – that these impossibilities should not be understood as "factual" ones – is now expressed by the more concrete saying that assertions like

(a) One cannot order the real numbers into a series

or

(b) The set ... is not countable

are dangerous, are deceptive, because, as Wittgenstein writes, they create the impression that one is dealing with facts of nature.

In the case of (a), one may think of facts concerning ourselves, as Hardy or Russell may have had in mind: both consider our inability to actually produce infinite sequences as a sort human weakness. But in the case of the real numbers even an overcoming of this weakness would not be of any use to us. In the case of the real numbers we still couldn't produce such a sequence. And with regard to (b), one may think of, so to speak, "mathematical facts of nature", as Wittgenstein mentions them in §40: as facts concerning "the natural history of mathematical objects themselves" which often reveal "the mysteries of the mathematical world". It is Hardy, of course, who is famous for this sort of view, explicitly stated, for example, in his essay "Mathematical Proof" (1929), where he says: "I have myself always thought of the mathematician as in the first instance an *observer*, a man who gazes at a distant range of mountains and notes down his observations".[61] In

[60] Also by what Wittgenstein wrote on p. 52 of MS 162a: "The appearance of impossibility (of the cannot) here arises through the way in which we introduce the concept. When a determination that forbids the ordering is subsequently introduced like a discovery about the already complete concept."

[61] Hardy (1929), p. 18.

both cases, (a) and (b), one veers away from the calculational core of the diagonal method.

As a corrective, Wittgenstein proposes that assertions like (a) or (b) be read as determinations or formations of concepts. Of course, Wittgenstein does not have a "theory" of concepts and of concept-determination/formation: here he is rather referring back to what he has said in former sections. With respect to (a), §16 is relevant, where he writes: "you want to show a difference between the *use* of 'root', 'algebraic number', etc. on the one hand, and 'real number' on the other" (my emphasis), a difference that shows itself in the fact that for ordering roots and algebraic numbers into a series, we have specific methods, whereas in the case of real numbers we have a totally different method to show that such an ordering is out of the question. Furthermore, there is no antecedent, superordinate concept, at least not in any substantial sense, that may be applied in both cases. Seen in this way, it doesn't appear inappropriate to call the introduction of new methods of this sort "concept-determinations/formations".[62] Case (b) has already been dealt with in §10, where Wittgenstein writes: "I *call* the number-concept X uncountable if it has been *stipulated* that, whatever numbers falling under this concept you arrange in a series, the diagonal number of this series is also to fall under that concept" (my emphases). Also in this case, it is the new method introduced by Cantor to deal with concepts like "real number", which allows us to talk now in a mathematically substantial way of the "uncountability" of them. Again, a "concept-determination/formation".

In §20, therefore, Wittgenstein proposes to replace (a) and (b) by, as he says, the following "humble" ["*bescheiden*"][63] statement:

> If one calls something a series of real numbers, the expansion given by the diagonal procedure is also called a 'real number', and one in fact says that it is different from all the members of the series.

The "If one calls something a series of real numbers", with emphasis on "to call", again refers back to §16, where Wittgenstein says that what one calls "a series" – in the sense of "ordered in a series" – rests upon certain paradigms and certain methods that let us say that this or that is a series or can be ordered into a series. But there is no antecedent general concept of such an ordering that can be referred to as the basis of this saying. And Wittgenstein's "the expansion given by the diagonal procedure is also called a 'real number'", again with emphasis on "to call", is reminiscent of the beginning of RFM II, where different methods to produce real numbers are considered. There, the diagonal method appears to be something totally new, in no way predetermined by the other methods or by our understanding of them, and it has the peculiarity of being a method of higher order, as explained above in my commentary on §3. Since Cantor's introduction of his diagonal method, one then subsumes under the concept "real number" also the diagonal numbers of series of real numbers. Finally, Wittgenstein's "and one in fact says that it is different from

[62] I will come back to §16 below.

[63] Anscombe's translation of Wittgenstein's "bescheiden" is "sober", which however doesn't appear adequate to me.

all the members of the series", with emphasis on the "one says", is a reverberation of §§8–9.

All this sheds light on the "it means nothing" of §13. But one might sense now a certain tension between §13 and §10. The person in the scenario of §13 can certainly conclude, after having seen the diagonal method: "Aha, *therefore* the real numbers cannot be put into a series", whereupon she abandons her undertaking. But doesn't "cannot be put into a series" mean the same as "uncountable", and hasn't Wittgenstein emphasized in §10 with respect to this "therefore" that it *means nothing*? The situation can be clarified, however, by recognizing that the "therefore" in §10 has a different meaning from the "therefore" that matches §13. In the scenario of §13 it expresses the action-guiding insight that one's undertaking should be abandoned, but there is no reference to the *basis* of this insight. Maybe this "insight" has the character of a decision rather than that of an insight concerning existing mathematical connections. What is important is its relevance to action. The "therefore" in §10, on the other hand, points to something deeper, indeed to a mathematical connection between the diagonal method and the "cannot be put into a series" regarded as having been uncovered by Cantor. It is this idea of a deep connection against which Wittgenstein turns in §10. His thought is – or at least §10 can be interpreted in this way – that before Cantor's presentation of his method, the phrase "cannot be put into a series" had no respectable mathematical meaning; and when Cantor eventually offers his method and thereby gives this phrase mathematical substance, the "therefore" becomes empty. Thus, seen in this way, there is no contradiction between §10 and §13; rather, §13 looks like an amendment to §10.

What is meant when, at the end of §13 and put in brackets, Wittgenstein adds that "another picture" may get the person to resume her undertaking? – Perhaps one may think of the picture that the diagonal number of a given series of numbers may subsequently be put at the beginning of the series and so would now belong to this new series. Perhaps someone *may* become unsettled by pictures of this sort, and an empirically given scenario, as described by Wittgenstein, allows such reactions (even if "intelligent" reasoning, as we would like to add, would show them to be improper).

However, Wittgenstein goes too far, I think, when saying that it might be "the whole and proper purpose" of the diagonal method (understood in the down-to-earth way as presupposed in §13) to get people to abandon the undertaking to search for a series of all the numbers of a certain sort. What about the purpose referred to at the very beginning of RFM II, namely, to produce a number that is different from all square roots? – One might say that, because of its artificiality, this certainly does not belong to the purpose of the diagonal method. But what about the purpose of constructing a transcendental number, as in fact done by Cantor himself? Why call this purpose "improper"? Wittgenstein's own scepticism towards any sort of essentialism should make him hesitate here. Maybe he wasn't sufficiently aware of the constructive nature of this Cantorian application of the diagonal method, and so he wouldn't be willing to consider it a "method" in the down-to-earth sense he had in mind. That is, his "it strikes me as if" may have some justification in view of his own restricted knowledge. But from a systematic point of view his talking of "the whole and proper purpose" of the diagonal method does not seem to be acceptable.

In §14 Wittgenstein himself uses the term "demonstration" ("Demonstration" in the German original) with respect to the purpose of the diagonal method described in §13:

> The procedure exhibits something – which can in a very vague way be called the demonstration that *these* methods of calculation cannot be ordered in a series. And here the meaning of "*these*" is just kept vague.

Again, in order not to come into conflict with §10, "demonstration" here, like the "therefore" appropriate in the scenario of §13, should be understood in a strictly action-guiding sense as explained above. What Wittgenstein concentrates on in §14 is not how "demonstration" should be understood, but rather the "these" of "these methods of calculation". The methods of calculation he means are methods of producing real numbers, and according to his non-extensionalist standpoint Wittgenstein allows himself to identify them with the numbers themselves.

Already in the first sections of RFM II, very different methods of this sort are presented, and he thinks that we do not have a precise overview of them. Therefore the meaning of "these" is kept vague, as he says. Not only is it vague, but it even may appear to be *necessarily* vague, at least when one argues as follows: 'Methods' in mathematics certainly should be understood in such a way that each of them can be characterized by finitely many words and symbols. If then their extent is not kept vague, all of them should be orderable in a series, and this ordering itself should be understood via a certain method. But then the diagonal method turns up as a new method of constructing a real number, one that has not yet been recorded .... .

Wittgenstein should have been familiar with an argument of this sort because it can be found in Turing's "On computable numbers, with an application to the *Entscheidungsproblem*" (1936), which he knew. In this paper, Turing famously presents his own precise definition of "calculation" or "computation", devising what afterwards has been called a "Turing machine", and he evades the contradiction just delineated by accepting computations – i.e., Turing machines – that may get stuck at certain points, for example when loops occur, so that at these points they do not have any outcome. When permitting computations of this sort, Cantor's diagonal method can no longer be applied and all is well.[64] In contrast to that, as we now see, Wittgenstein adopts the view that the extent of what may be called "method of calculation" is vague, and in RFM II there is no mention at all of Turing's approach.

[64]See Floyd (2012c), Sect. 2.2, Chap. 8. In the theory of recursive functions that is standard today, instead of considering expansions of real numbers one deals with functions from $\mathbb{N}$ to $\mathbb{N}$, and Turing's manoeuver then means that not only total functions are allowed – that is, functions that are defined for each element of $\mathbb{N}$ – but also partial functions; see, for example, Rogers (1967), pp. 10–12. – I completely ignore the recent developments of "computable real numbers" and "computable analysis"; see, e.g., Weihrauch (2000). They are not relevant in our present context.

This is no wonder because the aims of Turing and Wittgenstein are very different.[65] For Turing, considerations about our actual methods of calculation, which are in the foreground of Wittgenstein's thinking, merely function as the starting point in order to develop a precise concept of computability, which is precisely his concept of a Turing machine. In the course of this process, calculations become purely mathematical objects. They may be considered to represent idealized human computers,[66] or idealized computing machines, but they are mathematical objects nevertheless. The concept of a Turing machine is a mathematically precise concept; the same is true of its extension, even in such a way that, appropriately understood, this extension can be represented by a single Turing machine itself, the so-called *universal* Turing machine.[67] Moreover, Turing accepted what was later called "Church's thesis" – and what is now often called the "Church-Turing thesis" – which says that the concept of computability that was non-rigorously used before the time of Turing (often called "effective computability"; see the title of Rogers (1967)) has the same extension as the concept of Turing-computability, i.e., computability by Turing machines. For this reason, Turing could claim to have found *the* one, natural concept of computability which, moreover, has lost all the *vagueness* of its non-rigorous predecessors.[68] In the sequel, this view, the "Church-Turing view", as it were, has been adopted by many mathematicians and philosophers of mathematics, maybe even by their majority.

All this, of course, is well-known, and I only recapitulate it in order to contrast it with Wittgenstein's perspective adopted in §14, according to which the "*these*" in "*these* methods of calculation" is kept vague. Wittgenstein here simply refrains from going beyond our actual practice of calculation. This live practice, which should be in the foreground when we try to solve or dissolve philosophical problems, is not delimited as seen by the Turing-Church view, but is always open to potentially new methods of calculation. It is Cantor's diagonal method that serves as Wittgenstein's paradigmatic example here, and other new methods will be developed and accepted in the future. From this perspective, Church's thesis appears not as a "thesis" but as a conceptual decision, which simply restricts the notion of computation to what is

---

[65] Which doesn't mean, however, that Wittgenstein ignored the idea of a Turing machine in all contexts. There is, for example, his famous statement in RPP I §1096: "Turing's 'machines'. These machines are *humans* who calculate", which originates in a passage on p. 18 of MS 135, written on 30 July 1947. The Introduction of Floyd (2012c) gives a concise overview of the different ways in which this statement may be understood, and the issues are revisited in Sect. 8.8 below.

[66] The word "computer" was in fact originally used by Turing to refer to humans who calculate. In order not to confuse this with what is nowadays called a computer, the artificial word "computor" was later introduced for what Turing meant; see "Human computability" in the Index of Copeland et al. (2013).

[67] See Rogers (1967), pp. 22f., and the semi-popular exposition in Penrose (1994), Chapter 2.

[68] Around the time of Turing's definition of computability, there were other precise concepts of computability on the market, and one may raise the question of historical priority here. I will not go into them, however. It is important and worth mentioning, though, that many of the other precise concepts of computability proposed in the course of time turned out to be extensionally equivalent to Turing-computability – an astonishing fact of mathematical concept building.

encompassed in the concept of a Turing machine. It does not alter the fact, however, that this notion as actually used in the practice of mathematicians is an essentially vague one. A view of this sort was actually adopted by many at the beginning of the discussion of Church's thesis, for example by Kleene, who in his classic treatise *Introduction of Metamathematics* (1952) writes that the role of Church's thesis "is to delimit precisely an hitherto vaguely conceived totality" (Kleene (1952), p. 318). This is certainly a reasonable view, and it is in accordance with what we find in §14. However, the discussion about Church's thesis is many-faceted and controversial to the present day, and I cannot go into it further, in particular since I am unaware of any Wittgensteinian statement about it.[69]

7.6 The Concept of "Ordering in a Series" and Its Negation

In §16, Wittgenstein goes on with his critical reflections in the vicinity of Cantor's diagonal method by examining a further aspect of it. Suppose someone says

(O) The diagonal number of a series real numbers is a real number as well; therefore the real numbers cannot be ordered in a series.

In §10, Wittgenstein remarked that the "therefore" in this statement "means nothing" because the proper, mathematically substantial concept of "cannot be ordered in a series" concerning X-numbers is just the concept "the diagonal number of any ordering of X-numbers in a series is an X-number itself". According to §14, then, it is the "all" that Wittgenstein would call "vague". To be precise, what in §14 is called "vague" is actually the "*these*" of "*these* methods of calculations", but what Wittgenstein has in mind are methods of calculating real numbers and, as already said, he allows himself to identify these methods with the numbers themselves. Consequently, the talking about *the* real numbers in (O), meaning *all* of them, must be called "vague" as well.[70]

Now in §16 the "can be ordered in a series" is critically examined. Wittgenstein discerns a mistake in it that already shows itself in the case of the cardinal numbers. This is what he says in the first two paragraphs of §16:

---

[69]The literature about Church's thesis and its possible functions is practically endless; see, e.g., Olszewski et al. (2006) and Copeland et al. (2013).

[70]The question as to what Wittgenstein would say about conceptions of real numbers that are not oriented towards methods for calculating them, like those of Dedekind or Cantor, is extensively discussed in Parts I and II of this book. Wittgenstein's result was, very roughly speaking, that conceptions of this sort provide a certain "phraseology" for talking about real numbers, but that the real mathematical substance of the concept "real number" has to be seen elsewhere, especially in our methods of calculating them. So, despite this seemingly precise phraseology, Wittgenstein can insist that, measured against the notion "real number" in its substantial sense, our talking about "all" real numbers remains vague.

> The mistake begins when one says that the cardinal numbers can be ordered in a series. For what concept has one of this ordering? One has of course a concept of an infinite series, but here that gives us at most a vague idea, a lodestar for the formation of a concept. For the concept itself is *abstracted* from this and from some other series; or: the expression denotes a certain analogy between cases, and it can e.g. be used to delimit for the time being a domain that one wants to talk about.
>
> That, however, is not to say that the question "Can the set R be ordered in a series?" has a clear sense. For this question means e.g.: Can one do something with these formations that corresponds to the ordering of the cardinal numbers in a series? Asked: "Can the real numbers be ordered in a series?", the conscientious answer might be: "For the time being I can't form any precise idea of that". – "But you can order the roots and the algebraic numbers for example in a series; so you surely understand the expression!" – To put it better, I *have got* certain analogous formations, which I call by the common name 'series'. But so far I haven't any bridge from these cases to that of 'all real numbers'. Nor have I any general method of trying whether such-and-such a set 'can be ordered in a series'.

When he uses the term "cardinal number" here, Wittgenstein means, as always, finite cardinal numbers, that is the natural numbers 0, 1, 2, etc. The "mistake" then consists in believing that there is a general concept of "ordering in a series" that we have in advance and according to which, in view of any given totality – be it the natural numbers, the real numbers, points of a line or a surface, or similar things – it makes sense to ask whether or not its elements can be ordered in a series. In the case of finite totalities (at least if they are not too large) this may be no problem: if we have a heap of flowerpots, we can take out one pot and place it before us, then another one of the rest and place it alongside, etc., until the heap is completely transformed into an ordered series.

A process of this sort can be seen as the core of the idea of *well-ordering*, and one may think that the notion of a well-ordering, now applicable to any set whatever, may in fact be the sought-after general concept of ordering. This concept, however, is *too* general because one can show, at least with the help of the Axiom of Choice, that *every* set can be well-ordered, while in the present context we are interested in the difference between, say, $\mathbb{N}$ and $\mathbb{R}$. As a *prima facie* better candidate for the desired general concept, one may offer the set-theoretic notion of the successor of a set and its interplay with the Axiom of Infinity.[71] For each set x we define the *successor* of x to be the set $x' := x \cup \{x\}$. The *Axiom of Infinity* then says: There is a set I that contains the empty set $\emptyset$, and if x is an element of I, then so is x'. A set with this property is called an *inductive set*. So the axiom states that at least one inductive set exists, and all of them obviously contain the following sets: $\emptyset, \{\emptyset\}, \{\emptyset, \{\emptyset\}\}, \{\emptyset, \{\emptyset\}, \{\emptyset, \{\emptyset\}\}\}, \ldots$. These sets are named then "0", "1", "2", "3", …, and they are considered to be the natural numbers in set theoretic guise, with the successor-relation as the natural ordering. Furthermore, the set $\mathbb{N}$ consisting of only the natural numbers is then defined as the disjunction of all the inductive sets; i.e.: $\mathbb{N} := \{x \mid x \in I \text{ for every inductive set } I\}$.

[71] See, for example, Halmos (1960), pp. 43f., and Hrbacek and Jech (1999), pp. 39–45. What in the following, like Hrbacek and Jech, I call an "inductive set", Halmos calls "successor set".

Hasn't one now found the general concept of ordering-in-a series that is appropriate in the present context? One may want to say that this general concept is in fact based on the concept of succession in $\mathbb{N}$, as just explained, because then the ordering-in-a-series of the elements of a set M can be defined via a function f from $\mathbb{N}$ onto M which engenders the infinite row $f(0), f(1), f(2), \ldots$. Isn't this the sought-after super-general concept of ordering in a series? – This point of view, however, would be totally unacceptable from Wittgenstein's perspective because it relies on an abstract, theoretical modeling of our actual mathematical practice, while Wittgenstein is interested in this practice itself. What he wants is a surveyable representation of our actual use of symbols and not a set theoretic *ersatz* of it. Without doubt our actual view of ordering-in-a-series takes its orientation from certain paradigms with which we become acquainted already as children, and the paradigm of counting "1, 2, 3, 4, ..." is certainly the most prominent one.[72] But, as Wittgenstein says, "that gives us at most a vague idea, a lodestar for the formation of a concept", and if we accept the ordering 1, 2, 3, ..., as such a lodestar, as the standard against which we measure the "can be ordered in a series", the statement that the numbers 1, 2, 3, 4, ... themselves can be ordered in a series loses all substance.

Oriented by this lodestar, we then see, for example, how to order the rational numbers and the transcendental numbers in a series. The corresponding general concept of ordering-in-a-series would then be "abstracted" from these cases, or better: it would be oriented towards certain analogies between them. As Wittgenstein says, these analogies can "e.g. be used to delimit for the time being a domain that one wants to talk about", i.e., to delimit a certain domain of numbers, but also of other objects (linguistic objects, for example), that can be ordered in such a way. Wittgenstein's emphasis here is on the expression "for the time being" – "*vorläufig*" in the German original – which also occurs in the second paragraph of §16.[73]

There are two very different cases to distinguish here. First, the delimitation *from the inside* is not definitely settled because we remain open to ever new methods representing what we would call "ordering in a series". If we do not cling merely

[72]One may ask whether "1, 2, 3, ..." is really *the* paradigm for ordering-in-a-series or whether there may be different ones (e.g.: iterations of certain linguistic operations). This, however, doesn't seem to be really relevant here. Maybe every person has her paradigm, and that's enough as long as the paradigms of different people are compatible with one another.

[73]Characteristically, Anscombe has translated Wittgenstein's "*vorläufig*" in the first paragraph as "provisionally" and in the second as "for the time being", thereby destroying, as in so many similar cases, a connection that Wittgenstein hints at by using the same word. It is true, though, that Wittgenstein's "*vorläufig*" isn't easy to translate, and it *can* mean both, "provisionally" and "for the time being". But one has to decide here which to choose, and I would plead for "for the time being". Furthermore, it is worth noting that in his manuscript – see MS 117, p. 105 – Wittgenstein actually struggled with the word "*vorläufig*"! He first had written "*beiläufig*" ("casually" in English), then crossed this out and wrote "*vorläufig*" above it and "*ungefähr*" ("approximately" in English) below, but in the end also crossed out "*ungefähr*" so that "*vorläufig*" remained. (Interestingly, in Austrian German the word "*beiläufig*" can not only mean "casually" but also "*ungefähr*", i.e., "approximately", making the situation even more complicated.)

to the extensional concept of countability as defined in my commentary of §10 above, then the concept of "can be ordered in a series" certainly appears to be settled only for the time being – until new methods are developed. Secondly, there is the problem of a delimitation *from the outside*, i.e., concerning cases where no ordering-in-a-series is available. Again, if not blinded by the purely extensional concept of uncountability defined above, we should say that *before* Cantor there were no methods at all, not even any ideas at all, that would have given the concept of not-orderable-in-a-series mathematical content; and *after* Cantor we again should be open, as in the case of countability, to the development of new methods.

With respect to the time before Cantor, or with respect to people who are unaware of Cantor's method or other methods to show uncountability, one can doubt, as Wittgenstein does in the second paragraph of §16, that a question like "Can the set R be ordered in a series?" has a clear sense. If, when trying to answer this question, one is oriented only towards the ordering of the natural numbers or towards other successful cases of ordering that one has come to know, such as the rational or algebraic numbers, a feeling of helplessness arises because, as Wittgenstein writes at the end of the second paragraph, we "haven't any bridge from these cases to that of "all real numbers"", nor "any general method of trying whether such-and-such a set 'can be ordered in a series'".

For this reason he beautifully remarks:

> Asked: "Can the real numbers be ordered in a series?", the conscientious answer might be: "For the time being I can't form any precise idea of that".

This form of conscientiousness is an essential characteristic of his philosophical method, which seems particularly apt with regard to the question about countability versus uncountability. Wittgenstein's conscientious attitude is shown in an especially appealing way at the beginning of the third paragraph of §16, where he is bothered by our saying that with the diagonal procedure we now "have the proof that this ordering can't be done here". The whole third paragraph reads as follows:

> Now I am shown the diagonal procedure and told: "Now here you have the proof that this ordering can't be done here". But I can reply: "I don't know – to repeat – what it is that *can't be done* here. Though I see you want to show a difference between the use of 'root', 'algebraic number', etc., on the one hand, and 'real number' on the other. Let's say as follows: we call the roots 'real numbers' and so *too* the diagonal number constructed out of them. And likewise for all series of real numbers. For this reason it makes no sense to talk about a 'series of all real numbers', just because the diagonal number of the series is also called a 'real number' ". – Would this not be somewhat similar to the case if one ordinarily called any row of books a book itself and now said: "It makes no sense to speak of 'the row of all books', since this row would itself be a book".[74]

[74]The published version of this passage is unclear with respect to the quotation marks that Wittgenstein has placed before the sentence "I don't know – to repeat – what it is that *can't be done* here" (in German: "Ich weiß – wie gesagt – nicht, was es ist, was hier *nicht geht*"), because these quotation marks are not closed, and this is the case not only in the published German version in BGM II but also in the original manuscript: see MS 117, pp. 106f. In the Anscombe translation this missing quotation mark has been inserted at the end of the sentence "I don't know – to repeat – what it is that can't be done here", but it seems to me that Wittgenstein's quotation marks should go

The expression "this ordering" suggests a concept that is not only delineated for the time being, but defined once and for all, and Wittgenstein had denied that before. If I understand him correctly, this is what we means by saying, "I don't know – to repeat – what it is that *can't be done* here".

One can say, of course, that one doesn't know how this or that specific method of ordering – for example, one of the methods applied to the rational or the algebraic numbers – may work in the case of the real numbers. But with the claim that the diagonal method shows "that this ordering can't be done here", one wants to say more, namely, that every conceivable method of ordering-in-a-series doesn't work in the case of the reals; that every such method of ordering somehow doesn't 'fit' in this case, that the reals somehow 'balk at' such methods. What this may mean, however, remains unclear, and for that reason Wittgenstein in §16 offers the following statement as the more conscientious one:

> [The diagonal method shows] a difference between the use of "root", "algebraic number", etc., on the one hand, and "real number" on the other. Let's say, as follows: we call the roots "real numbers" and so *too* the diagonal number constructed out of them. And likewise for all series of real numbers. For this reason it makes no sense to talk about a "series of all real numbers", just because the diagonal number of the series is also called a "real number".

In contrast to the roots, i.e., the roots of natural numbers as presented in §1, or the algebraic numbers, etc., we assign to the real numbers the property that for *each* series of these numbers its diagonal number is a real number as well. This is precisely the substantial concept of *uncountability* as already proposed in §10. And according to this concept it in fact *makes no sense*, as Wittgenstein now says in §16, to talk about a "series of all real numbers" because it is ruled out by this concept from the outset.

Precisely this point is underpinned by the analogy with the "books" at the end of §16, which Wittgenstein introduces slightly clumsily with the words "Would this not be somewhat similar" ("Wäre das nicht etwas ähnlich" in German): Given an arbitrary row of books, then this row itself does not belong to the books of which it consists, but if one now calls this row a "book" itself then one has something outside this row that is a book itself, and according to this sort of use of the word "book" it *makes no sense* to speak of the "row of all books". This situation is obviously similar to that of our use of the word "real number" according to which we consider the diagonal number of any series of real numbers as a real number that doesn't occur in the given series.[75]

---

further, that they should reach until "is also called a 'real number'". In other words: I think that the whole package from "I don't know – to repeat –" to "is also called a 'real number'" is presented as an answer that Wittgenstein accepts ("But I can reply"), similar to the sentence at the end of this section, "It makes no sense to speak of 'the row of all books', since this row would itself be a book", which Wittgenstein has put in quotation marks as a whole.

[75]It is also structurally similar to the case of what recursion theorists call "productive sets", a notion that, as Rogers writes, "is basic in applications of recursive functions theory to logic" (Rogers (1967), p. 90). A *productive set* is a subset A of $\mathbb{N}$ such that there is a recursive procedure which to each recursively enumerable subset W of A assigns an element of A that is not in W

What is the difference between Wittgenstein's view just explained and the usual set theoretic understanding? – To this I have already said enough in my commentary on § 10: According to set theory, a set M is called "countable" iff there is surjection from $\mathbb{N}$ onto M, and M is called "uncountable" iff it is not countable. One then proves *by means of Cantor's diagonal method* that if $\mathbb{R}$ were countable, i.e., if there were a surjection f from $\mathbb{N}$ onto $\mathbb{R}$, the corresponding diagonal number would not be in the range of f, in contradiction to the presupposed surjectivity of f. So, according to the set theoretic view, the uncountability of $\mathbb{R}$ would not reside in the *concept* of uncountability, while in Wittgenstein's view precisely this is the case. That is the reason why he says that *it makes no sense* to talk about a "series of all real numbers".

Wittgenstein's understanding of "uncountable", as explained in § 10 and elaborated on in § 16, is aimed at the computational core of the diagonal method, which is so simple that it might even be taught to school-children. This is the content of § 17:

> Here it is very useful to imagine the diagonal procedure for the production of a real number as having been known long before the invention of set theory, and familiar even to school-children, as indeed might very well have been the case. For this changes the aspect of Cantor's discovery. This discovery might very well have consisted *merely* in the new view of this well-known, elementary calculation.

When saying that this "might very well have been the case", Wittgenstein isn't *so* far away from the actual historical truth, for in the first paper, in which the uncountability of the real numbers is proven, the paper "Über eine Eigenschaft des Inbegriffs aller reellen algebraischen Zahlen" (1874), Cantor begins by constructing a transcendental number in precisely such a computational way.[76] Moreover, the whole paper is written in a manner that was even acceptable to Leopold Kronecker, the editor of the journal in which it appeared, who afterwards became one of the fiercest enemies of set theory.[77] Wittgenstein's next remark is insightful and comes close to historical reality: "For this changes the aspect of Cantor's discovery. This discovery might very well have consisted *merely* in the new view of this well-known, elementary calculation." In Cantor's 1874 paper the actual set-theoretic result – that the set of real numbers is uncountable – is displayed by showing that each series of real numbers determines a real number in form of its diagonal number that doesn't occur in the series, so that the assumption of having all the real numbers ordered in a series immediately leads to a contradiction. One can call this sort of presentation, as Wittgenstein proposes, a "new view of this well-known, elementary calculation",

(see Rogers (1967), pp. 84f. and 90–104). According to this concept, a productive set cannot be recursively enumerable itself, or, as Wittgenstein might say, it makes no sense to call a productive set "recursively enumerable", because according to its definition the concept of productivity explicitly precludes this. This concept was introduced by James Dekker in the 1950s, and so it could not have been known by Wittgenstein; but as we see, the sort of structural property expressed by it was familiar to him.

[76] See Gray (1994), pp. 820f. As already said in a former footnote, I here ignore the fact that, strictly speaking, the diagonal method is only implicitly contained in Cantor's (1874) paper.

[77] See Gray (1994), p. 828.

a view that can lead into set theory proper, into which, however, we are in no way forced.

What Wittgenstein imagines in §17 is certainly useful, but it is not what really happens when *we* get to know Cantor's method. This, however, is sensitively described on p. 49 of MS 162a:

> [I]t seems as if Cantor *taught* me no new technique; he only needs to show me the picture and I immediately have mastered the technique; he just tells me one that I already knew. But the picture is *new* and suggests something new, even if we understand the suggestion straight away.

The question that is in the air, then, concerns the mathematical substance of the new view presented by set theory: does it add something respectable to the diagonal method, understood as a "well-known, elementary calculation"? In the sequel, precisely this was disputed by Kronecker and others, and the quarrel broke out four years later when Cantor, in his paper "Ein Beitrag zur Mannigfaltigkeitslehre", proved the seemingly astonishing theorem that the set $\mathbb{R}$ of all real numbers and the set $\mathbb{R}^2$ of all pairs of real numbers are equipotent, i.e., have the same cardinality. The actual core of this proof, however, can again be seen in a sort of "elementary calculation" that is anything but astonishing: Simplifying the situation, let us consider not the entire set $\mathbb{R}$ but only the real numbers from 0 to 1, and instead of $\mathbb{R}^2$ the set of all pairs of real numbers from 0 to 1. Then the totally trivial function $0.a_1a_2a_3a_4\ldots \mapsto (0.a_1a_3a_5a_7\ldots, 0.a_2a_4a_6a_8\ldots)$ maps the first set one-to-one onto the second, and the equally trivial function $(0.a_1a_2a_3a_4\ldots, 0.b_1b_2b_3b_4\ldots) \mapsto 0.a_1b_1a_2b_2\ldots$ is obviously the inverse of the first[78]; hence these sets are equipotent and therefore, as is easily seen, also $\mathbb{R}$ and $\mathbb{R}^2$. However, if this triviality is the actual core of Cantor's proof one can rightly ask to what extent its set theoretic clothing is more than a stagey gown that lacks mathematical substance. Cantor seems, then, to be offering a "puffed-up proof" ("ein prahlerischer Beweis"), as Wittgenstein writes in §21. By using this expression Wittgenstein isn't actually commenting on the proof just given, but on the proof of the uncountability of the real numbers; but this latter proof as well raises the suspicion of being "puffed-up". It is the suspicion that the set theoretic pepping up of elementary situations may be no more than a sort of bluff.

Wittgenstein pursues this suspicion, and it is true that time and again he gives the impression of considering it confirmed (for example at the end of §22, to be quoted in a moment). But I would like to propose a more cautious interpretation that is more adequate to the spirit of his later philosophy: that he *conscientiously investigates* this suspicion with the aim of understanding the actual mathematical substance of the set theoretic way of thinking and talking, without prematurely anticipating the result of this investigation.

---

[78]Because of the non-uniqueness of the decimal representation of the real numbers, certain small alterations of the definition of these functions are necessary which, however, hardly change their elementary character; see Mühlhölzer (2002), p. 309.

The next section, §18, is too insignificant to be treated here, and the §§19–21 have already been dealt with above. §22, however, is important. It reads as follows:

> The usual expression creates the fiction of a procedure, a method of ordering which, though applicable here, nevertheless fails to reach its goal because of the number of objects, which is greater even than the number of all cardinal numbers [i.e., natural numbers].
>
> If it were said: "The reflection about the diagonal method shows you that the *concept* 'real number' has much less analogy with the concept 'cardinal number' [i.e., 'natural number'] than we, seduced by certain analogies, are inclined to believe", that would have a good and honest sense. But just the *opposite* happens: in that the 'set' of all real numbers is allegedly compared in magnitude with the set of cardinal numbers [i.e., natural numbers]. The difference in kind between the two conceptions is represented, by a skew form of expression, as difference of extension. I believe and hope that a future generation will laugh at this hocus pocus.

This section, which is the very last remark of the "Ansätze" in MS 117, before RFM II then goes on with passages from MS 121, begins with a modified recapitulation of some of the previous considerations. A usual expression such as the one mentioned in §19, that "one cannot order the real numbers into a series", suggests a universally applicable method of ordering which reaches its goal in the case of the set of natural numbers but not with respect to the set of real numbers. For, with respect to any series of real numbers Cantor's diagonal method presents a real number not included in it. Furthermore, this can lead us to say that the set of real numbers is "greater" than the set of natural numbers – as if one had discovered a deep fact in the realm of sets. In contrast to that, Wittgenstein now proposes, as already in §20, a more modest description, one that has "a good and honest sense": "The reflection about the diagonal method shows you that the *concept* 'real number' has much less analogy with the concept 'cardinal number' ['natural number'] than we, seduced by certain analogies, are inclined to believe".

What are the "certain analogies" Wittgenstein may have in mind here? He doesn't explicitly say, but his use of the word "set" in his subsequent sentence, put in single quotes, suggests the analogies presented by set theory – with all the highly abstract properties which the sets of natural and real numbers are considered to share with all sets, and which are registered in, or which follow from the usual axioms of set theory (like those of Zermelo and Fraenkel). As already foreshadowed in §10, however, and explained in detail in my commentary on §10, these properties are too abstract to give us substantial notions of universally applicable methods of ordering and of corresponding comparisons concerning "magnitude". To put it another way: the set theoretic phraseology alone does not possess sufficient substance to make comparisons of this sort really useful in our normal mathematical practice. This is Wittgenstein's point with respect to set theory throughout.

In particular, it is the extensionalist view conveyed by this phraseology that conceals our common non-extensionalist approach to numbers, which is based on different methods and techniques. The corresponding critique is beautifully expressed in the following statement in §22 that already constituted the motto of Chap. 2 of Part I of this book:

> The difference in kind between the two conceptions [that of the natural numbers in contrast to that of the reals] is represented, by a skew form of expression, as difference of extension.

As for the drastic final sentence of §22 – "I believe and hope that a future generation will laugh at this hocus pocus" –, what is remarkable about it is not its vehement rejection of set theory but the year it was written: the year 1938. For in that year, set theory was already a flourishing and well-established part of mathematics, at least in the following two respects: it provided a useful language for mathematics, and it had developed into a self-contained special field of mathematics with its own questions and methods, as impressively shown in precisely the year 1938, when Gödel published his groundbreaking note "The consistency of the axiom of choice and the generalized continuum hypothesis".[79] Wittgenstein's rejection as such, however, is nothing extraordinary. It can be found in the writings of Kronecker, Poincaré, Brouwer, Weyl (at least during his constructivist phase) and many others,[80] and it can be buttressed by a multitude of arguments. All the same, Wittgenstein's dismissive statement had been deleted in the first edition of *Bemerkungen über die Grundlagen der Mathematik*,[81] as if it were something to be ashamed of – for Wittgenstein himself and for those who want to take his philosophy of mathematics seriously.

Incidentally, 1938 is the year a wager came to a close that was made in Zürich in 1918 between Hermann Weyl and George Pólya, in which Weyl predicted that within twenty years Pólya himself and the majority of representative mathematicians would admit that the concept of real number as conceived in set theory is totally vague and should be abandoned.[82] There is no doubt that Weyl lost this wager, but one can at least say that Weyl was in good company when criticizing set theory. As for Wittgenstein, the "I believe" in his hocus-pocus statement is wrong, as was Weyl's prediction, but the statement as such is nothing to be ashamed of. However, the extensionalist point of view in standard set theory can be worked out in the form of axiom systems which, by and large, are sensed as satisfying by most mathematicians. In this way hocus pocus can be avoided. At the same time, the hocus-pocus avoidance by means of axiom systems has its price: most of the mathematical methods and techniques actually used – those that are already in the center of our learning of mathematics, which are then developed further and in which our mathematical competence lies – are not covered by set theoretic

---

[79]Gödel (1938) only presents Gödel's results and some of his ideas concerning the independence of axiom of choice and continuum hypothesis. Detailed elaborations followed in the years 1939–40; see Gödel (1990), pp. 1–101, and also Kanamori (2012a), Section 3.4.

[80]In the majority of cases, however, the rhetoric of rejection is less drastic than Wittgenstein's. It has often been said of Poincaré, for example (even lately in Havil (2012), p. 204), that he had called set theory a disease from which mathematics should be cured, but to my knowledge there is no evidence of such an utterance. Poincaré was certainly no friend of set theory, but his statements about it that are known to me sound relatively mild.

[81]See BGM_I, pp. 56 and 57, and RFM_I, pp. 56e and 57e, where the text of §22 of Part II of the second edition appears in §3 of Appendix II of Part I, without this last sentence!

[82]See van Dalen (2013), p. 310.

axiomatics, or merely in a sort of *ex post* modelling far away from our practice. It is precisely at this point where the value of Wittgenstein's non-extensionalist point of view becomes obvious: that it is mainly concerned with this practice and aimed at raising our awareness of it. Furthermore, in their accompanying prose mathematicians rather often slip back into hocus pocus, especially when talking about infinity, and Wittgenstein is particularly sensitive to this tendency.

It is not difficult to find examples of this tendency in pronouncements made by mathematicians. Cantor, for example, after having proved that the line and the plane, represented as $\mathbb{R}$ and $\mathbb{R}^2$, have the same cardinality, exclaimed to his pen-pal Dedekind: "There you see what wonderful force lies in the ordinary rational and irrational numbers!"[83] This "wonderful force" especially reveals itself in the uncountability of the real numbers, whereby an abyss seems to be opening up. Cantor has seen it in precisely this way, at least if the following anecdote that was told by Emmy Noether and transmitted by Felix Bernstein is to be believed:

> *Dedekind* remarked, with regard to the concept of a set: he imagines a set as a closed sack that contains specific things which one cannot see, however, and of which one doesn't know anything except that they exist and are determined. Some time later, *Cantor* revealed his image of a set: He straightened up his colossal physique, made a grandiose gesture and said with a gaze directed at something indefinite: "I imagine a set as an abyss."[84]

With this, of course, Cantor hadn't in mind finite sets, and also not countable ones, but the uncountable $\mathbb{R}$ and all the sets of even higher cardinality. They can very well evoke the image of an abyss. And shouldn't this abyss also be the key to understanding the astonishing fact, mentioned in my commentary above on §11, that the irrational numbers of the interval [0,1] have measure 1, the same measure as the interval itself? Doesn't this show the almost incomprehensible magnitude of the uncountable set of irrational numbers in [0,1], which seems to swallow up, as it were, the rationals? Something like this has been expressed, for example, by Harro Heuser who, in his textbook *Lehrbuch der Analysis, Teil I* (2009), says the following:

> The set $\mathbb{Q}$ of rational numbers is in a certain sense "small": It is infinite, although *merely countable*. In another sense, however, $\mathbb{Q}$ is "great": $\mathbb{Q}$ "lies dense in $\mathbb{R}$", i.e., in each ϵ-neighborhood of each real number there is a rational number. [...] Now we want to show that, according to a third point of view, $\mathbb{Q}$ in a quite surprising manner is "small" again. [...] Speaking intuitively – but this intuition is difficult to accomplish – the set $\mathbb{Q}$ [...], although lying dense in $\mathbb{R}$, *can be completely covered by countably many intervals with arbitrarily small "sum of lengths"*.[85]

---

[83]See Cantor and Dedekind (1937), p. 34; my translation.

[84]See Becker (1975), p. 316. Dedekind's characterization of sets seems to me phenomenologically rather adequate: in the way as described by Dedekind we very often treat sets in our set-theoretic practice. Perhaps Dedekind's characterization is even acceptable from a Wittgensteinian point of view.

[85]Heuser (2009), p. 193; my translation; emphases by the author.

In the sense just explained, $\mathbb{Q}$ is, as Heuser says, "small", and correspondingly the rest, i.e. the set of irrational numbers, is "big", big to such an extent – or better perhaps: "deep" to such an extent – that the image of an abyss again offers itself.

When looking at the corresponding proofs, however, dramatizations of this sort appear inappropriate. The sameness of the cardinalities of line and plane is essentially based on the trivial manipulations

$0.a_1a_2a_3a_4\ldots \mapsto (0.a_1a_3a_5a_7\ldots, 0.a_2a_4a_6a_8\ldots)$

and

$(0.a_1a_2a_3a_4\ldots, 0.b_1b_2b_3b_4\ldots) \mapsto 0.a_1b_1a_2b_2\ldots$[86];

the uncountability of $\mathbb{R}$ on Cantor's diagonal method which may be taught already to school-children; and the case of the rational and irrational numbers in [0,1] (or in $\mathbb{R}$, as described by Heuser (2009), but this makes no real difference) can be easily de-dramatized by clarifying the simple measure-theoretic mechanisms that are used, especially that the measure of the union of disjoint measurable sets is the sum of the measures of these sets, together with the trick to cover the rationals, which are considered as ordered in a series, by intervals "with arbitrarily small 'sum of lengths'", as Heuser says. Wittgenstein is looking at mathematical situations in precisely such a sober way, far from all verbal mystifications, concentrated only on the actual mechanisms used in the proofs.

There is a beautiful, but unfortunately unpublished passage in Wittgenstein's *Nachlass*, where he compares the mystery-producing behavior of mathematicians with the case of a magician "who performs his conjuring tricks in front of a mirror, such that he sees them like his audience, and who now is astonished himself that he could do *this*", and in parentheses Wittgenstein adds: "In such a way the proposition fools the mathematician."[87] In the light of observations of this sort, one should distinguish between those views of set theoretic results that involve a hocus pocus and the more sober ones that avoid it.[88] I would like to understand the last sentence of §22 as primarily emphasizing this sort of distinction.

§22 completes the part of MS 117 that Wittgenstein captioned "Ansätze", and from the remaining sections of RFM II, which are a selection of remarks on pp. 27r-92v of MS 121, I will now discuss the §§28–34. These are the only passages concerned with the subject of the present chapter, Cantor's diagonal method, and the rest of RFM II, interesting as it may be in itself, can be readily ignored.

---

[86]As already said in a former footnote, this is the main idea of the proof. It needs some technical corrections which, however, do not alter the proof's essential triviality.

[87]MS 162a, p. 92; my translation. A magician is already mentioned on p. 87, but in a less pointed way. Compare p. 251 below.

[88]A paradigmatic investigation in this vein is Kanamori (2012b).

7.7 The Multiplicity and Openness of the Concept "Real Number"

In §28 Wittgenstein makes a new start concerning the question of why the real numbers are non-orderable:

> Why should we say: The real numbers cannot be ordered? – We have a method by which to disrupt any order ["jede Ordnung zu stören" in Wittgenstein's German].[89]

Wittgenstein's answer, "We have a method by which to disrupt any order", *prima facie* sounds a bit strange. What is meant, of course, is Cantor's diagonal method, and one might wish rather to say that this method does not "disrupt" any order but shows that *there isn't* a way to order all the real numbers and that, therefore, they *cannot* be ordered. This sort of parlance, however, is the typical extensionalist one, according to which the totality of real numbers is a complete one, so that the question makes sense whether it *can* or *cannot* be ordered. Cantor's method then allows us to give the negative answer. The value of §28, however, may be seen precisely in how it irritates us with this strange "a method to disrupt any order", to make us think about its possible sense from a non-extensionalist perspective.

Seen non-extensionally, the real numbers do not exist as a complete totality but are produced in diverse ways that are not defined in advance. These ways *may* typically bring about certain orderings of the real numbers, but any commitment to such an ordering is then undermined by Cantor's method. In this way Cantor's method in fact "disrupts" any ordering one may think of.[90] A point of view of this sort is beautifully described in the following scenario invented by Wittgenstein four years later[91]:

> Think of someone who has a patent for a rule to produce rules according to which series of cardinal numbers [that is, natural numbers] can be produced (for the purpose of numberings, for example). He says he has found a rule according to which *all possible* endless series can be produced and that no rival would be able to construct a rule that does not contain all the rules produced by his rule. And now Cantor shows him that this is not

[89]Instead of "real numbers", Wittgenstein actually wrote "irrational numbers". This, however, is inappropriate, as I explained above in my commentary on §13, and I simply change it here, and later again, without further explanation, when commenting on §§29–33.

[90]In the manuscript one can see, however, that instead of "stören" – "to disrupt" in my translation – Wittgenstein actually wrote "(zer)stören" which, when understood without brackets, means "to destroy" (see MS 121, p. 35v). This is already neglected in BGM, where one merely reads "stören", and the "stören" then has been taken over to the English edition. To make things even more complicated, in the original Anscombe translation "stören" is translated as "to upset". But "to upset" neither means "stören" nor "zerstören". In my presentation I also decided to ignore the "(zer)", i.e., the "destroy", because I don't think that it is very important. I actually do not know why Wittgenstein wavered between "stören" and "zerstören", that is, why it was important to him to ponder this distinction.

[91]On October 23, 1942; see pp. 10–13 of MS 126. MS 126 is the basis of the first half of RFM V and therefore important with regard to Wittgenstein's annotations to CPM. The text of RFM V, however, starts on p. 24 of MS 126, that is, after the passage that I now quote.

> possible. This proof without doubt changes his concept of the endless sequence of numbers. Previously, having taken so much trouble to include *all* rules in his system, he had believed that he couldn't have omitted any. Now, he thinks *quite differently* about the matter.
>
> Like someone who did not know that the construction of a partition of an angle might be impossible. Now, he looks at this quite differently.

This inventor of a rule produces rules for obtaining endless sequences of natural numbers. We can, of course, imagine him also as someone who produces rules for obtaining decimal expansions of real numbers and then applies for the patent for a sort of super-rule, with the claim that in this super-rule *all possible* decimal expansions of real numbers are included. Furthermore, since offering it as a proper rule, he presupposes that it leads to an ordering of all these decimal expansions, and therefore of the real numbers themselves, in a series.

A scenario of this sort may be suggested from a non-extensionalist point of view, according to which the concept "real number" is determined by such a super-rule and the totality of all real numbers is not considered as something simply being there but as produced by this rule. The question raised from the extensionalist standpoint as to *whether* this rule encompasses all the real numbers may then be considered as being *without sense* because to say "*all* the real numbers" simply *means* to be produced by this rule. Imagine, for example, a society in which our inventor is an undisputed mathematical authority who is regarded as having *defined* what "all real numbers" means: it means a number produced by his super-rule. In fact, definitions of this sort are nothing extraordinary in mathematics. But now Cantor turns up with his diagonal method, and he thereby acts indeed as a *disrupter* of our inventor's idea, however this idea might be elaborated in detail (as long as his super-rule claims to give us all the real numbers in the form of a series). From this perspective, Wittgenstein's "disrupt any order" in §28 appears justified.

Wittgenstein then imagines that our inventor, confronted with Cantor's argumentation, eventually abandons his idea of ordering all the real numbers in a series, and with respect to this situation Wittgenstein certainly has good reasons to say that now our inventor's *concept* of an endless sequence of numbers – or, in our variant, his *concept* of real numbers – is changed and that "he thinks *quite differently* about the matter". And isn't this sufficiently analogous now to *our* situation that we can say the same about ourselves: that *before* Cantor we thought quite differently about the matter – or maybe even not at all – and that our change of attitude prompted by Cantor has the character of a *conceptual* change?[92]

[92]In this respect Cantor's letter of 2 December 1873 to Dedekind is interesting, in which, among other things, he writes: "Incidentally, I want to add that I never seriously occupied myself [with the question of whether all real numbers can be counted] because it doesn't have any practical interest for me, and I totally agree with you when you say that it doesn't deserve much effort. Nevertheless it would be nice if it could be answered; for example, presupposing that it would be answered with *no*, a new proof of Liouville's theorem would be given that there are transcendental numbers." (Cantor (1891), pp. 32–34; my translation) This passage sounds as if even the *question* concerning the countability of all real numbers had not crossed anybody's mind before Cantor, not to mention an answer to it. (Quite remarkable is Cantor's phrase "no practical interest". I think what he means

The end of the passage from MS 126 just quoted is also instructive, where Wittgenstein, as already in §2, draws the comparison with the construction of a trisection of any given angle,[93] in the present case, however, with another direction of thrust. What he has in mind is again a construction within the construction game presented in Euclid's *Elements*, but now with the thought that in the framework of this game the impossibility of such a construction – i.e., a construction applicable to any angle whatsoever – is not comprehensible in a mathematical way. It was an empirical fact that despite much effort the mathematicians had found no such construction, from which they might have ventured the hypothesis that it will never be found, but precise *mathematical* arguments were lacking. This changed with the advent of Cartesian analytic geometry, which allowed a transformation of the geometric problem into an algebraic one, and thereby a *proof* of the non-constructability of the angle-trisection was possible. Without a doubt, Descartes' new view of Euclidean geometry involved a totally different geometric thinking, making it appropriate to speak of a *change of the concepts* involved.[94] In an analogous way Wittgenstein now sees the question concerning the countability of the real numbers and the negative answer to it by means of Cantor's diagonal method.

If one understands, as just explained, the diction of §28 as hinting at the view that Cantor's diagonal method should be regarded as a change of concepts and not as a means to discover new mathematical facts, §29 is an elucidation of it:

> Cantor's diagonal procedure does not show us a real number different from all in the system, but it gives sense to the mathematical proposition that the number so-and-so is different from all those of the system. Cantor could say: you can prove that a number is different from all the numbers in the system *by* proving that it differs in the first place from the first number, in its second place from the second number, and so on.
>
> Cantor is saying something about the multiplicity [*Multiplizität*] of the concept "real number different from all the ones of a system".

With the first statement of §29 Wittgenstein expresses nothing new. It can be found in §§19–22, which I have already commented on. The second statement – "Cantor could say: you can prove that a number is different from all the numbers in the system *by* proving that it differs in the first place from the first number, in its second place from the second number, and so on" – then says that Cantor should proffer his method as a new method of proof that is constitutive of the *concept* "real number different from all the ones of a system".

At the end of §29 this is put in more specific terms with the help of the word "multiplicity": "Cantor is saying something about the multiplicity of the concept 'real number different from all the ones of a system'". "Multiplicity"

---

by this is the practice of the *pure* mathematician and not a Wittgensteinian orientation concerning what is practical in a down-to-earth sense.)

[93] Wittgenstein writes rather sloppily: "the construction of a partition of an angle", but I interpret him as meaning the construction of a trisection of any angle.

[94] This is obviously analogous to the case of the non-constructability of the regular heptagon, extensively discussed in LFM; see the reconstruction in Mühlhölzer (2001).

– "Multiplizität" in German – as understood here is one of the few technical terms of Wittgenstein's philosophy. It originates from the 4.04 remarks of the *Tractatus* (with a short resumption in 5.475), where Wittgenstein, however, writes "Mannigfaltigkeit" instead of "Multiplizität". But by this he means the same as by "Multiplizität". In the English translations of the *Tractatus*, beginning with the first one by C.K. Ogden (to which Wittgenstein gave his blessing) "Mannigfaltigkeit" is always rendered as "multiplicity", and Wittgenstein also used this word when expressing himself in English.[95] Wittgenstein made use of his originally Tractarian *termini technici* "Mannigfaltigkeit" and "Multiplizität" until the 1940s,[96] and this usage may be seen as representing an important aspect of continuity between his early and later philosophy.

In TLP 4.04 the term "multiplicity" is introduced as follows:

> In a proposition there must be exactly as many distinguishable parts as in the situation that it represents. The two must possess the same logical (mathematical) multiplicity. (Compare Hertz's *Mechanics* on dynamical models.)

Here, Wittgenstein speaks of a "logical multiplicity" and only in brackets of a "mathematical" one, but it is the latter term that he uses in the further course of the *Tractatus*. This sort of diction, and in addition the hint at Hertz's *Mechanics*, show that this term is meant as a technical one. However, Wittgenstein gives no explicit definition of it, either in the *Tractatus* or in later texts, and one has to gather its meaning from the way he uses it, especially from the examples he gives.

In his first texts after the *Tractatus* that mention "multiplicity", in RLF and the lectures of 1930–1933 reported by Moore (MWL), Wittgenstein only uses the combination "logical multiplicity". Moore looks at this expression rather critically, and on two occasions he cannot resist adding in brackets: "whatever

[95]"Mannigfaltigkeit" is used by Wittgenstein also in its normal colloquial meaning, for example in §§23f. of PU, but then it has only very superficial points of contact with the technical term as used in the *Tractatus* and also in our present context. Unfortunately, G.E.M. Anscombe translated the occurrences of the word "Mannigfaltigkeit" in PU §§23f. as "multiplicity" throughout, thereby suggesting a connection with the *Tractatus* that doesn't exist. In the new translation by Peter Hacker and Joachim Schulte, the "Mannigfaltigkeit" of PU §§23f. is translated, much more adequately, as "diversity" or "variety".

[96]On p. 109 of MS 109 Wittgenstein at first writes "Multiplizität", but then crosses it out and replaces it with"Mannigfaltigkeit", perhaps with the thought that "Mannigfaltigkeit" is his word in the original German version of the *Tractatus* – and why should he conform to the English translation when writing in German? Later, however, he became less scrupulous, and on p. 7 of MS 112, p. 25 of MS 113, pp. 31r-31v of TS 155 and p. 576 of TS 213 (the *Big Typescript*) both words, "Multiplizität" and "Mannigfaltigkeit", occur in close proximity to one another without crossing-out. If I see matters correctly, the last time in Wittgenstein's texts when the word "Multiplizität" turns up is on p. 391, in remark No. 1536, of TS 229, written in 1947 (adopted from a remark on p. 30v of MS 123, also written in 1947, published as No. 868 of RPP I), and the word "Mannigfaltigkeit", in its technical sense, occurs for the last time on p. 752, in remark No. 613, of TS 323 written in 1948 (adopted from a remark on pp. 52b-53a of MS 137, also written in 1948, and published as No. 614 of RPP II and as §437 of Z).

that may mean".[97] What does it mean? – In the *Tractatus* and also for a while thereafter, what Wittgenstein had in mind by "logical/mathematical multiplicity" is the structure of a proposition that makes the proposition represent a state of affairs by faithfully *picturing* the structure of it, in the *mathematical* sense of "picturing", i.e. in the sense of structure-preserving mapping. In other words, it refers to a structural identity between proposition and state of affairs. Precisely this is the essence of Wittgenstein's 'picture theory' of the proposition, according to which the representation achieved by a proposition consists in the fact that the proposition and what it represents share the relevant structure. Later, however, he not only omits the words "logical" and "mathematical" but also – and this is an essential step to his later philosophy – refrains from his picture-theoretic stance.

What remains, then, is merely the multiplicity of propositions and, furthermore, of signs per se, of concepts, rules, "systems" and proofs, but no longer of states of affairs or of phenomena with which the propositions, etc., could have the same multiplicity. Nevertheless, the conception of "multiplicity" is still an important one because it can now be deployed in order to compare propositions, etc., *with each other* as to their multiplicity. The basic idea is to point to the respective *conceptual structure*, as one might say. Making this idea precise, however, is a difficult task, and if I see it correctly there are no attempts in this vein in Wittgenstein's texts. One has to look at the examples he gives.

Let me mention three of his examples.[98] The first one, discussed in RLF of 1929, concerns phenomena of continuity, especially the continuity of space and time. Although Wittgenstein still clings to the picture-theory of the *Tractatus* at this stage, the example appears especially instructive because it reveals characteristic images in the back of his mind that he associates with the term "multiplicity". This is what he says on pp. 165–168 of RLF:

> We meet with the forms of space and time with the whole manifold of spatial and temporal objects, as colours, sounds, etc., with their gradations, continuous transitions, and combinations in various proportions, all of which we cannot seize by our ordinary means of expression. And here I wish to make my first definite remark: it is this, that for their representation numbers (rational and irrational) must enter into the structure of the atomic propositions themselves. I will illustrate this by an example. Imagine a system of rectangular axes, as it were, cross wires, drawn in our field of vision and an arbitrary scale fixed. It is clear that we then can describe the shape and position of every patch of colour in our visual field by means of statements of numbers which have their significance relative to the system of co-ordinates and the unit chosen. Again, it is clear that this description will have the right logical multiplicity, and that a description which has a smaller multiplicity will not do. [...] The system of co-ordinates here is part of the mode of expression; it is part of the method of projection by which the reality is projected into our symbolism. [...]
>
> I maintain that the statement which attributes a degree to a quality cannot further be analysed [...]. [T]he atomic statement must have the same multiplicity as the degree which it attributes, whence it follows that numbers must enter the forms of atomic propositions.

[97] Moore's critical discussion of "logical multiplicity" in MWL is on pp. 83–85 of Moore's original publication in *Mind* of 1954/55; the addition in brackets on p. 84.

[98] Others are mentioned in Black (1964), p. 176.

This vivid idea, however – that the characteristic multiplicity of a continuous phenomenon should be mirrored by a corresponding multiplicity of the proposition representing it – is completely abandoned by Wittgenstein immediately after 1929.[99] But the view just presented remains interesting because it shows where the notion of "multiplicity" can lead. Even more vivid, and particularly appealing for the straightforward way in which it is written, is the following passage on p. 55 of MS 105, presumably also noted down in 1929, where Wittgenstein writes:

> In order to represent space, one needs – as it appears to me – an elastic sign.
> Perhaps a sign that allows an interpretation analogous to the decimal system.
> The sign must have the multiplicity [in German: "die Mannigfaltigkeit"] and the properties of space.[100]

The second example I want to mention is on pp. 27f. of the same manuscript:

> When I know the rules of elementary trigonometry I can control the sentence sin $2\alpha = 2$ sinα cosα, but not the sentence sin $x = 1 - x^3/3! + \ldots$. But this means that the sin of elementary trigonometry and the sin of the higher one are *different* concepts. If we give them, however, the same name, this has the good reason that the second concept includes the multiplicity [in German: "die Mannigfaltigkeit"] of the first; but for the system of elementary trigonometry the second sentence has *no sense* and the question whether sin $x = 1 - x^3/3! + \ldots$ is of course senseless as well.[101]

In elementary descriptive geometry the trigonometric functions sin and cos are defined for any angle α in such a way that α is taken to be an angle in a rectangular triangle and then sinα:= (side opposite α)/hypotenuse, cosα := (side adjacent to α)/hypotenuse. By elementary means one can then easily prove – "control", in Wittgenstein's parlance – that $\sin 2\alpha = 2\sin\alpha\cos\alpha$. However, in order to show that for the sin function as just defined it also holds that $\sin x = 1 - x^3/3! + x^5/5! - x^7/7! + \ldots$, one has to expand the rules of elementary descriptive geometry considerably: one has to devise the theory of infinite series, and the geometric notions of triangle and angle have to be connected to this theory, which is normally done by integration. The new rules introduced in this way are "higher" than the elementary ones to such an extent that it appears very apt to say, as Wittgenstein does in the passage quoted, that we now have two *different concepts* of sin.[102] As long as

---

[99] And his general verdict is that the entire essay, "Remarks on Logical Form", is worthless; see the corresponding remarks by the editors in PO, p. 28.

[100] Wittgenstein used this naïve manner of speaking – "an elastic sign" [in his German: "ein dehnbares Zeichen"] – not only in this manuscript but he took it even over into his TS 208: see there on p. 5.

[101] Again this passage is taken over, with minor alterations, in TS 208 (p. 53) and TS 209 (p. 73).

[102] In "higher" mathematics the trigonometric functions are no longer defined geometrically, but from the outset *via* their standard series: $\sin x := 1 - x^3/3! + x^5/5! - x^7/7! + \ldots$; $\cos x := 1 - x^2/2! + x^4/4! - x^6/6! + \ldots$ (see, for example, Grauert and Lieb (1967), pp. 137–142). In Heuser (2009), pp. 274f., the functions sin and cos are at first introduced *via* axioms that describe certain characteristic properties of them – for example the property $\sin(x + y) = \sin x \cos y + \cos x \sin y$, wherein Wittgenstein's formula $\sin 2x = 2 \sin x \cos x$ is automatically contained! –, and on p. 358 Heuser shows that the standard series then follow. However, according to Heuser the question as to the existence of the sin and the cos function isn't settled with this move, and in the

it is not expanded in the way described, in the system of elementary trigonometry, as Wittgenstein says, the sentence $\sin x = 1 - x^3/3! + x^5/5! - x^7/7! + \ldots$ has no sense. And his statement that the *multiplicity* of the second concept includes the multiplicity of the first appears appropriate, for on the "higher" level one can in a clear sense express, distinguish and show what is possible on the "lower" level, but also much more besides. Precisely this constitutes the "greater multiplicity" of the second concept.

My third example of Wittgenstein's use of "multiplicity" is on p. 70 of MS 162a, written on 12 January 1939. In its down-to-earth simplicity it is a typically Wittgensteinian example and needs no commentary: "*Multiplication* and *division*. Division has a greater multiplicity ["Mannigfaltigkeit"]; it also shows that $a \times b$ is *not c*."

The use of "multiplicity" in our second and third example need no longer to be seen as having a *Tractarian* picture-theoretic component, and what Wittgenstein says in their case can easily be transferred to the "multiplicity" at the end of §29, for when Cantor introduces his diagonal method, the multiplicity of the concept "real number different from all the ones of a system" is increased, too. I understand the word "system" here as referring to a given totality of numbers together with a rule – a method or a technique – to order these numbers in a series.[103] A "system" of this sort is presented by Wittgenstein immediately at the beginning of RFM II in form of the square roots of natural numbers: $\sqrt{1}$, $\sqrt{2}$, $\sqrt{3}$, etc. Before Cantor one could bestow mathematical content on the concept "real number different from all the ones of this system" by referring, as Wittgenstein himself says, to the number $\sqrt{2}$, or to the number 1.3142... that agrees with the decimal expansion of $\sqrt{2}$ at every place except the first one which is diminished by 1; or, to mention another natural possibility, by constructing the arithmetic mean of two consecutive numbers of the system. But with his diagonal method, Cantor then presented a quite new, hitherto unheard-of possibility, the peculiarity of which can be seen in diverse ways. The most important of these lie in the fact that it presents a real number different from all those of the system by referring to *all* the numbers of the system, which was not the case with the "natural" methods considered before. Furthermore, with this idea Cantor extended our conceptual possibilities in a totally unexpected way. So, for example, one can now most easily show that in the case of the system of all algebraic numbers, a "real number different from all those of this system" emerges which we then consider to be transcendental. Obviously, this can justly be called an increase of the respective conceptual "multiplicity".[104]

---

end, after this quasi-axiomatic detour, he feels forced nevertheless to define sin and cos explicitly *via* their standard series.

[103]Presumably the word "system" as used here, like the word "multiplicity", has its origin in Wittgenstein's thinking in the *Tractatus*. It would certainly be desirable to thoroughly discuss the connections of the present context with the *Tractarian* context, but this would lead me too far away and must be postponed to a later occasion.

[104]In a rather long passage near the end of MS 121, on pp. 89v-91v, amidst the notes that have been published as the last section of RFM II §62, the subject "multiplicity" is readopted. This passage,

The next two sections, §§30–31, make it clear that Wittgenstein does not reject the notion "extension", which in the present context is always meant not in the set-theoretic sense as expressed by the "axiom of extensionality", but in the sense of regarding decimal expansions of real numbers extensionally, i.e., by abstracting from possible methods of expansion and by considering only their results. The two sections belong together and they read as follows:

> 30. Cantor shows that if we have a system of extensions it makes sense to speak of an extension different from them all. – But with that the grammar of the word "extension" is not yet determined.
>
> 31. Cantor gives a sense to the expression "extension which is different from all the extensions of a system", by proposing that an extension should be so called when it can be proved of it that it is diagonally different from the expansions of a system.[105]

As may already be seen in §4, Wittgenstein's non-extensionalist standpoint does not amount to a rejection of our talking of "extensions"; rather it says that extensions shouldn't be regarded as *primary*. In the foreground of Wittgenstein's considerations are the different methods for expanding, but then he is interested also in the comparison of these methods and especially the comparison with respect to their *results*. Such a result is not regarded as given in advance, to be grasped by this or that method, but it is of methods themselves that we ask whether they should be considered as having the *same results*. Such results may then be called "extensionally equivalent" (although this is not a term that Wittgenstein himself uses) and Wittgenstein accordingly allows himself to talk about *extensions* as such. In the §§30–31 he then expresses in the language of extensions what he had just said in §29.

This is relevant with regard to Cantor's diagonal method, which is commonly understood as mainly oriented towards extensions. It is the decimal expansions as such that seem to be in the foreground, and not the methods for expanding. Of course, when applying Cantor's method to a computational ordering of the algebraic numbers, with the result that *via* an appropriate diagonalization one gets a method for constructing a transcendental number, the methods of expansion are important as well, all the more so in the application of Cantor's method in recursion theory; but in the usual set theoretic context methods of expansion play no role. There, the real numbers are considered as defined by Dedekind cuts of the set of rational numbers or by equivalence classes of Cauchy sequences of rational numbers, and it can then be proved that these real numbers are representable by decimal expansions.

---

interesting as it may appear, cannot be easily interpreted, however, especially because of its opaque references to the *Tractatus*, and I will not go into it here.

[105] In the original German version of these sections, in MS 121, pp. 36v-37r, Wittgenstein in fact uses the word "Extension", i.e. "extension", as presented in my translation, and I am sure that what he means are decimal expansions extensionally conceived. This, however, cannot be seen in Anscombe's translation in which "Extension" is rendered as "expansion" throughout, as if a Wittgensteinian slip had to be corrected. Or is "Extension" perhaps a slip in MS 121? – No, this can be ruled out because similar considerations, and again with the word "Extension", can also be found in MS 162a, pp. 40–42.

It is even possible to define the real numbers in an extensionally acceptable way by using decimal expansions from the outset.[106] The totality of real numbers and of their decimal expansions then appears to be a precisely delimited one, and the classical picture of this totality is the number line, about which Wittgenstein will write in §33. This picture in particular gives the impression that the real numbers are a well-delimited totality and, moreover, a thoroughly homogeneous one.

Instead of this picture and by transferring the considerations of §29 to extensions, Wittgenstein in §§30–31 proffers the view that with his method Cantor does not presuppose a definite sense of "extension", thereby revealing mathematical facts concerning extensions, but gives a sense to the expression "extension which is different from all the extensions of a system". This sense can be captured by the expression "diagonally different", according to which a real number with extension $\ldots .b_1b_2b_3\ldots$ (i.e. with this decimal expansion, irrespective of the method of expansion) is called "diagonally different" from those of a system of extensions $\ldots .a_{i1}a_{i2}a_{i3}\ldots$ (i.e., a series of respective decimal expansions, irrespective of the methods of expanding) iff $b_i \neq a_{ii}$ for all i. And who knows how our talking of "extensions" may gain sense in further new ways? Seen in this way, the grammar of the word "extension" does not yet appear to be determined, as Wittgenstein says in §30.

Using this new concept – "diagonally different" – one can now formulate specific tasks as mentioned in §32: "Find a number whose expansion is diagonally different from those in this system." I have already discussed this, together with the interesting *Nachlass* passage immediately following it, in my commentary on §2 and need not go into it once more. As it turns out, the main point of §32 is that tasks of this sort can be too trivial to be called "tasks" at all – like performing the task of constructing a number that does not agree with $\sqrt{2}$ precisely at every second decimal place simply by adding 1 or -1 at every second decimal place (where one should add 1 if this place is unequal to 9 and -1 if it is 9) – and that we should be intent on understanding them, or on formulating them, in a way that gives them mathematical substance. After this remark, which appears to be a belated insertion that comes back to the problem alluded to in §2, Wittgenstein not only emphasizes the possible variety of concepts like "extension which is different from all the extensions of a system" but also of the fundamental concepts "irrational number" and "real number" themselves. This is the content of §33:

> One might say: Besides the rational points there are *diverse systems* of irrational points on the number line.
>
> There is no system of the real numbers – but also no super-system, no 'set of the real numbers' of an infinity of higher order.[107]

[106]See Gowers (2017b) and Tao (2006), pp. 380–388.

[107]In this second paragraph of §33, instead of "real numbers" Wittgenstein actually writes "irrational numbers" ("Irrationalzahlen" and "irrationale Zahlen", respectively). This is inappropriate, as in several similar cases before, and I have corrected it accordingly.

In keeping with his non-extensionalist view Wittgenstein emphasizes the diversity of systems of irrational numbers – and therefore also the diversity of systems of real numbers – which he had already done in the *Big Typescript*. There, on p. 751, he says that "constructs that are logically quite different are called 'irrational numbers' without any clear limits given to the concept", and on p. 762, "it is incorrect (or misleading) to speak of irrational numbers by contrasting them as a type of number with cardinal numbers [i.e., natural numbers] and rational numbers; because what are called 'irrational numbers' are actually different types of numbers – as different from each other as the rational numbers are different from each of these types". Cantor's diagonal method now directs our attention to what Wittgenstein calls in the present context *systems* of irrational numbers, i.e. (as I understand him), to series of irrational numbers, and these series can be produced by quite diverse means. Hence, there are also quite *diverse systems* of irrational numbers.

The statement from the *Big Typescript* just quoted – that what are called "irrational numbers" are actually different types of numbers, as different from each other as the rational numbers are different from each of these types – then has to be changed, because it gives the impression that at least the rational numbers are all of the same type. But the diagonal number of a system of irrational numbers can in certain circumstances indeed be a rational number itself (see my commentary on §10), and shouldn't Wittgenstein hold that this rational number is of a different type than the numbers simply given by our ordinary rules of decimal expansion that rely on our usual procedure of dividing integers? Seen in this light, perhaps Wittgenstein should have written that on the number line, besides the rational points given in this elementary way, there are diverse systems of irrational points *and of rational points*. The decisive issue concerns the diversity of systems of *real* numbers, i.e., of irrational and rational ones.

It may seem puzzling that Wittgenstein talks here about the number *line* and *points* on this line, whereas normally our use of the picture of this line is meant as representing the *homogeneity* of the continuum considered as a complete totality, which Wittgenstein is just disputing in §33. It may be, however, that with this sort of geometric parlance he wants to sensitize us with respect to this picture: the picture in fact suggests a "super-system" of real numbers with an "infinity of higher order", whereas through Cantor's diagonal method we have only learned that there is *no system* of all real numbers, i.e., no series of them. However, I regard Wittgenstein's claim that *there is no* super-system of the real numbers, no "set of the real numbers" of an infinity of higher order, not only as somewhat strange but also as too coarse. It is not only too dogmatic but also smacks of "ontology", which is far removed from Wittgenstein's intent in his later philosophy. I interpret him here as merely saying that at the present status of his investigations, *he cannot see* the legitimacy of this idea of a super-system representing an infinity of higher order.

Of course, this does not mean that he would call the real numbers "countable". Already in §10 he proposed his own version of the concept of uncountability: "I call the number-concept X uncountable if it has been stipulated that, whatever numbers falling under this concept you arrange in a series, the diagonal number of this series is also to fall under that concept", and according to this concept the real numbers *are* uncountable – "unabzählbar", in Wittgenstein's German. However, he doesn't call them, as is common in German mathematical language, "überabzählbar", which means "excess-countable" or "trans-countable". This latter term suggests the notion of an infinity of higher order, of which Wittgenstein is wary. Instead, with respect to the diagonal numbers referred to in the definition just quoted, he prefers to talk about a *difference* of higher order. This is proffered in the next section, §34, the last section of RFM II that is explicitly concerned with Cantor's diagonal method:

> Cantor defines a *difference of higher order*, that is to say a difference of an expansion from a *system* of expansions. This definition can be used so as to show that a number is in this sense different from a system of numbers: let us say π from the system of algebraic numbers. But we cannot very well say that the rule of altering the places in the diagonal in such-and-such a way is thereby proved different from the rules of the system, because this rule is itself of 'higher order'; for it *deals with* the alteration of a system of rules, and for that reason it is not clear from the outset in which cases we shall be willing to declare the expansion of *such a* rule different from all the expansions of the system.

The thought expressed in §29, that Cantor's diagonal procedure doesn't show us a real number different from all those in the system, but that it rather *gives sense* to the mathematical proposition that the number so-and-so is different from all those of the system, is turned over and over again by Wittgenstein with different formulations, before and after §29, and in §34 he refers to it in the expression "a difference of higher order". In saying this he is not alluding to an "infinity of higher order", of which §33 is critical, but instead to a *conceptual* difference with respect to types of real numbers. In §34 this is explicated in such a way that the Cantorian rule for altering the places in the diagonal of a system of rules (i.e. a system of rules to expand real numbers) *treats of* the alteration of this system and is therefore something totally new, *not yet* regulated before, and that consequently, as Wittgenstein writes, "it is not clear from the outset in which cases we shall be willing to declare the expansion of *such a* rule different from all the expansions of the system". This declaration is only given by Cantor's method itself, and we shouldn't think that with the help of this method we would prove something the content of which had already been explained before.

This is, as I understand it, the central point of Wittgenstein's reflections on Cantor's method in RFM II. Wittgenstein makes these reflections a bit shady, however, by his example using the number π.[108] Maybe it is possible to manage the arrangement of the algebraic numbers in such a way that π results as the diagonal number, but this is certainly not what Wittgenstein meant. His idea seems to be that we present a certain series of algebraic numbers, without initially aiming at π, and then prove afterwards that it is possible to construct the diagonal number of this series in such a way that π results. I do not know of such a proof, and it would be rather strange to see the value of Cantor's method in its capability to produce proofs of this sort. At this point Wittgenstein should cease to think of the diagonal

[108]The same questionable reference to π also occurs on p. 24 of MS 162a and a variant concerning $\sqrt{2}$ on p. 32.

number as being one of the prominent transcendental numbers, like π or e. It is enough to speak of "the diagonal number of a given system" without expectations concerning its prominence, and as we have seen Wittgenstein does not object to that. His criticism concerns the idea that with Cantor's method things may be shown with regard to concepts already settled beforehand, against which he develops the view that what this method achieves is actually the determination of new concepts.

7.8 Résumé

As already said in my commentary on §2 above, and in what I have in a way just repeated, the central issue of Wittgenstein's remarks on Cantor's diagonal method in RFM II is the disparity between the simplicity of Cantor's idea, on the one hand, and the supposedly deep and amazing mathematical results gained by it, on the other. These results are often said to reveal, for example, the breathtaking magnitude of the set of real numbers, so that – to speak now in Cantor's words – this set seems like an abyss. When thinking and talking that way, however, philosophers and mathematicians appear to forget the simple technical means by which the respective results are actually proven. In §17 Wittgenstein rightly says that the diagonal procedure for the production of a real number could have been known long before the invention of set theory and taught even to school-children; and on p. 49 of MS 162a he observes that, strictly speaking, it seems as if Cantor did not even *teach* us a new technique: "he only needs to show me the picture and I immediately have mastered the technique; he just tells me one that I already knew." Nonetheless, mathematicians seem to draw the deepest consequences from it. This is a puzzling situation that should be clarified.

Wittgenstein solves, or dissolves, this puzzle by diagnosing in the relevant cases that we merely form new concepts, concepts pertaining to the simple means used in the proofs, and do not reveal mathematical facts described by concepts that we had sufficiently understood. This is beautifully expressed in §22, and the relevant passage is worth quoting again:

> If it were said: "The reflection about the diagonal method shows you that the *concept* 'real number' has much less analogy with the concept 'cardinal number' [i.e., 'natural number'] than we, seduced by certain analogies, are inclined to believe", that would have a good and honest sense. But just the *opposite* happens: in that the 'set' of all real numbers is allegedly compared in magnitude with the set of cardinal numbers [i.e., natural numbers]. The difference in kind between the two conceptions is represented, by a skew form of expression, as difference of extension.

In precisely this lies the essence of Wittgenstein's investigations concerning Cantor's diagonal method in RFM II.[109]

[109] I am grateful to Juliet Floyd and Marianne Mühlhölzer for valuable discussions about the material of this Chapter.

Chapter 8
The Generality of Cantor's Diagonal Procedure (Juliet Floyd)

Abstract This chapter explores the non-extensionalist notion of "generality" in connection with the real numbers, focusing on diagonal argumentation. The notions of "technique" and "aspect" are distinguished in the development of Wittgenstein's philosophy. Then the textbook of Hobson on functions of a real variable (the standard presentation of the real numbers in England through the 1940s) is considered, with a focus on Hobson's non-extensionalist point of view and his reaction, both to the extensionalist point of view and Cantor's diagonal argument in particular. Next Turing's reference to Hobson in his 1936 "On computable numbers" is considered in connection with Hobson's idea of the novelty and open-endedness of our linguistic capacities in his response to the Richard paradox, which is related directly to Turing's use of diagonal argumentation to resolve the *Entscheidungsproblem*. The emergence of rule-following in Wittgenstein's work 1936–1937 is characterized, along with the notebooks he kept while teaching his seminar on the foundations of mathematics at Cambridge 1939, attended by Turing. Finally, Wittgenstein's own rendition of Turing's diagonal argument from 1947 is canvassed with an eye toward his non-extensional point of view.

8.1 Introduction

In what sense did Cantor provide, with his diagonal procedure, a *general* procedure or method of proof? In what sense, to put this in Wittgenstein's own terms, a *technique* of proof? What we call a "technique" must be what we call "general" in the sense that it gives us more than a local logical stratagem for proving just one result. It must be robust enough to be transported into new contexts. As Felix Mühlhölzer has just explained in his Chap. 7, it is the surprising combination of novelty and simplicity attaching to the procedure in its initial application by Cantor that Wittgenstein sought to clarify in his RFM II remarks on the diagonal argument. In fact, as I shall argue in what follows, Wittgenstein wants to explore Cantor's procedure of argumentation *as* a technique partly by laying it beside *other* techniques we have for handling the real numbers, and weighing differences and

J. Floyd, F. Mühlhölzer, *Wittgenstein's Annotations to Hardy's Course of Pure Mathematics*, Nordic Wittgenstein Studies 7,
https://doi.org/10.1007/978-3-030-48481-1_8

similarities. The purpose is to clarify, not only Cantor and the real numbers, but the very notion of a mathematical "technique". This is necessary, for as we have seen our techniques in handling the real numbers are, from Wittgenstein's non-extensionalist perspective, various.

This chapter argues that we should more precisely understand the central role played by the *notion* of technique in Wittgenstein's later philosophy, and that we can do so if we analyze the connections between his uses of it and his grappling with Cantor's diagonal procedure. A quick anticipatory point of translation is in order here: the German *Technik* can mean both "technique" and "technology". We are moving into an exploration of the "technology", that is, the techniques, that are constructed for our handling of real numbers. Recall that the notion of a "technique" appears more than once in Wittgenstein's annotations to Hardy (on CPM pp. 28 and 119). His non-extensionalist conception of the real numbers associates them above all with everyday techniques of calculation and proof, as we have argued.

Surprisingly, the term "*Technik*" first enters Wittgenstein's writings only in 1937, both as a notion and as an object of reflection – it does not occur, for example, in BB or any earlier writings. When it enters, it is marked by an allusion to his summer 1937 discussions with Alister Watson and Alan M. Turing.[1] This important fact connects the ideas of diagonalization and "incompleteness" to his deepening reflections on the ideas of *following a rule* and *Regelmässigkeit*, or regularity. After this point these terms fill the pages of his writings and lectures, becoming signature notions in his mature philosophy. The notion of a *technique* was duly entered into the first version of *Philosophical Investigations* (the *Frühfassung*) [FF] in 1937–1938 and appears in all subsequent versions, deployed across wider and wider areas of philosophy in connection with the notions of *real number*, *meaning*, *understanding*, *intention*, *mastery*, *seeing an aspect*, and so on.[2]

Alongside this shift is the fact that the notion of a *form of life* [*Lebensform*] also becomes central to Wittgenstein's philosophy only after 1937, entering as a substitute for the idea of *culture* [*Kultur*] in the fall of 1936. In BB to imagine a language is to imagine a culture [*Kultur*], but in PI it is to imagine a form of life [*Lebensform*], and this is something different from a "language-game"', a full-fledged "culture", or even a life-world [*Lebenswelt*] in the phenomenologist's sense: it means instead the activity of embedding words, concepts *in* life.[3]

This history provides a clue to the role of Wittgenstein's idea of a *technique* [*Technik*] in his later writings – including those aimed at Hardy's presentation of the

[1]The first two occurrences of "Technik" are (in 1937) in MS 118, pp. 45–6 and 87r, the second one containing the reference to Watson. For a characterization of the summer 1937 discussions among Turing, Wittgenstein and Watson see Floyd (2001b).

[2]See *Frühfassung* [FF] §229 and §322 in PUK, the origins, respectively, of RFM I §143 and §133, quoted and discussed below on p. 199 and in Sect. 8.2.

[3]See Floyd (2016, 2018b), and our discussion above in Sect. 1.3.

real numbers. After the fall of 1937 we see the idea of technique appearing often in his manuscripts, notebooks and lectures. Most notably, in the spring 1939 LFM lectures at Cambridge – the ones Watson and Turing attended – the term occurs 117 times, an indication that the discussion among Wittgenstein, Watson and Turing was focussing on what role it should play in the foundations of logic and mathematics. Above all, they are discussing the idea of the *limits* of mathematical techniques – something brought out by diagonalization, as a procedure, quite explicitly.[4]

In order to tether Wittgenstein's condensed and often confusing thought-trains to something more comprehensible, this chapter will give, first, an overview of the emergence of the notion of *technique* in his later philosophy (Sect. 8.2), contrasting it with the notion of an *aspect* of a number or technique (Sect. 8.3). Next (Sect. 8.4) will be rehearsed a brief analysis of Cantor's diagonal procedure as it was redeployed in presentations of analysis in England during the first decades of the twentieth century, when debates about the Axiom of Choice, the paradoxes, and the new, more abstract methods of analysis were grappled with by the mathematical community. In particular we shall consider the non-extensional perspective advocated by the Cambridge mathematician Hobson, whose interchanges with Russell were known to the young Wittgenstein and whose *The Theory of Functions of a Real Variable and the Theory of Fourier's Series* was the standard Cambridge textbook in presenting the subject before Hardy's book was published. Throughout this chapter our focus will be the notion of *definability*, the idea of using language to characterize a real decimal expansion or other concept, and how the notion of a technique of characterization may be put to work. We fasten on Hobson's treatment of the Richard paradox – the paradox *par excellence* dealing with the idea of what it is to give an "adequate" finite characterization of a real number – to provide us with a thread through our discussion.

Our analysis of different conceptualizations of Cantor's procedure of diagonalization in the context of real analysis will allow us to shed light on Turing's novel deployment of what he calls "the diagonal process" in §8 of his famous (1936) (Sect. 8.5). We shall see that Turing uses Cantor's diagonal method as a technique, fitting it into a new context where it behaves differently from its context in Cantor's original argument. Finally, we shall revisit how Wittgenstein approaches the method of diagonalization as a technique, reconstructing what I have called "Wittgenstein's Diagonal Argument", a variant or adaptation of Cantor's and Turing's (1936) diagonal arguments penned in 1947 (Sect. 8.8).

Wittgenstein's diagonal argument was already expounded above by Felix Mühlhölzer (p. 144). Mühlhölzer takes its significance to lie in the fact that already in the formation of a diagonal sequence the heart of Cantor's argument may be found: the "switching" of digits is a secondary matter. In this chapter we shall draw out the philosophical significance of this in light of Wittgenstein's overall point of

---

[4]It is thus almost a scandal, as Felix Mühölzer pointed out to me, that the term "Technik" ["technique"] does not appear at all in the Register or Index of BGM II [or RFM II], while the term itself occurs repeatedly in these remarks.

view as it developed in notebooks he wrote during 1939, up to the point where he annotated Hardy's CPR. In a regression to what is most basic or simple, in these remarks Wittgenstein finally formulated a use of diagonalization evincing what is central to his own non-extensional point of view. Considered in the light of the background in Hobson and Turing, the argument shows that his emphasis on our everyday techniques with real numbers is neither an idiosyncratic philosophical obsession nor a dispensable ideological add-on, but, rather, a logical *must*. This deepens our understanding of his critical reactions to Hardy's presentation of the real numbers in CPM in 1941–1942.

Wittgenstein's remarks on the real numbers hearken back to the problematics of an earlier era, before Cantor's argument was fully articulated against the backdrop of a successful axiomatic theory of sets. As we have seen, Wittgenstein's major objection to Hardy's presentation in CPM is that it fails to clearly convey the importance of the distinction between the non-extensionalist and the extensionalist perspectives. Hardy makes it seem as if the transition in perspectives is smooth, requiring no discussion of the prices paid for each or the subtlety of the transitions between them.[5] Wittgenstein criticizes this, insisting on emphasizing cross-cutting shifts in perspective.

The narrative in what follows will offer a selective, targeted, historical reconstruction, designed to show that there is a sense of *necessity* in Wittgenstein's turn toward ordinary mathematical practice with real numbers, toward rules and techniques utilized in everyday life. The stress on the objectivity located in *these* – as opposed to the objecthood per se of the entities discussed – is a theme of the whole later Wittgenstein.

8.2 Technique in Later Wittgenstein

For Wittgenstein the idea of technique is not one of cold unthinking roteness or automaticity. Instead, a technique must provide us with something that may be taught and mastered, something purposeful and useful, something capable of being

[5] Another good example of this (though not singled out by Wittgenstein in his annotations to CPM so far as we know) is §101 of CPM, where Hardy proves the Intermediate Value Theorem directly by application of Dedekind cuts. The theorem says that any continuous function f on an interval [a,b] must take on every value η between $f(a)$ and $f(b)$ within the interval. But an appeal to Dedekind cuts is unnecessary for the proof: the existence of such a value may be shown for any arbitrary η by the bisection method, the process of taking the midpoint $f(x_1) = (a+b)/2$ and if $f(x_1) \neq y$, considering $[a, x_1]$ if $f(x_1) > y$ and $[x_1, b]$ if $f(x_1) < y$, and iterating this to get $x_1, x_2, \ldots$ converging to an x such that $f(x) = y$. Kreisel (1989) p. 135, reminiscing about discussing this issue and Hardy's CPM with Wittgenstein in 1942, credits Wittgenstein with having sparked this experience with he later came to know as "constructive content". Wittgenstein mentions different methods of proving the intermediate value theorem in both 1940 (117, p. 163) and 1944 (127, p. 189). In both places he remains interested in the mixing of geometrical, extensional, and calculative attitudes embodied in different proofs. See Kanamori (2018b) for a discussion of Kreisel's shifting attitudes toward his teacher Wittgenstein.

carried forward in a variety of ways and "fit" to new circumstances. A technique in the relevant sense yields a *way of thinking* about *how* to operate within this or that proof context, and it shows itself *as* a technique through its applicability in more than one kind of situation. The kind of generality it carries is not then quantificational, but purposive: it widens our environment of affordances.

American mathematicians sometimes speak of a "trick" in an initial proof, which later on becomes a method.[6] We shall not use the idea of method, but, following Wittgenstein, that of technique. A technique is something that may be adapted and re-embedded in new situation by different people, re-applied: this is the meaning of the notion in everyday parlance and in working mathematics.

In terms of the later Wittgenstein's philosophy, we may think of a mathematical technique as something that plays an initial role in a single proof mechanism, but then may be removed, as a working part, and fit into a new situation, perhaps newly integrated with other techniques. A gear box solving a certain mechanical problem may be seized upon in other machines, moved and transported into a new context (see PI §12). It is thereby adapted, perhaps somewhat altered, in the process: it then *itself* becomes a technique for solving problems. – Wittgenstein's own later philosophical techniques involve ways of comparing and contrasting uses of sentences and words conceived of *as* techniques. Techniques reveal their forms of generality, their reality, their potentialites, in their uses.

To give but one example in foundations of mathematics, we might say that the method of forcing is a technique in present-day set theory for generating models of certain kinds – although when it first appeared, in Cohen's proof of the independence of the Generalized Continuum Hypothesis from the other axioms of Zermelo-Fraenkel set theory (Cohen 1963, 1964), it appeared to be a trick. The arithmetization of syntax, in Gödel (1931), was at first a trick, but its general import as a technique or proof, a kind of device, is what made it matter. As a matter of fact, the interweaving of the techniques of diagonalization and arithmetization *itself* would become a kind of technique in later argumentation in logic. A final example, closer to Wittgenstein's own history, is Russell's theory of descriptions (Wittgenstein praises it at TLP 4.0031). This "theory" rests upon a technique of rewriting definite descriptions with quantifiers: it is a contextual device.

To count as a technique, something must prove itself to be more than just a proof or argument element in one context. One must be able to deploy it in a new way, stitching it to a new context where it shows its worth. And so it is with Cantor's diagonal procedure. As we shall see in this chapter, it is regarded by Wittgenstein as a technique for handling techniques of characterizing real decimal expansions, and it may be applied in a variety of contexts.

The idea of "technique" is there in Wittgenstein, as we have seen, to counterbalance other, perfectly legitimate but sometimes unbalanced ideas: that of the "depth" of a theorem, the "size" of an extension, or the "genius", "creativity" and "insight"

[6]An example is "Scott's trick", which allows one to define set representatives for ordinal and cardinal numbers; see https://en.wikipedia.org/wiki/Scott%27s_trick.

of an individual mathematician like Cantor. Earlier we saw him questioning the idea of an "eternally valid proof" machinery in connection with indirect proof of the irrationality of the square root of two and Gödel's incompleteness proof (see p. 58 above). Ideas of "eternal validity" shape our feelings and impressions of logical laws, and of proofs such as Cantor's as giving us a picture unveiling a gapless abyss. It is not that Wittgenstein will deny us the ideas of insight, depth or genius. But he wants their potential for misuse to be understood soberly, in a down-to-earth way. As in musical performance and criticism, uninformed notions of surprise and depth can blind us, encouraging us toward inarticulate, "puffed up" effusions (see RFM II §21), ones that, unmoored from their pedestrian roots in the day-to-day mastery of specific techniques, spread enthusiasm unreflectively around a community.

Here we run up against the large issue of where techniques begin, and where they end, and the problem of what is "indescribable" or "uncharacterizeable in words". Words, as the later Wittgenstein sees them, take on their meanings with human techniques of use, and these have their limits. Moreover these uses themselves shape further uses and the development of further techniques. The whole is dynamic, not static. In Wittgenstein's mature philosophy the situation is, he argues, not free of gaps and jumps: it does not take place in one fixed or formally unified space that is beyond or orthagonal to human affordance, as the extensional point of view pictures things.

It is not that Wittgenstein is reducing everything *to* mere techniques.[7] Rather, he seeks to characterize the respective roles of different techniques and terms of criticism, weighing and balancing them against one another.

An important remark – rather a breakthrough, actually – occurs in his pocket notebook 160, in 1938 (p. 27), where a contrast is drawn between an available technique and the idea of community agreement or consensus of opinion:

> I do not say, "Do not believe that the logical laws must be valid everywhere, there may very well be a case in which the law of the excluded middle does not hold" - as if this were an empirical proposition and one had to warn against blindly assuming that he agrees! – But I do say: get used to seeing a variety of techniques of sign usage (i.e. of thinking). I do not want to eliminate a prejudice of opinion but of technique. For example, do not be dismayed in principle by a contradiction.

Wittgenstein admonishes himself here to "get used" to a different approach. Appreciating the use of the law of excluded middle *as* a technique, or the generation of a contradiction (for example, in diagonal proofs) *as* a technique is not a matter of opinion – of (for example) generating majority consensus or agreeing to disagree. Rather it is a matter of conveying how a technique may be fit into a novel situation, how it contrasts with the variety of other fittings of the same technique to a variety

[7] It is sometimes held that the fact that Wittgenstein was trained as an engineer *explains* his remarks on mathematics *tout court*, as if only "applications" of mathematics in engineering were to him the heart of things. But this cannot explain the entrance of the notion of "technique" into his mature writings. For an insightful criticism of readings of Wittgenstein that are either reductive about techniques (Rorty) or too dismissive of the idea of a technique (Putnam) see Diamond (1999).

of different situations. It is a matter of seeing, not an actuality, but a possibility of application.

This is the context in which the more general issue of rule-following appears in this notebook (160, p. 26r):

> How is the application of a rule fixed? – Do you mean "logically" fixed? Either through further rules or not (at all). – Or do you mean: How does it come about that all the applications agree so, and are not applied otherwise? Through training, drill and the form of our lives.
>
> It is a matter, not of consensus of opinions, but of forms of life [*Lebensformen*].

The important distinction drawn here – which becomes more and more explicit as Wittgenstein goes on – is between an opinion and a technique. One may choose this or that way of fixing a system of rules, using or not using the law of excluded middle, for example. But one also develops distinct techniques. And there will, characteristically at least, be agreement that in one system one proposition follows by logic, and in another system not. There may even be mathematical techniques developed to show us transitions that may be made from one system to another.[8] This is not only a matter of opinion, but also of technique. To adapt an important precursor remark from the TLP (2.003), techniques are used to embed words and proofs as "forms", or "*possibilities* of structure" *in* life (*Lebensform*).[9]

The emphasis on the *variety* of techniques and the generality of logical laws is also to be found in the very first place where the notion of technique first enters into Wittgenstein's writing, in the fall of 1937. As has been mentioned above, Wittgenstein associates this turn of thought with Alister Watson, with whom Wittgenstein and Turing had discussed undecideability results that summer in Cambridge:

> The propositions of logic are 'laws of thought' "because they express the essence of human thought" – but more correctly: because they express or show the essence, technique (Watson), of thinking. They show what thinking is and also ways of thinking.[10]

The allusion here is to the Preface to volume I of Frege's *Grundgesetze der Arithmetik* (1893/1903) where Frege, drawing a sharp distinction between logic and psychology, insists that the laws of logic are essential and internal to what counts as thought as such. In countenancing the idea of logical aliens who would pursue different laws of thought from ours, Frege contrasts his conception of the laws of truth with the idea of human thinking as 'our thinking', "the thinking of humanity up until now". He argues that the laws of logic are "boundary stones of thought anchored in an eternal ground, which our thinking may wash over but yet cannot displace".

---

[8]See, e.g., Gödel's interpretation of intuitionistic logic in classical logic, Gödel (1958).

[9]See Floyd (2019a) for a discussion of the importance of this remark in relation to Wittgenstein's early views on truth.

[10]MS 118, p. 87r; see *Frühfassung* [FF] §322 in PUK, the origin of RFM I §133. The allusion to Watson is not in the published version of RFM I §133.

Wittgenstein does not disagree, in the above-quoted remark, that logic gives us "the laws of thought" and that these "express the essence of human thought" *as* thought. But he writes that what these "laws of thought" show is the essence of thinking *as* a matter technique – and here he alludes to Watson. His idea is that the different ways we have of thinking and proceeding (plural "ways") *are* thinking. This inverts the structure of Frege's expressed picture. A technique, in being a technique, is not forever and everywhere the same in its uses: it is internal to its *being* a technique that it may be redeployed in different ways.

This is why in a related remark Wittgenstein explores an "alien" tribe of woodsellers who sell wood, not according to the weight of a lot, but by some other measure "alien" to us (the area of ground covered; how much labor it took to cut the trees, etc.). Wittgenstein writes, of the idea of "logical aliens", that

> Frege says in the Preface to the *Grundgesetze der Arithmetik* "…here we have a hitherto unknown form of madness" – but he never said what this "madness" would really be like.[11]

From very early on in his writing of the *Tractatus* and ever after, Wittgenstein held that essential limits of thought cannot be given by true laws or absolutely certain propositions or principles, but instead show themselves as we formulate tautologies or contradictions in putting words together in our own tongues. The limits of thought show *themselves*, in other words, in our efforts to think with them, and seeing that we are not making claims of a usual kind. These are not matters on which we can hold an opinion, or demand consensus. An opinion (or "prejudice") may be contradicted, justified, agreed to, or not. But the law of excluded middle is not just one more opinion: it is a technique put to work in a variety of ways.

As a matter of technique there remains the living question, in Wittgenstein's as in our time, of the mathematician's ability to develop techniques that allow one to go one way or another way, smoothly and naturally *in* logic, thereby overcoming certain clashes of opinions *through* the development of different possibilities, different parameters and procedures. Different techniques may be titrated with one another, contrasted, and compared. On Wittgenstein's view in 1937, this is fundamental to appreciating logic itself. He brings the notion of technique into his later philosophy in order to steer clear of a reductive operationalism or verificationism about meaning: now there is an evolving plasticity, a multi-coloredness, to logic. The price of this perspective is that now a single sentence is thought of as a technique, as bearing multiple dimensions of meaning. And the entanglement and separation of these dimensions cannot be charted in advance, any more than it can for any collection of techniques.

In the right kind of situation (and it is clear that Wittgenstein has in mind here Cantor's diagonal argument as a paradigmatic example, as well as Gödel's later implementation of its method of diagonalization[12]), a "contradiction" or "paradox"

---

[11]RFM I §152, from 1938; originally in 117, p. 49; FF (§240).

[12]Actually that Gödel is at stake is not obvious in 160, but in pocket notebook 161 Wittgenstein mentions provability (p. 6), and in pocket notebooks 162a and 162b the connection with Gödel is

may show us the possibility of *differing* techniques, *differing* ways of going on – while within each there is a real sense of *being* correct or incorrect.

The point is that we should not think of sentences as describable without our being able to engage, even if only imaginatively, in an investigation of the variety of their possible applications in differing situations. The modalities of possibility and necessity are part and parcel of our counting something *as* a mathematical or logical technique. And here we see the connection with Wittgenstein's annotations to Hardy's CPM. He wants to emphasize the *variety* of differing ways we have of conceiving (working with, seeing) what the real numbers *are*.

In the case of the contradiction generated by Cantor's indirect form of argument, Wittgenstein asks (160, p. 28f.):

> Are you not gripped by a prejudice? Could you really not have gone in any other way? Do you have to say that the diagonal number is different from all those by which it was formed? Are there not other ways that we can recognize on an equal footing? Or: Are you fully aware of the unusual nature of this mathematical situation? Do you realize how far you have moved away from the original conditions of familiar technique? Have you failed to survey anything in the surroundings of a calculation? Can I not move you, in the altered circumstances, e.g. not to regard a contradiction as a sign of a deadly logical disease of the calculus?

A "contradiction" means in Wittgenstein's writing more generally a "paradox".[13] A "paradox" may lead to a new technique being invented. Wittgenstein mentions as an example the issue of whether Cantor's diagonal number is or is not "the same" as the numbers from which it is formed (160 p. 28r). He is fully aware of the difficulty of saying when precisely a new technique begins, and an old one ends, writing that "every [real] number – we could say – creates a new situation and every such situation could be taken as an opportunity to deviate from the usual application of the rule" (160, 25v): that is, to invent a *new* technique for handing it.

Wittgenstein's idea of the importance of a variety of techniques to the objectivity of logic and mathematics is not an easy one. And it did not come to him easily. In notebook 160 (from 1938) between pp. 18 and 27 we see a long train of thought in which at first he tries to associate just one technique of applying a statement to each statement, or just one application of a description. Only by p. 27 do we see that he has reached the point of telling himself to "get used to seeing a variety of techniques of sign usage (i.e., of thinking)".

---

made clear. On Cantor see LFM p. 253; on unprovable propositions see LFM pp. 47, 56, 188–9 and Wittgenstein (2017).

[13]The terms "contradiction" and "paradox" are treated interchangeably by Wittgenstein, as they were by Russell; see Russell (1903), §79 and *Principia* and compare Ramsey, who does the same (1926). This usage has confused many of his readers, who assume that when Wittgenstein says "contradiction" he only or primarily means formal contradictions in a formalized system of logic; some readers do so in order to associate Wittgenstein's later remarks on "contradictions" with non-classical approaches to truth in logic, e.g. Priest (2004). This was true in the *Tractatus*, where the term assumed the narrower technical form as part of Wittgenstein's effort to clarify and streamline *Principia*; but Wittgenstein came to reject this narrowing in his mature philosophy.

Moreover, he adds just here (160 p. 27):

> This whole issue hangs together with: Are you clear about the unuseable character [*Unbrauchbarkeit*] of the usual conception of reference [*Bedeutung*], meaning [*Sinn*] and truth [*Wahrheit*] (more on this later).

He means not only the usual views of these notions among other philosophers, but also his own, as held, e.g., in the *Tractatus*. What he aims to do is to work out a different view of these concepts, one that builds the *plasticity* of technique into their natures and interrelations.

It would take a long time for Wittgenstein to rework the "usual" conception of the relations among our concepts of reference, meaning and truth in light of his ideas of technique and form of life. As we have seen, his annotations to Hardy formed part of this development, with their emphasis on the ways in which our special techniques with real numbers are not less primary than the general set-theoretic or extensional ones.

His most mature formulation of these remarks is the famous PI §§240–242, not drafted until 1944:

> 240. Disputes do not break out (among mathematicians, say) over the question of whether or not a rule has been followed. People don't come to blows over it, for example. This belongs to the scaffolding from which our language operates (for example, yields descriptions).
> 241. "So you are saying that human agreement decides what is true and what is false?" – What is true or false is what human beings *say*; and it is in their *language* that human beings agree [*stimmen die Menschen überein*]. This is agreement [*Übereinstimmung*] not in opinions, but rather in form of life [*Lebensform*].
> 242. It is not only agreement in definitions, but also (odd as it may sound) agreement in judgments that is required for communication by means of language. This seems to abolish logic, but does not do so. – It is one thing to describe methods of measurement, and another to obtain and state results of measurement. But what we call "measuring" is in part determined by a certain constancy in results of measurement.

These passages are crafted to form a transition to the whole discussion of "private" language, where Wittgenstein is also aiming to avoid reducing meaning to social consensus, investigating the importance of first- and second- person technique *in* life to first-person expressions of "inner" states.[14]

In general we can say this. Wittgenstein has moved in a kind of regression downward: from the idea of language-game in BB (1933–1935) to the ideas of technique and form of life (1937–1941), according which to imagine a language is to imagine, not a culture, but a specific embedding of words *in* a form of life, and, finally, to the centrality of the grammar of speech and occasion of speech by

---

[14]Pocket notebook 160 also contains lengthy discussions of the need for technique in the "direct" description of "immediate experience". See 119, p. 104v (November 13, 1937), also embedded in a discussion of 'private' language:

> I do not want to say that "communication" is what you call our language-technique [*Sprachtechnik*]. Thus, and in *excessively* simplified cases, it is difficult to say what we should and should not call it.

particular speakers (1944–1951).[15] These are the ideas that frame his conception of logic as a kind of "scaffolding" for our thinking (see PI §240): something that is always available, no matter what structure or theory we are building; something providing a variety of techniques for human accessibility and use; something that supports nothing, and is supported by itself; something modular and portable, whose elements may be arranged in a variety of ways, depending upon the needs or circumstances. Scaffolding does not *need*, because it does not *have*, either eternal boundary stones limiting its uses or human consensus on its fundamental content and scope of use. It is piecemeal, fit together for a variety of situations utilizing a variety of techniques. Yet without it, tall and complex buildings could not be erected by human beings.

It is of course true, as T.S. Eliot wrote about the idea of technique in poetry, that we cannot always "say at what point 'technique' begins or where it ends".[16] Here, like Wittgenstein, Eliot was warning against a certain danger that faces critics: we might imagine we can use a "technique of feeling", mere emotional or psychological association, to analyze their content.[17] In exploring different tendencies in our impressions of Cantor's procedure, Wittgenstein wants, just as Eliot and Frege did, to get *past* mere feeling or techniques of idea-association, to genuine, thoughtful criticism.

And so Wittgenstein must explore the sometimes murky area where technique begins, and where it ends. The diagonal procedure instantiates, in his view, such a murky place. He will interject, for example, in the midst of his reflections on diagonal arguments, "How do you know that it makes sense to say you are describing a feeling?" (MS 162b, p. 59r). He wishes to clarify the grammar of the *notion* of technique as opposed (for example) to that of feeling, for these are notions he himself is increasingly using in philosophical criticism. And such criticism should not be a matter of feelings and psychological responses, but of grammar. In the end Wittgenstein is grappling with the issue of what it *is* for a definition, description or characterization of something (e.g., a particular technique, feeling, taste, etc.) to be accurate and/or adequate, inaccurate and/or misleading.

What he wants to say, as Eliot said of poems, is that a proof, like a poem, "has its own life" within our ways of comparing and contrasting it with other proofs (poems), that "its parts form something quite different from a body of neatly ordered biographical data" and that it is "something different from the ... mind" of the

[15]Floyd (2016).

[16]Eliot (1928), p. 414.

[17]Eliot (1928), Preface. This work, seemingly influenced by Eliot's study with Russell, was titled with an allusion to Frazer's *The Golden Bough*, which Wittgenstein also made remarks about, first in 1931 and then again in 1937–1938, when he was putting together the remarks on Cantor's diagonal procedure and the *Frühfassung* of the *Investigations* (Wittgenstein 1931); see PUK. Wittgenstein critically explored the limits of a "technique of feeling" for describing an experience of poetry in his 1940 "Lectures on Description" (SWL, p. 170); the difficulty that without a technique we may not know what to *call* a proper description is raised, in connection with I.A. Richards' *Principles of Literary Criticism* (1928), at SWL p. 149 and LA p. 37.

mathematician (critic).[18] What we have to work with are our inclinations to say certain things about Cantor's method, our "impressions" of the proof; but these can be sharpened and clarified through an investigation of its various applications. For example, labeling something "indescribable", like labeling it "uncountable" or "unimaginable", requires an investigation of what it is, specifically, that we "cannot" do with the techniques that are our words.

As we have said above (p. 53), Wittgenstein compares a philosophical investigation of proof (e.g. going through a proof such as Gödel's "from A to Z") with an aesthetic exploration of fittingness or purposiveness.[19] This may be compared with a remark in 162a (from 1939, pp. 67–68):

> But what can one answer to the question: "How many cardinal numbers are there?" - Why not give the answer: "This question does not mean anything"? – But does not the question make sense? Not if you do not give it any sense. And if you give it one, you can give it one or the other.

The point here is not that the determination we choose will be arbitrary, or that answers to mathematical questions are merely stipulated conventions, but rather that there is a necessity of making a determination one way or the other *through* the development of techniques. It is this point which, I believe, he convinces Turing of in LFM.[20]

The point is made more generally as he continues in his 1938 notebook (162b, p. 68):

> In mathematics concept determinations [*Begriffsbestimmungen*] are made, not characteristics of concepts found [*gefunden*].
>
> There should then be characteristics such as purposiveness (amenability) [*Zweckmässigkeit*].
>
> A technique "treats" of nothing. However, it may be taught with a view toward this and that application.

When we are dealing with Cantor's diagonal method as a *technique*, we require an investigation of what we are doing with it.

8.3 Aspects vs. Techniques

Wittgenstein's contrast between an *aspect* and a *technique*, as I shall explain in this section, is designed to articulate a response to philosophical debates about discovery vs. invention in mathematics. As he sees it, there are different ways of looking at "the same thing", but sometimes the "thing" is a feature of a concept, sometimes a

---

[18] See Eliot (1928), Preface.

[19] See a letter he wrote to Schlick July 31, 1935, responding to Schlick's question whether Wittgenstein believed in "true but unprovable" propositions (in Wittgenstein 2005b), and my discussion of the letter in Floyd (2012a).

[20] This would need further argument that I cannot engage in here; but see LFM p. 168.

property of an extension, sometimes of a technique. Exploring examples carefully can help us resist the tendency to allow the different ways – here, the extensional and the non-extensional perspectives – to be characterized in *ideological* terms, as a matter of right and wrong. The ideological terms express a refusal to hold in mind the tensions among different aspects of the real numbers to which Bernays, as we saw above, called attention.[21] But it is these tensions among differing aspects, the differing dimensions and perspectives from which they may be considered, that lend the real numbers their true richness and depth.

From Wittgenstein's non-extensionalist viewpoint Cantor's diagonal procedure gave us of a way of thinking about certain problems. Its generality and distinctiveness *as* a technique are *shown* in the multiplicity, or diverse variety, of its actual and possible employments. And these novel applications of the diagonal technique entangle us in questions about the method itself: what belongs to it properly, and what belongs to our artful development of it as reflected in its myriad uses.

Evidently what is to count as a technique is not given by immediate grasp of an overarching extension given in advance: the hallmark of a technique is that it is open-ended in its range of applications, while remaining robust enough to be recognizably re-applied. Indeed, technique is a classic example of what Wittgenstein called a "family resemblance" notion: one bound together by a variety of partial fibers reaching through contiguous and analogous applications, each connected directly to others, but without necessarily *one* thread or technique reaching all the way through (see PI §67). – The point of the extensional perspective, by contrast, is to eliminate altogether appeal to techniques.

Specific implementations of Cantor's diagonal procedure may be adapted to demonstrate this in specific contexts, and Wittgenstein explores this in his 1937–1947 writings. The issue, as he poses it in his notebooks 162a–162b (from 1939–1940), are the contours and complexities of the idea – differentially instantiated by the diagonal method's applications in Cantor, Gödel, Turing and elsewhere – that "I cannot teach you all techniques through a technique (162b, p. 1)", a remark he investigates at length.

His approach underscores once again the importance of not confusing the extensional and the non-extensional perspectives on the real numbers, as is so easily done in philosophical rationalizations and justifications. There is the need for a systematic theory of extensions, as a way of thinking. Set theory, which structures the notion for purposes of handling the infinite, is perfectly cogent mathematics, and in fact essential for representing explicitly (in an axiomatic context) certain aspects of the real numbers. Of course there remain outstanding questions today – just as there did in Wittgenstein's time – about how far and in what manner these multifarious aspects may be summarized in one comprehensive theory.[22] But in any case there is no question of set theory representing *all* the aspects, or features, or

---

[21] Bernays (1957), p. 4, quoted above on p. 33.

[22] See the remarkable program of Woodin, described in Woodin (2017).

guises under which the real numbers appear to us as we do mathematics. It is the latter point that interests Wittgenstein.

It is crucial to understand that we have the notion of extension in mathematics for a reason: to *suppress* the diversity of actual human techniques and procedures of proof used in informal mathematics of the real numbers. We do this in order to handle the infinite, as well as to separate the idea of order from that of collection. And yet, at the same time, this in no way erases the importance and interest of the techniques. The study of the characteristics of these techniques form a different subject matter for investigation, but one interpenetrating with our whole interest *in* the extensional point of view.

Thus a debate about whether we can quantify over "all" real numbers leads ineluctably toward a skewed understanding of Wittgenstein's aims in his remarks on Cantor's procedure.[23] Wittgenstein is interested, not in establishing an extensional range, but instead in carefully calibrating and attending to different ways Cantor's method might be deployed, in different contexts, all the while investigating and exploring the similarities and differences among cases that shed light on the technique itself. This is of course a very different, non-extensional conception of the "generality" of Cantor's technique.

Recall that Wittgenstein's aim is neither to justify nor to overturn any proofs. Nor is it to resist or advance a systematic point of view of techniques, putting them into one overarching perspective. He engages instead in an exploration of differing partial "techniques", an investigation of the generality of our very concept of a proof technique itself, in order to know our way about.

By "generality", then, I do not mean – as in the view of extensionalism about the real numbers I *should* mean – a sharply expressed quantificational understanding of how far we may generalize over "all" real numbers, in terms of scope. This idea, if what has been argued in the last chapter is right, can lead to a "skewed" understanding of the difference between concept-formation and extension in the context of Cantor's diagonal proof that the real numbers are uncountable. It is not that Wittgenstein denies that we can speak about "all" real numbers, on grounds that any such talk is necessarily indeterminate. Instead, the idea is to promote determinacy of expression about the real numbers, but insist that this comes only with an exploration of how the non-extensional and the extensional points of view about "all reals" occlude one another in complex, sometimes unanticipated ways in our thinking. We must reflect upon and choose our standpoint, according to a purpose: this is not something done for us by the real numbers themselves, as they are viewed from within this or that standpoint.

For Wittgenstein it is a matter of refusing to look at the samenesses and differences of thoughts about the real numbers *one-sidedly*, to invoke an important metaphor from Frege.[24] In stressing the way in which the same thought may be analyzed wholly differently, depending upon which dimensions of generality one

[23] Compare, for a different view, Putnam (2007) and the response in Floyd (2020).

[24] Frege (1879), VII; (1897), pp. 158–9; (1892b), p. 30.

wishes to emphasize, Frege was careful never to fall into psychologistic ways of speaking about appearances of thoughts: he held that a thought "divides itself" or allows for its analysis in a variety of different ways.[25] And this is not a psychological or an ideological matter but a matter of logical and philosophical investigation. The extensionalist tends to flatten our perspective out. But so does the non-extensionalist if she fails to confront and explore the details of the interaction – or failure of interaction – between the two perspectives, extensional and non-extensional.

As we have argued at length in this book, Wittgenstein's difficulty with the extensionalist perspective is its tendency to suppress the contrasts among particular techniques in favor of what the extensionalist regards as a primary level of fundamentality, one that is designed to reduce references to human procedures and practices of representation and proof to efforts to somehow encompass or uncover a huger and huger extension, or insist on the properties of the real numbers as lying there ahead of time, to be discovered. In this way the extensionalist perspective comes to occlude the relevance of any other perspective, and even that it itself *is* a perspective, while not fessing up to its limited ability to shed light on – or even connect in any intelligible way with – our everyday practices with particular real numbers. In contrast, throughout his manuscripts Wittgenstein is investigating, usually very tentatively, how one might design a way of thinking that will connect Cantor's diagonal procedure to these, but without falsification.

There are of course a variety of forms of constructivism, pursued in logico-mathematical programs designed to extract the constructive content from arguments (including Cantor's) and calibrate this precisely, bringing human or computational (idealized) procedures into view. Wittgenstein's remarks are by contrast neither intended to systematically develop any new mathematics, nor to reject non-constructive approaches – although as has been noted, Cantor's diagonal argument as a *technique* is wholly constructive, and this is something we may regard RFM II as drawing out.[26] Instead, Wittgenstein explores the interstices of the diagonal method itself, how it may be fitted to procedures in a variety of possible ways, how it occludes and reveals different aspects, not only of the real numbers, but of our efforts to analyze and present them systematically.

The importance of these issues to Wittgenstein's 1938 remarks on Cantor's diagonal argument in RFM II has been detailed by Felix Mühlhölzer in Chap. 7 above. This analysis of these writings, situating them (for the first time) against the backdrop of Wittgenstein's surrounding manuscripts, sheds light on just what Wittgenstein meant by technique, by rule of procedure, and by method. The accent here is on the multiplicity of the diagonal procedure: its many faces, in terms of the

---

[25]English translators struggle with Frege's language, often psychologizing it by drawing in the idea that we use our *imagination* to analyze a thought into components, drawing out *aspects* of thoughts. Frege's language bears the seeds of this understanding, but not psychologistically, only logically, and the same is true of Wittgenstein's conception. See Frege (1879), §10 in Bauer-Mengelberg's translation (improved in Beaney's translation in Frege 1997); Frege (1891), pp. 5, 6, and 17 in Geach's translation. Frege himself does not speak of logical aspects, as Wittgenstein does.

[26]See Kanamori (2012a).

aspects of the real numbers that it reveals, and the differing kinds of proofs it helps us concoct and execute in a variety of differing contexts.

This also explains the dynamic character of Wittgenstein's remarks on the real numbers: his emphasis on the lack of easy or smooth connections between the two perspectives – extensional and non-extensional. As we shall see in the following sections of this chapter, his investigations of different contexts of diagonal argumentation runs them through a series of stress tests, seeking a range of variation to reveal aspects of the real numbers and our work with them. In the end, his remarks suggest that he regarded these as forming a family of *Gestalten*: one that can be investigated and tethered to procedures and techniques.

We have just argued that in the self-reflective way that logic and philosophy often demand, Wittgenstein's deployment of his non-extensionalism required him to critically investigate the very *notion* of a "technique" or "method" of proof. He does not begin with a theory of concepts or proof-techniques which is then applied. Instead he critically probes the possible "twisting and turning" of meanings required to push the idea of the "same" proof-method forward.[27] "We treat", as he writes, "the grammar of the word 'technique'" (MS 162a, 45).

In attempting to make some progress here, Wittgenstein developed a distinction between *aspects* and *techniques*. This distinction is designed to do justice to the rhetoric surrounding the extensional and non-extensional points of view. Most simply put, as a slogan: aspects are *discovered* or noticed or revealed, whereas techniques are *invented*.[28] Techniques, including proof techniques and calculative procedures and modes of indexing argumentation, help *reveal* the character of the real numbers to us.

This distinction answers the objection, framed by Ramsey (perhaps with the middle Wittgenstein in mind), that although analysis of our experience may seem always to invent new complexity that we impose in interpreting, we also want to say that it is complexity that is already there that is revealed in any adequate analysis. This is known more generally as the "Paradox of Analysis": if we analyze x, then we transform it; but then we no longer have x. The difficulty we face, according to Ramsey, is due to the vagueness of our notion of understanding.[29] Wittgenstein argues that if the open-ended notion of technique, embedded in life, frames the backdrop to analysis itself, then there is no such intrinsic difficulty.[30] We can admit the vagueness of our notion of understanding, but achieve clarityin particular cases.

[27]See RFM II §4, analyzed above in Sect. 7.4.

[28]BT, §134; RFM II §38, RFM III §§46ff; MS 122, pp. 15, 88-88, 90; PI §§119,124–129,133, 222, 262, 387, and 536; xi, p. 196; PPF xi, §130. Floyd (2018a) analyzes this distinction, while Kanamori (2018a) applies it to the real numbers.

[29]Ramsey (1929), pp. 263ff.

[30]At 162b, p. 9 Wittgenstein writes that "Ramsey was therefore completely right in what he wrote about [the problem] in his Last Papers [(1929, see pp. 263ff.)] . . . ; but he failed (I believe) to see that the problem lies in the connection between that vagueness and logical exactitude".

An analysis may not be more "fundamental" than a technique, but that does not take away the interest of the analysis or make it useless (compare PI §63).

Wittgenstein used the notion of an aspect to allude to features revealed in logical or mathematical analysis, whereas technique stands in for the procedures we devise to be able to survey and proceed with an analysis. We have already seen that his conception of "technique" is open-ended and not surveyable or foreseeable in general. His aspect-analogy is between the facial features of a human being that express, show, and contain our grasp of her inner state *in* the display of aspects, and logical features or the grammar of sentences or mathematical procedures that help us grasp what they say, mean, or imply.

Wittgenstein's master simile between facial features and logical features of sentential expressions[31] reworks Russell's epistemological notion of "acquaintance": Wittgenstein substitutes for Russell's idea of infallible epistemic contact between mind and logical form (say, of a proof) a more fallible, everyday idea of being familiar with, able to recognize, an individual person or psychological state. The key to his conception is that it draws modality – possibility and necessity – as well as the issue of our techniques of comparison and proceeding into view.

A face can assume all the usual features of fear, and yet be deceiving us, just as a sentence can assume all the usual features of grammatical meaningfulness and truth, and still be false or misleading. A proof, a sentence, may not be fully understood, and yet we may come to understand it better by looking at it in a specific, different way. Whether or not a sentence does or does not actually carry a particular logical valence requires its situation against the backdrop of an environment, in life; here it is truly like a human face. It is multiply analyzeable, in differing dimensions.

Examples of "discovered" aspects include the famous duck-rabbit of Jastrow[32]:

and the Necker Cube, with its corners that may be seen pointing upward to the right or downward to the left:

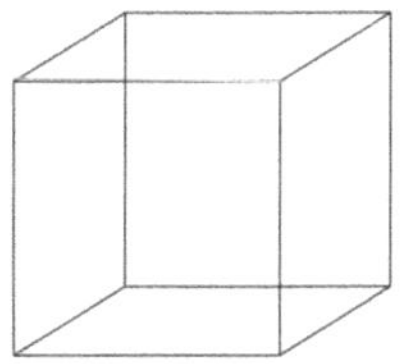

[31] See Floyd (2018a) for a discussion.

[32] See PI PPF xi §118.

Each of these figures give rise to alternative possible experiences. Techniques involve uses to which such pictures may be put, in illustrating this or that point about cognition, concept, or experience.

In Wittgenstein's early philosophy he held a rather unstable view of these figures. This may be seen in the *Tractatus*, where he uses a pair of labels, or parameters, to keep track of the two aspects. He expresses here two different techniques – in the parlance of the later, but not the earlier Wittgenstein – and says that we perceive two different facts. These alternative parametrizations, written in the form of two different pairs of labels or parameters a and b, are used to distinguish the two differing ways of perceiving the "front" edges of the figure (see TLP 5.5423):

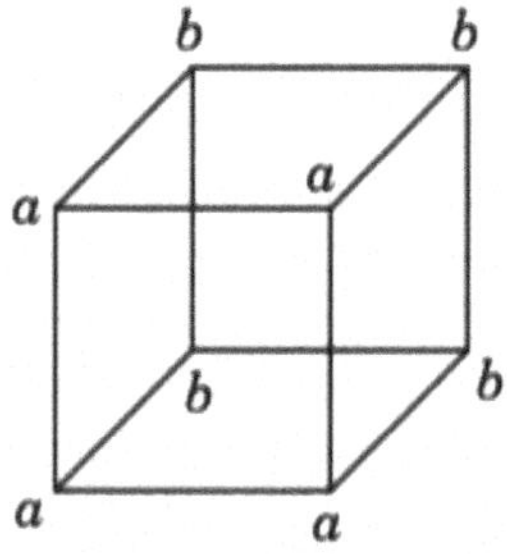

Given the *Tractatus* insistence that there is only *one* logical space, the specific role of these parameters is troublesome. Wittgenstein wrote that here we see "two different facts" (TLP 5.5423). Yet there is only *one* fact that is seen, the diagram on the page; our experience of shifting aspects shows that there are multiple manners of analysis here, multiple ways of drawing the figure into three dimensions.

What we "see" with the charactery, the use of the parameters, are two different possible ways of projecting the same configuration. With this we open up two differing fields of necessity and possibility.[33] Going back to the duck-rabbit: if *this* is an ear, then *that* is a nose, and *that* a place where a whisker might go; but if *this* is a beak, then *this* is the back of the head of a duck, and *this* a beak. And if *this* is the front of the cube, then *that* is a point in its rear, and *that* a point above it, and so on and so forth.

There are many examples of aspect-shifts in mathematics. We haven't the space to pursue an exploration of examples further here, but Wittgenstein investigates many different ones in his remarks. The fact is that figures, numbers and calculating procedures have differing characters, or faces that they show to us, that we draw out from them.[34] These are familiar shifts in the ways we see the possibilities of structuring in a mathematical situation: think of proofs in Euclid (cf. Sect. 3.4 above) or a sum generalized in algebra, or an empirical problem embedded in a theory, e.g.,

[33] See Floyd (2018a), p. 372.

[34] On aspects in Wittgenstein's philosophy generally, see Floyd (2018a).

Euler's famous problem of the thirty-six officers, embedded in the theory of Latin Squares.[35]

Wittgenstein places a great deal of weight on this idea throughout his discussions of mathematics, for it affords him a way of making sense of the difference between, and the complicated relations among, illustrations and applications (see Sect. 3.1 above). In RFM I the notion of aspect receives a great deal of investigation in connection with alternative analyses of geometric figures and notations as models, or pictures; the notion of differing techniques is connected to shifts in aspects. In RFM II, in connection with Cantor's "discovery", the language of "aspects" and "techniques" is also appealed to. In RFM III Wittgenstein regards mathematics as a many-colored collection of techniques of proof (RFM III §46), emphasizing that these techniques draw out new aspects: for example when an unabbreviated piece of notation is abbreviated (RFM III §§46–68). He emphasizes by contrast that the extensional point of view on $\sqrt{2}$ is a matter of seeing it "in terms of the picture of a cut somewhere along the straight line, *hence extensionally*" (RFM V §34).

By helping ourselves to this language of aspects – and adding to it the idea of a parameter used to fix or mark off a technique – we do not have to worry that there are two modes or kinds of truth, "duck-truth" and "rabbit-truth", or "a-truth" and "b-truth", or "extensional truth" and "non-extensional truth". We are not marking off possible *worlds*. Nor do we have to decide which way of regarding the figure we prefer (though we might need or wish to on a given occasion). What we are able to do is to mark a difference in experiences, perspectives, in what is seen, and proceed. We may say that these truths are there to *be* seen *in* the figure, though the articulation of different projections or techniques is necessary to appreciate what their truth conditions *are*. As Wittgenstein wrote in PI, "the substratum of this experience [of aspects] is the mastery of a technique" (PI PPF xi §222).

In the history of mathematics there are cases of something being newly seen: the mathematician may as it were glimpse, as Cantor said, an "abyss", or new depth, or an analogy, or a problem, or a structure. This would correspond to the example of seeing two faces and suddenly *noticing* an aspect that they share (see PI PPF xi). For example, one can come to see the real numbers under addition as a group; one can see the real numbers as a structure or set, with relations imposed upon it.

In life, of course, there may be many occasions in which we fail to communicate what it is that we see. In a sense, when it comes to people and what they "see" what we ultimately have are only sketches of larger landscapes (see PI Preface). And there may be no proofs, only the aspiration to sharing an experience or way of experiencing, for example, what a particular demeanor expresses (see PI PPF §355, on *Menschenkenner*, who are expert at judging the genuineness of emotion). But in mathematics, the technique establishes a procedural hold, a way of revealing the aspects through proof and calculation, modes of definition and characterization – through, in short, the development of a *system*.

[35]I discuss these and a number of other such examples in Floyd (2012a).

To take a very simple example, the ratios 1/2 and 13/26 are very different, if we think in terms of techniques of dividing things into parts; 1 is not the same as 13, nor is 2 the same as 26. All the same, once embedded in a system of calculations where we know how to *establish* samenesses (cross-multiplying, adding, multiplying and so on) the notion of being "the same" (or not) makes perfect sense, and standards of correctness and incorrectness are clear. And these obtain quite independently of this or that person's or community's opinion. – We have seen in Wittgenstein's reactions to Hardy's CPM an emphasis on the importance of techniques to our very idea of the "sameness" of real numbers. Wittgenstein will extend the point, in his rule-following reflections, even to the sameness or recurrence of a symbolic configuration – something he jokes about with Turing in 1939.[36]

Aspects are there to be seen. But only if embedded in life by techniques. I shall carefully adhere to the later Wittgenstein's distinction between aspects and techniques in what follows. Discovery of new aspects may of course lead to the invention of new techniques, and applications of well-grasped techniques may take place in wholly novel circumstances, revealing new aspects of things. Eliot's point about the difficulty of seeing at what point technique begins, and at what point it ends, remains in place. What matters for Wittgenstein, however, is that the range of generality of a concepts extension is something quite different from the generality of a technique, which in turn differs from the generality of consensus, of more or less widely held opinions.

Investigating sentences or proof methods as to their modes of possible employment – i.e., treating them as offering us techniques for saying or proving things – helps get at the "multiplicity" of a variety of genuine aspects – say, of the real numbers. Aspects that are revealed by way of our applications of sentences and procedures in proofs are not invented: they are there to be seen. But what is of interest are the ways in which these "extensions" (extrapolations) of concepts are revealed in the deployment of techniques of different kinds. – Of course, a face can deceive us about what it is really expressing: we can find a method of projecting its evincing that is fully in order, and yet misses the mark. Wittgenstein's view is that the same holds of sentences, even in mathematics.

---

[36] In LFM, Lecture I Wittgenstein's pointed joke is targeted directly at a feature of Turing's analysis of logic that we discuss below (see Footnote 82). He imagines teaching Turing what the letter "σ" is:

> What is a "representative piece of the application"? Take the following example. Suppose I say to Turing, "This is the Greek letter sigma", pointing to the sign σ. Then when I say, "Show me a Greek sigma in this book", he cuts out the sign I showed him and puts it in the book. – Actually these things don't happen. These misunderstandings only immensely rarely arise – although my words might have been taken either way. This is because we have all been trained from childhood to use such phrases as "This is the letter so-and-so" in one way rather than another.
>
> When I said to Turing, "This is the Greek sigma", did he get the wrong picture? No, he got the right picture. But he didn't understand the application.

A technique is only a technique insofar as it finds transportable uses in a variety of situations: it must be detachable. A sentence or proof structure may be considered in terms only of the structure or deductive configuration of a particular proof or proof type, or an alphabetical arrangement of letters. But then it is not regarded as serving in the manner of a technique, but rather as a kind of extension in an alphabet. Sentences themselves, and proof structures too, may be seen to contain a variety of different aspects, as possible features there to be seen. It requires work to draw these out. Wittgenstein's investigations of possible manners of employment of words, sentences, and characterizations are intended to do just this.

If I understand Wittgenstein, an invented technique is more diffuse in its applications and loose than a method of proof: the latter may be treated as something of interest for mathematicians as a genuine subject – for example, in explorations of purity of method.[37] Wittgenstein's notion of a technique registers something else, akin but not reducible to the notion of a rule, always conceived of as something we can use and master. Because he is interested in the nature of logic, he focusses on comparing different realizations of logical and mathematical modes of argumentation with sentences so as to gain insight into their ranges of applicability *as* techniques.

As he writes in 1939, in a notebook preparing for his Cambridge lectures on the foundations of mathematics,

> When you look for a proof you don't try to find, or to produce, a certain object, the written proof e.g.; but you try to find or (what here says the same) to produce a technique of producing certain objects. We discover a new way of finding. We extend our idea of finding.[38]

Rather than ascending to a metalevel, in the manner of Tarski and the mathematical logicians, Wittgenstein has in mind that, as he puts it, "the technique of use of the words we describe in words" (see 160 p. 17r).[39]

---

[37]For an informative discussion of this concept, see Detlefsen and Arana (2011).

[38]161, p. 7v; the original is in English. As I argue in Floyd (2017), the impact of this idea on Turing's thinking appears to have been considerable, for he takes "intelligence" to consist in appreciation of the variety of different kinds of searching and finding (see Turing 1948).

[39]This passage is reproduced in CE, p. 415, excerpted, however, from its original context. At this point, in 1937, it seems to me Wittgenstein has not yet gotten to full clarity about the idea that there may be more than one technique associated with a description or explanation (see p. 198 above). He does however emphasize that "to give a definition is not more important than to realize that we do not know how to give one" and also that "if we want to understand the use of our language as it is, then we must be interested in what questions we *do not* know how to give an answer to". Rhees' failure to print either of these remarks in his excerpting from 160 in CE is a typical case in which Wittgenstein's own diffidence and conscientiousness is papered over in the presentation of his thoughts. These excerpts are misleading for our purposes for another reason: they suppress all references to diagonal argumentation and the limitative logical results Wittgenstein had in mind at the time.

8.4 Definability: Hobson

In this section we shall clarify the resilience of the non-extensional point of view by threading our way through extensions and applications of Cantor's diagonal procedure inspired by the Richard Paradox.[40] The Richard lies on the cusp of the linguistic and the logical, presenting a structured challenge to our thinking about a variety of notions of *definability* in relation to truth and denotation. In this way, it is located at the crossing of the non-extensional and the extensional points of view, as well as the crossing of language and what it characterizes. We have already seen how Wittgenstein presses the issue of Cantor's diagonal method toward problems about inexpressibility and characterization; in this section we explore how the Richard illuminates that pressing. It evinces clearly how, given its simplicity and constructive core as a procedure, Cantor's diagonal procedure may be "fitted" to new situations, "generalized" to afford us a variety of different forms of proof. The diagonal procedure is *itself* a technique. This deepens Wittgenstein's non-extensionalist point that diagonalization need not be taken to apply an antecedently clear general notion of *can be ordered in a series* to an antecedently clear extension, but may instead be regarded as a technique for revealing new aspects of numbers and concepts (e.g., in our uses of such notions as "decimal expansion").

The notion of an "adequate" characterization of "the" real numbers is problematical territory, filled with traps and potentially misleading byways. In the Richard, Cantor's diagonal method is applied to the digits of an array of decimal expansions, but only indirectly, *via* the very linguistic apparatus by means of which the characterizations are framed. This is clearly an extension of Cantor's procedure into a novel setting (it invents a certain new use or application of Cantor's diagonal procedure, revealing a new *aspect* of our concept of *definability*) by turning the argument upon the activity of listing out decimal expansions given through "suitable definitions". With this new use the diagonal technique comes to reveal two different ways of regarding sentences and words: the extensional one (sentences and words as constructible sequences of letters in a fixed alphabet) and the non-extensional one (sentences and words as techniques with different physiognomic "faces", revealing aspects). This double perspective on sentences and words is of course a major theme for the mature Wittgenstein.

Cantor's original proof of the uncountability of the real numbers (1874) utilized what might retrospectively be conceived to be an application of the diagonal method. However, it did not force or even allow us to abandon altogether previously recognized techniques of working with particular real numbers and their varieties: he relied on the conception of reals as Cauchy sequences. In comparison to the later diagonal argument (Cantor 1891), the 1874 argument may be therefore be regarded as appealing to merely *ad hoc* contrivances of bijection.[41] In the seventeen years

[40] Richard (1905).

[41] See Kanamori (2012a) and the discussion on pp. 76ff above.

between the papers Cantor came to see a new, more general *aspect* of his original proof: the collapsing of two variables into one. This grammatical move, made in a footnote *en passant* in the original paper,[42] is transformed into a simple procedure in the later diagonal proof through the representation of a 2-dimensional array of decimal expansions in tabular form. In the later proof, the indexing used in the representation of particular real numbers as decimal expansions "$a_1a_2a_3, \ldots$" is identified, grammatically speaking, with the parameters used to index the process of writing out such sequences in a list (thereby generating the diagonal expansion a_{nn}). Bringing the two different uses of "1,2,3,..." together in this way shows us how in general to merge two different dimensions of use, or techniques, into one.

This transformation was not extracted before Cantor's (1891) *as* a technique. Now seen to apply to *any* listing of sequences, the complex grammar of our notion of a *sequence* is taken a step further toward its contemporary extensional construal. A "sequence" may now be regarded, not merely as a matter of succession in time or procedure of calculating – the root meaning of "sequence" derives from rules of succession in a royal office – but extensionally, as a function from the ordered series of natural numbers onto a set, a function with a fixed, ordered, infinite "domain" and "range". As Felix Mühlhölzer argued above (p. 126), "function", "domain" and "range" are thus absorbed into an extensional understanding, even if nowadays the notion of an "assignment" of an output to an input draws us back towards a procedural understanding in the midst of this extensional language. In particular, it unfolds a new way of regarding our everyday uses of "series" with indices indicating processes of calculating a function's behavior (e.g. $\sum_{n=1}^{\infty} 2^{-n} = 1$). In a sense, the idea of a sequence *eclipses* the earlier, process-of-calculation understanding, but without destroying it.

This new way of looking at series and sequences reveals a new picture of what "a technique for ordering" a collection of numbers *is*, as was argued above in Sect. 7.6. Cantor's distillation of a new technique re-organized our understanding of the earlier, particular techniques of parametrization, affording a new perspective in which the concept of *uncountability* may be seen. The key was his developing applications of the extensional perspective, the idea of a covering [*Belegung*] of a set by a function and the exploitation of 1-1 surjection as a criterion of sameness of cardinality applied to finite sets. This opens up new dimensions of the concept of *countable*.[43]

The Richard Paradox is set out in Russell's Introduction *Principia Mathematica*, Chapter II, where we can be sure that Wittgenstein saw it as a young man (this

[42]See Cantor (1874), note 1 and the analysis in Kanamori (2020, in press), §2.1; the connection with diagonalization is explained in Kanamori (2012a), p. 8.

[43]See Jourdain's remark at Introduction, p. 82 to his translation of Cantor's papers on the founding of the theory of transfinite numbers, Cantor (1915):

> The introduction of the concept of "covering" [*Belegung*] is the most striking advance in the principles of the theory of transfinite numbers from 1885 to 1895.

presentation also set Gödel in motion, on the way to his understanding of the undefinability of "truth" and his incompleteness proof.[44]) Russell writes:

> Richard's paradox [Note: see Poincaré (1906b), sections VII and IX and Peano (1906), pp 149 ff.] is akin to that of the least indefinable ordinal. It is as follows: Consider all decimals that can be defined by means of a finite number of words; let E be the class of such decimals. Then E has $\aleph_0$ terms; hence its members can be ordered as the 1st, 2nd, 3rd, ... Let N be a number defined as follows. If the nth figure in the nth decimal is p, let the nth figure in N be $p + 1$ (or 0, if $p = 9$). Then N is different from all the members of E, since, whatever finite value n may have, the nth figure in N is different from the nth figure in the nth of the decimals composing E, and therefore N is different from the nth decimal. Nevertheless we have defined N in a finite number of words, and therefore N ought to be a member of E. Thus N both is and is not a member of E (Whitehead and Russell 1910, p. 64).

If we take all finite strings of all combinations of letters of the English alphabet – *Or is it any possible alphabet?* (here is an issue to which we must return in assessing the argument's generality *as* a technique of proceeding) – we may order all the permutations of sequences of letters, by length and lexicographically, and so produce an exhaustive list of all means of characterizing real decimal expansions in English. Striking off of the list all those definitions that fail to define such an expansion, call the remaining collection of definitions E. Lexicographically ordering them, we can present this set in a definite order, and by its means a series of decimal expansions $u_1, u_2, \ldots$ that are "definable in a finite number of words". We next apply Cantor's diagonal procedure to this list, defining another decimal expansion N in a finite number of words by altering the digits along the diagonal. Call this definition, given by Russell above, "G". N does not belong to the set E, but N is defined in a finite number of words. So we face an apparent contradiction.

In his original version of the paradox Richard proposed the following solution. Although the string of words in G must have appeared on the original list, in *that* particular place the definition has no meaning, because it mentions the set E, "which has not yet been defined" (Richard 1905, p. 144). Richard offers a kind of step-by-step picture of defining that invokes the idea of an activity or process, rather than an extension. Viewed procedurally, N can be inserted into the enumeration of the set E, $u_1, u_2, \ldots$ at any rank k, and we can then increase the indices i for all u_i with $i > k$ by 1. This yields a new diagonal definition G' that will define another number N', distinct from N – although of course this gives us the wherewithal to diagonalize again, with G''. In this way what was conceived of as static is now re-construed as dynamic and procedural.[45]

[44]The explicit reference to the Richard Paradox is in Gödel (1931), Introduction; see Floyd and Kanamori (2016), where Gödel's excavation of his incompleteness argument from these pages in *Principia* is reconstructed.

[45]This reconstrual is taken to be fundamental by Poincaré, in his presentations of the Richard; see Poincaré (1905, 1906a,b, 1909) and compare Luna (2017) for a detailed and informative discussion of Poincaré and the Richard Paradox. In the *Tractatus* Wittgenstein proposed such a procedural approach with his notion of an operation, indexing applications; see TLP 3.333f.

The set of numbers "definable in a finite number of words" is "self-reproducing", as Russell would have said,[46] or "indefinitely extensible", as Dummett would have said.[47] As we have seen above, Wittgenstein would have admitted that a listing of decimal expansions of real numbers is "uncountable" [*unabzählbar*] in the sense that it draws in a new concept of "ordering in a series" *via* the diagonal method (see the discussion by Felix Mühlhölzer above in Sect. 7.5). The series N, N', N'', ... continues to fall under the same concept as diagonalization proceeds, while the defining capacity of the language fundamentally "expands", and as we go on diagonalizing into new "systems", new aspects of the "definable" are revealed. We have here an indefinitely extensible *technique*, as the mature Wittgenstein would say.

At just this place in *Principia*, having outlined the Richard Paradox, Russell added a footnote to draw in discussions pertaining to the real numbers: to König (1905), Dixon (1906) and, most important for us, Hobson (1906).

Author of the standard Cambridge textbook *The Theory of Functions of a Real Variable and the Theory of Fourier's Series* (three editions 1907, 1921, and 1927), E. W. Hobson inaugurated a large debate over the Axiom of Choice – and the extensional point of view generally – in England in his (1907).[48] His influence on the theory of functions of a real variable extended, as we shall see, well into Wittgenstein's time, and his differing presentations of Cantor's argument in the different editions of his influential book resonates with Wittgenstein's concerns.

In fact, Hobson's obituary was written by Hardy. He described Hobson as

> ...a keen philosopher. Philosophy, indeed, was his strongest external interest, as one could judge from passages of his great book. He was not a 'mathematical logician', but he was attracted by fundamentals, and was the first English mathematician to see the point of the discussions of the 'antinomies' and to recognize the importance of 'Zermelo's axiom'...
>
> Hobson's book ...is, if any book ever was, a 'standard treatise', and is probably the most important book written by a modern English mathematician (Hardy 1934, p. 939).[49]

Hardy stresses that Hobson was a pioneer and a victor in bringing the modern abstract study of theories of measure and integration into England, making the theory of functions of a real variable "the most highly developed mathematical discipline at Cambridge, a subject recognized even as a 'good Tripos subject', the most popular and paying subject in 'Schedule B'." However, Hardy adds,

> Hobson never quite understood how completely he had won his fight. He always retained something of the air of the protagonist of an unpopular cause; he was a little too old to understand fully that everything that he had been fighting for had been achieved (Hardy 1934, p. 939).

[46]Russell (1905/6).

[47]Dummett (1993); see Shapiro and Wright (2006) for a formulation of "relative indefinite extensibility".

[48]Moore (1982), Chapter 2.7 describes this debate in fine historical detail.

[49]Hardy also mentions that Hobson had a close friendship with philosopher James Ward, whom Wittgenstein also knew; see Monk (1990), p. 537.

Given Wittgenstein's severe reactions to Hardy's *laissez faire* attitude about mixing the non-extensional and extensional perspectives – as well as the rise of other, non-set-theoretic approaches to the foundations of mathematics in the wake of Turing's analysis of "computable" – there is more than a little irony here. Wittgenstein does not wish to reject or defeat the more abstract, extensional perspective in favor of the non-extensional perspective: he does not see modern, more abstract methods involving us in a "fight". Rather he is interested in comparing and investigating the differing approaches to the real numbers. He is pressing the question, "But *where* and *how* have we settled? And what has been 'achieved'?" As we know, he is especially critical of Hardy's breeziness in mixing up and glossing over contrasts between the non-extensional and the extensional perspectives in CPM.

Given that he was a leading pioneer in England developing expositions of abstract topological methods, it is interesting that Hobson from early on – never recanting – resisted using the Axiom of Choice. Despite what Hardy says, Hobson always emphasized the non-extensional perspective's importance in the presentation of the theory of real numbers. In fact, what Hobson argues from a non-extensionalist perspective will nicely engage with what we have said about Wittgenstein's insistence that our ordinary techniques for working with particular real numbers are central to our understanding of the concept of "real number".[50] In the end we will turn to look at Turing's mention of Hobson in his "On computable numbers" (1936) (Sect. 8.5). This will show that concern about the interplay between the extensional and the non-extensional perspectives formed part and parcel of a larger Cambridge tradition in mathematics and philosophy in the twentieth century.[51]

About Hobson Russell writes (p. 64):

> The solution [to the Richard Paradox] offered in [Hobson (1906)] depends upon the variation of the "apparatus of definition," and is thus in outline in agreement with the solution adopted here. But it does not invalidate the statement in the text, if "definition" is given a constant meaning.

What is meant here by a "definition"? What by an "apparatus"? In *Principia* Russell's solution to the Richard is that the very notion of "definable in a finite number of words" is ambiguous, and must be typed. "Finitely definable", is seen to suffer, like the notions of *truth* and *falsity*, from "systematic ambiguity" (Whitehead and Russell 1910, Introduction, Chap. II, Sec. III). If this is granted, then the notion

[50] We have no proof that Wittgenstein had studied any of Hobson's works, and there is no reference to Hobson in his writings. However, through Russell's references (including his 1905/6), Wittgenstein would have known of Hobson's role in the disputes. Hobson also authored (1913), a survey of impossibility proofs of construction. I note in Floyd (1995) that the trisection of the angle was one of Wittgenstein's favorite examples of an impossibility proof throughout his career; this and other problems of construction are the primary examples he used in the early lectures of LFM in 1939, and Hobson's books were those used by Wittgenstein's students (and possibly by him as well).

[51] For an in-depth discussion of Wittgenstein and Turing as part of this wider Cambridge tradition, see Floyd (2017).

of "definable" can be replaced by a notion "definable$_{\alpha\beta}$", which is indexed as to type and order within Russell's ramified hierarchy. This implements Poincaré's idea of avoiding vicious circles,[52] and also takes on board Peano's idea (1906) that the way to avoid paradox is to avoid loose, everyday language by formalizing.[53] Because every formal definition – being of finite length and occurring at a specific step – fails to contain a term referring to what comes after it in the enumeration, there can be no formulation of the paradox.[54]

Hobson (1906) is a rejection of König's introduction of a sharp dichotomy between those elements of the continuum which are "capable of being "finitely defined", and those which are not. König had argued (against Zermelo 1904) that the set of real numbers cannot be well-ordered.[55] His argument was that because the real numbers definable in finitely many words form a countable sequence of alphabetical arrangements, whereas the continuum is itself uncountable, there must exist real numbers that are not specified by any definition. But this generates a contradiction. For there is, for example, the finite definition "the least real number not in the enumeration", which appears to define a real number.[56] König concluded that the irrational numbers capable of being finitely defined form an "enumerable" (countable) collection, just as the rational numbers do; but the remainder of the irrational numbers, being "ideal" elements, are neither "countable" nor "finitely definable".

Hobson rejected König's absolute dichotomy between real numbers that are, and those that are not, "capable of being defined in finite terms". König takes this to reflect something "inherent" in the particular real numbers themselves, and Hobson rejects this idea, drawing in a varying sense of "definition" by making our *language* dynamic and open-ended.[57] Hobson's strategy was to uncover ambiguities in König's *use* of the notion of "finitely definable" to avoid the idea of a number that is "inherently" undefinable. He applied Cantor's diagonal procedure to do so, taking it to demonstrate that the "apparatus of definition" is dynamic, not fixed.

---

[52]Poincaré (1905, 1906a,b, 1909).

[53]Peano regards the diagonal number as indeed finitely definable, but only in a contradictory manner, as it must assert of at least one digit of the list, depending upon where it is *on* the list, that it both is and is not identical with itself.

[54]Actually Russell collapses ramification within types by means of the "Axiom of Reducibility", which asserts that there is a predicative, minimal level propositional function equivalent to any one specified by ramification within each type.

[55]On Zermelo (1904) see Kanamori (2004). It would follow from König's argument that the Continuum Hypothesis is false, for his argument could be applied to the collection of finitely definable, countable ordinals, thereby showing that Cantor's second number class is not a set in a proper sense.

[56]See Cantini and Bruni (2017) and Franchella (2016) for detailed discussion of König's Paradox, the latter treating Hobson's objections in detail and discussing the whole arc of König's work on it.

[57]See Hobson (1921), p. 87.

Hobson admitted that there is a way to list all irrational numbers such as $e, \pi, \sqrt{2}, \ldots$ "which are in common use in arithmetical analysis". Yet, he argued, this is not enough to ground König's dichotomy:

> ... by means of a discussion of the possible modes of formal definition of irrational numbers ... the distinction between the definitions applicable to the special class, and to other irrational numbers, is not of such a character as to justify our speaking of some irrational numbers as capable of finite definition, and of others as not so (Hobson 1906, p. 22).

First, Hobson claimed,[58] *no* irrational number is inherently capable of being defined in a finite number of words and symbols each of which has a definite and unique meaning:

> [Even the] simplest possible definition of any of the ordinary irrational numbers, such as $\pi, e, \ldots$ involves the use of a symbol n (or of some form of words equivalent to the use of such symbol), to which no unique meaning is attached, but which is capable of denoting the numbers of the integer sequence 1,2,3 ..., of type ω. Thus, for example, a definition of the number e may be given, in which the expression $1 + \sum_{i=1^n} 1/n!$ occurs, and no arithmetical definition of the number e can be given which dispenses with the use in some form of the "variable" n, of which the meaning is not unique (Hobson 1906, p. 22).

This reflects an implicit idea: a variable expresses a rule of some kind, and cannot in general be reduced to its range, or extension, of variation. Hobson's application of the diagonal technique is not framed from within the extensional perspective, where the rows of the two dimensional array are conceived to be arbitrary extensions. Instead he applies Cantor's technique to "adequately defined" decimal expansions in order to clarify his non-extensional point of view. He will find a way to *attach* Cantor's diagonal technique to his non-extensional conception of functions of a real variable.

Hobson granted that an aggregate E_i could be imagined that consists of all such one "variable" definitions of irrationals using n (1, 2, 3, ...) – though the ordering by listing such definitions would be, as he also says, "difficult to carry out". Putting E_i together with the aggregate of all the rational numbers, E_r, to get E, this at least gives a provisional characterization of König's aggregate of "finitely definable" real numbers.

Hobson then points out – just as we have stressed in our discussion of Wittgenstein – that while an irrational number in *general* may be defined to be a Dedekind cut, this does not give us a way to characterize *particular* irrational numbers. He puts this in terms of definability. A real number is only "*adequately* defined", he writes, when we

> ... are supplied with a means of deciding, in respect of any arbitrarily assigned rational number, whether the irrational number ordinally precedes or ordinally succeeds the assigned rational number... [which implies that] a definition of a particular irrational number [in a

[58] We shall have to return to this claim. As B. Jack Copeland has pointed out, a real number *can* be adequately defined without the explicit use of variables in a Turing-machine specification. This is connected with the absoluteness of Turing's definition of "computable", to be discussed below in Sect. 8.5.

> particular interval] must supply us with directions for calculating, for any prescribed integer n, by an arithmetic process, the first n figures of the decimal representation of the irrational number. The number of steps in the process must be definite for any assigned value of n, but has no upper limit for all values of n (Hobson 1906, p. 23).[59]

And he argues that, given *this* conception of an "adequate definition", Cantor's diagonal procedure may be applied to produce an adequately defined real number that cannot be formulated "in a form of words and signs which involves only the use of the one 'variable' n" (Hobson 1906, p. 24).

Assume that we have a list of numbers $x_1, x_2, \ldots$ in the interval (1,0) each of which has been given an "adequate definition" in Hobson's sense. The decimal representations of $x_1, x_2, \ldots$ may then be listed, adapting the Cantor diagonal technique:

$$x_1 = a_{11}a_{12}a_{13}\ldots$$

$$\mathrm{x}_2 = a_{21}a_{22}a_{23}\ldots$$

$$\mathrm{x}_3 = a_{31}a_{32}a_{33}\ldots$$

$$\cdot$$

$$\cdot$$

$$\cdot$$

Clearly for any a_{nr} we are in possession of a means of calculating the particular value of n and r "by a finite process dependent upon those assigned values" – and this includes our diagonal number a_{nn} for each n contained in the enumeration of $x_1, x_2, \ldots$. In other words, we are in possession of a procedure or technique of expansion, to use our earlier way of phrasing things.

We then may define, in the usual way, the diagonal number N: $a_n = a_{nn} + (-)1^{a_{nn}}$. It differs from a_{nn}, for if $a_{nn} = 9$, then $a_n = 8$; and if $a_{nn} = 0$, then $a_n = 1$; and so on. N is adequately defined in the relevant way, for "we have the means of calculating $[a]_{nn}$, and thence a_n, and thus $a_1 a_2 \ldots a_n \ldots$ can be calculated in a finite number of steps" for a_{nn} (Hobson 1906, p. 24).

The fact that it is not possible to enumerate all the "adequate" definitions of real numbers using one variable n naturally leads Hobson to an expanded conception,

59 See Hobson's remarks about "the method of postulation" (i.e., Dedekind's cuts); he holds (Hobson 1906, p. 23) that

> To have recourse to the method of postulation, in order to provide elements in the continuum [i.e., in Dedekind's way], would appear to involve the postulation of the existence of entities which are not clearly distinguishable from one another; for, if determinate ordinal relationship with the rational numbers is supposed to appertain to them, we are unable, in default of adequate definitions, to ascertain what that relationship in any particular case may be.

that of an "adequate" definition containing not only n, but also an ordered collection $\alpha_1, \alpha_2, \alpha_3 \ldots$ of numbers regarded as having been already defined with different parameters, arranged in the order type ω.[60]

But, as he argues, these specifications do not deserve the name "finite definition" in the same sense as before. And this exposes a kind of "ambiguity" in the notion of "finitely defined" (p. 26). For each definition in this more general sense "contains directions for carrying on a process which is essentially endless", even if the process of comparing the definition with any particular rational number is a finite one in the sense of employing only a finite number of parameters (p. 26).

Thus the Cantor diagonal method may be applied in all such cases to deny that when various collections of parameters are taken into account, the result is enumerable:

> For, let us suppose, if possible, that all the numbers so definable form an enumerable set which can be denoted by $x_1, x_2, x_3 \ldots, x_n, \ldots$; then the same reasoning as was applicable when $\{x_n\}$ was taken to be the set E suffices to shew that a number $\overline{N}$, not belonging to the $\{x_n\}$, admits of a definition of the type considered; and thus that there is a contradiction in the assumption that all the numbers definable in this manner are contained in the enumerable set x_n.

Hobson says that

> This proof is ... a modification of one of Cantor's proofs that the arithmetic continuum is not enumerable [viz., the (1891) proof]; and it completes that proof, by shewing that a number can be adequately *defined* which does not belong to the assumed enumerable set (Hobson 1906, p. 25).

In this way, language is not enumerable or "countable" in Hobson's view.

Hobson's idea is to exploit Cantor's diagonal procedure in order to attack König's idea of an "inherently" finitely undefinable real number. His real intent becomes, in fact, to insist on the creativity and open-endedness of our procedures of defining. This overturns König's idea of what is "inherent" to definable real numbers by insisting on the always-expandable nature of what belongs to our language.

For Hobson endorses a conception according to which there is a "norm" for every viable "object of thought" in mathematics. In fact Hobson never changed his mind about this. A Hobsonian "norm" is a rule- or grammar-like characterization of a collection which, in Hobson's words, "is of sufficiently precise character to make it

---

[60] In a footnote he shows how this idea would work with an example (p. 25):

> It is easy to construct directly definitions of irrational numbers in which the parameter cannot be taken to be the integers 1, 2, 3, ... For example, an irrational number M may be adequately defined as follows: —Let the decimal representation of M be such that the nth figure is identical with the nth figure in the decimal representation of $2^{1/P_n}$, where P_n denotes the nth of the prime numbers 2, 3, 5, 7, 11, ... Since P_n is not expressible in finite terms as a function of n, the parameters of the definition may be taken to be the set of prime numbers.

logically determinate as regards any particular object whatever, whether such object does, or does not, belong to the aggregate".[61]

In his (1906) Hobson agreed with König's rejection of the Axiom of Choice: that axiom, expressing the extensional point of view, allows us to regard a definition of a collection as "adequate" which nevertheless fails to allow us to "adequately" distinguish or define every particular member of that collection. But he rejected König's too-rigid idea of "finitely definable" as something "inherently" belonging to real numbers. Instead, for Hobson the "mind", in concert with "adequate" norms of specification, is able to creatively "single out" objects of thought, but its resources are not limited to a single, semantically fixed, countable alphabet.[62] The choice and expressive power of newfangled combinations of parameters is not determinable in advance by a set of rules.

Of course Hobson's demand for a non-extensionalist "norm" is vague, as Russell was quick to point out in his classic paper (Russell 1905/6). Hobson's notion of an "adequate definition" is not sufficient for showing the existence of a collection: as Russell's paradox shows, we have a norm, "$x \notin x$", which nevertheless does not secure the existence of any collection. Moreover, assuming that we *have* specified a collection, we cannot assume that there is a property, or "common mark" holding the set together that is denoted by what we express, as Hobson assumes (Russell 1905/6, p. 31).

In substituting his notion of a "propositional function" for Hobson's vaguer notion, Russell developed a more powerful non-extensional perspective, one that did not assume, as Hobson's did, that our "mental activity" of framing definitions and singling out objects of thought is required to assert existence or cardinality.[63] Russell was also able to disentangle the idea of "ordering", hence the difficulties about the Axiom of Choice, from the more fundamental problem of the "contradictions", operationalizing definability within the theory of types. (The whole idea of "merging" parameters into a definition would be further clarified through Tarski's and Turing's subsequent analyses of "definable" and "computable'.)

Yet Hobson made a Pilgrim's progress over time. In the first edition of his textbook (1907, p. 209f.) he was polemical, criticizing Zermelo's use of the Axiom of Choice, only briefly mentioning alternative perspectives on "inconsistent aggregates" such as Russell's, and relegating foundational issues about the contradictions to an Appendix. But in the much expanded second and third editions (Hobson 1921, 1926, 1927) he offers a far more balanced presentation, stating (in his Preface) that the controversial status of the Axiom of Choice raises such delicate issues that he will make no dogmatic pronouncements about it, but simply avoid its uses in proofs where possible, and point to its applications where these are necessary. He also added an extended discussion of debates concerning the philosophy of number

[61]Hobson (1905, 1905), pp. 172–3, (1907), pp. 196–7, (1921), p. 236.

[62]Hobson (1905), pp. 179–180; (1907) §§1, 158 at pp. 204n, 205; (1921) §§1, 190.

[63]Russell (1905/6), p. 41; see Hobson (1905), p. 180, (1907), p. 204.

where he canvassed differing views across the spectrum from Heine through Cantor, Frege, Husserl and Russell (Hobson 1921, pp. 9ff.).

In the end, Hobson never recanted his non-extensional presentation of the theory of functions of a real variable. He retained his notion of a "norm", insisting he had no need of taking an extensional viewpoint on the continuum, because his non-extensional conception sufficed to generate sufficient specifications of all relevant mathematical properties of the continuum studied with functions of a real variable.[64] He never retreated from the idea that for any adequate norm, a "common mark" holds together the class of numbers that is individuated in thought (Hobson 1921, p. 236). Nor, in keeping with his non-extensional perspective, did he accept Russell's reduction of cardinal numbers to classes of classes that may be put into 1-1 correlation with one another. Numbers, from Hobson's point of view – at least in the finite case – may be regarded as the "common characteristic of a family" of such classes, "created" in thought through abstraction from the nature of the objects in the collection and the order in which they are given.[65]

In short, Hobson regarded set theory as largely superfluous to the theory of functions of a real variable. And in his debate with Russell over norms, the contrast between the idea of our *creating* the singling out of properties of real numbers with norms, and *discerning* them, is made explicit. Russell held that properties are discerned, whereas Hobson always took it that they are created or at least fixed on and singled out in thought.[66] Against König's "inherent" idea of a real number's indefinability Hobson set the idea of our language. But here he meant "language" in the sense of a growing, indefinitely extendible apparatus.

As we have seen, after 1937 Wittgenstein attempted to fashion a way of accommodating both the idea of *inventing* (or *creating*) and the idea of *discerning* (or *seeing*) aspects of the real numbers, while rejecting the whole idea that an essential set of common marks is intrinsic to our idea of the generality of concepts (his so-called "family resemblance" idea, see PI §§66ff). His distinction between techniques and aspects does not track that between the non-extensional and the extensional points of view; nor does it postulate anything about activities of mind *per se*. But it aims to do justice to both Hobson's and Russell's philosophical concerns. Wittgenstein's point is that we must have both creating (techniques) and discerning (seeing what is there to be seen), but that we must align these two carefully. For if we do not then the very plasticity of our means of defining may collapse into mere convention or stipulation, losing thereby its connection with discovery and insight. Hobson's idea is that we may always add further names to our language, so that no language is enumerable. Wittgenstein, following Russell, resists this claim, while at the same time attending to the idea of a grammatical "norm", or technique, as central to the non-extensionalist point of view.

---

[64]See Hobson (1906), p. 27; (1921), p. 88; (1927), p. 91.

[65]Hobson (1921), pp. 10, 240. On Wittgenstein's rather different non-extensionalist rejection of Russell's claimed reduction of cardinality to sets see Floyd (2004), pp. 97–98.

[66]Russell (1905/6), p. 41 and Hobson (1907), p. 204n.

What now happens to our the idea of the continuum? Hobson also uses Cantor's diagonal procedure to raise the question of whether the set of real numbers can be characterized by a "norm", thus being legitimately considered "a single object of thought". As he puts the result of Cantor's argument,

> The set of real numbers cannot be tabulated in such a manner that no number fails to occur at some definite place in the table. No set of rules or specifications can be given which suffice to determine successively all the numbers of the set, and no finite set of symbols can exhaustively exhibit the numbers.[67]

Therefore,

> The only sense in which the numbers of the set are determinate is that each such number is the limit of a convergent sequence of numbers, taken from the unending table formed by the rational numbers. It may fairly be doubted whether such a negative specification of elements amounts to a valid synthetical definition of a determinate aggregate...

Hobson argues that what he calls "the arithmetic continuum", which is "obtained by adjoining to the set of rational numbers the set of all their limiting points" does have an order-type with sufficiently determinate characteristics: that of being *connex* (given two numbers, other numbers ordinally between the two can be defined) and *perfect* (if $x_1, x_2, x_3, \ldots$ be any defined convergent sequence of numbers, each of them adequately defined, the limit x of the sequence is definable adequately in a form involving the use of $x_1, x_2, x_3, \ldots$ as parameters, and conversely any adequately defined number can be exhibited as the limit of a convergent sequence of other numbers, in particular of rational numbers).[68] This makes of the continuum a sufficiently determinate object of thought, from the non-extensional point of view – though of course it does not pretend to "exhaust" it.

About a possible "intuition" of the continuum through idealized abstrations from space and time, Hobson is far less lackadaisical than Hardy would be in CPM. He writes that

> ...it is difficult to see how precision can be introduced into the intuitional notion of the continuum apart from some theory relating either to points or to infinitestimals; and the language employed in such a theory must be of a symbolical character amounting to the use of some kind of arithmetical notation. Regarding the geometrical continuum in this way as a single object of which we have a direct knowledge obtained from our intuitions of space and time, the reduction to a precise abstract form may be regarded as being made upon the assumption that the system of rational numbers, with their limits adjoined, is adequate to the analytical description of the continuum, in the sense that each point in the continuum is represented uniquely by a single real number, and that there is no point in the continuum which is not so represented. This amounts to a definition, in a certain sense, of the contents of the geometrical continuum.[69]

---

[67]Hobson (1907), p. 73; in (1921), p. 85 Hobson adds a reference to his §60, which treats the Cantor Argument.

[68]Hobson (1906), p. 27; see (1907) §59 and (1921) §63.

[69]Hobson (1921) §63.

Hobson comes in this way to acknowledge the use of the "Cantor-Dedekind Axiom" as an *axiom* holding that each point on the real number line "corresponds" to a Dedekind cut:

> From the point of view of Arithmetical Analysis the existence of the aggregate of real numbers, as a single definite object, possessing assigned properties, may be regarded as a fundamental postulate; the validity of such postulation being subject to the law of contradiction.[70]

And in fact, Hobson writes – echoing Dedekind – this postulation is a "free creation of our mental activity".[71]

Creative aspects of our mental activity are also demonstrated, according to Hobson, by the unending fertility of our possible instruments for characterizing the real numbers. Cantor's diagonal technique may be used to assure us that the whole, in terms of its particulars, cannot be represented. As Hobson argues,

> an enumerable aggregate may itself be denoted by a single word or symbol that can be created or assigned ad hoc, whenever the enumerable aggregate has been defined...[72]

And this dynamism evinces what Hobson calls the "essence" of Cantor's proof, namely that

> ...there exists, and can exist, at any time, no stock of words and symbols which cannot be increased for the purpose of defining new elements of the continuum. The theorem that the continuum is unenumerable is equivalent to the fact that the assumption of the existence of a *final and complete* stock of words and symbols, by means of which alone elements of the continuum can be defined by finite definitions, leads to a contradiction. Those elements of the continuum which are at the present time, or will be at a future time, capable of definition by means of the words and symbols in existence, may form an enumerable set; but this does not prove that there exists any element of the continuum that is inherently incapable of finite definition.[73]

Hobson is not attacking the actual infinite *per se*; he is only insisting on a non-extensional point of view. That is appropriate, because what he is attacking is *König*'s way of drawing a dichotomy between elements of the continuum that are

---

[70] Hobson (1921) §63, p. 86.

[71] Hobson (1905) p. 175. A less mentalistic idea, suggested in Kanamori (2012a), would be to treat the Cantor-Dedekind Axiom as something akin to a kind of "Church's thesis" for the number line; see Kanamori (2012a), p. 3, discussed in Floyd (2013a), p. 1029.

[72] Nowadays the Richard Paradox is put together with the so-called paradoxes of denotation, instanced by the puzzle of how we are to convey to whom we refer when introducing a new name. "That man's name is 'Smith'" uses the mentioned name, and so it appears to be impossible to informed about anyone's name, for what one hears is the name of the name, and not what it denotes (see Anscombe 1957, Read 1997, Gaskin and Hill 2013, Reach 1938). I was told that a copy of Reach's (1938) paper was in Wittgenstein's room at the time of his death, but I no longer remember who told me this. Notes of Newman's spring 1934 logic course (given the year before Turing sat in on it but likely following the same format) show Newman using a variant of the Richard, i.e., "the nearest star not mentioned in 1934" to produce a sample paradox to illustrate the diagonal procedure (Smithies 1934), May 10, 1934).

[73] Hobson (1921), pp. 87–88; see (1927), pp. 90–91.

finitely definable and those which are only ideal, i.e., as Hobson puts it elsewhere, "intrinsically undefinable" (Hobson 1921). Because of the dynamic character of our language, which grows, this dichotomy must be surrendered.

Now despite their differences over the issue of whether the objects of mathematics are discerned or created, in the exchange between (Russell 1905/6) there is a shared agreement that the limits of mathematical characterizations must be sought in the idea of being "logically determinate". Whatever is meant by a logically "adequate" specification of a real number or collection of real numbers – whether this is connected to human calculative procedures or not – it must not be "lawless", but somehow play the role of a determining norm or rule.[74] The idea of an object that is *inherently* incapable of being finitely defined is "an object ...regarded by many, if not most, mathematicians as not being an existent object for mathematical thought".[75] There are no "indescribables".

As Russell writes,

> The problem concerned in such cases [of uncharacterizeability] is like that of the "lawless" decimal, which reduces to the problem of the "lawless" class of finite integers. If we consider all the classes that can be formed of finite integers, it seems at first sight obvious that many will consist of a perfectly haphazard collection, not definable by any formula. But this is open to doubt. It would seem that, as Dr. Hobson urges [Hobson (1905), p. 182], an infinite aggregate requires a norm, and that such haphazard collections as seem conceivable are really non-entities. In the case of Cantor's "proof"[76] that there are more classes of finite numbers than there are finite numbers, it is shown that no one denumerable set of formulae will cover all classes of finite numbers; but the class shown to be left over in each case is defined by a formula in the process of showing that it is left over. Thus this process gives no ground for thinking that there are classes of finite numbers which are not definable by a formula.[77]

8.5 Turing on the Diagonal Process (1936–1937)

Hobson's claim that the "essence" of Cantor's diagonal proof leads us to a dynamic picture of our apparatus of definability is referred to by Turing at the beginning of §8 of his "On computable numbers, with an application to the *Entscheidungsproblem*" (Turing 1936).

This section, entitled "The Diagonal Process", is where Turing proves the unsolvability of the *Entscheidungsproblem* by applying his analysis of what a formal system *is* in terms of his "machines". Turing's proof helps delimit the range of

---

[74]For example, a number whose digits are generated by arbitrary throws of the dice will not do. Compare Hobson (1906), p. 27 and Wittgenstein's remarks on throwing the dice to form an expansion and whether this is "lawless", PR §§181ff.

[75]Hobson (1921), p. 87.

[76]The scare-quotes are in Russell's original paper.

[77]Russell (1905/6), p. 52.

"logic" in a formal sense, holding beside one another the extensional and the non-extensional perspectives in a subtle and novel way.

A "Turing machine" has an extensional face, as a proper mathematical object: it is nothing but a list of n-tuples. But it also has another, non-extensional face: it reveals a possible step-by-step process of writing down an expansion of a real number according to a rule followed step-by-step. In a sense, it is *we* who bring dynamism, the idea of steps unfolding in time, *to* this "process". Turing's model of a "computable number" is in fact founded on the idea of a human "computor" writing down digits step-by-step, reckoning according to a rule: a non-extensional perspective.

Turing's reference to Hobson has never been adequately explained, having been hardly noticed.[78] But why does Turing refer to Hobson? The answer is that he is interested in clarifying his particular application of Cantor's diagonal "process", or technique. He wishes to emphasize that in the context of the *computable* real numbers, which he first characterizes in this paper, Cantor's technique does *not* lead outside the collection, as it does in the familiar arguments of Cantor and Hobson. Turing has in fact *replaced* the idea of a real number with that of a computable real number, and this changes the situation *vis-à-vis* the diagonal technique and the issue of the apparatus of definition. Turing makes very few references in his paper: this one, we shall see, is of philosophical significance.

Turing opens §8 of his (1936) this way:

> It may be thought that arguments which prove that the real numbers are not enumerable would also prove that the computable numbers and sequences cannot be enumerable. [Footnote: See Hobson (1921), pp. 87,88.] It might, for instance, be thought that the limit of a sequence of computable numbers must be computable. This is clearly only true if the sequence of computable numbers is defined by some rule.

Turing's reference is to the striking passage quoted above (Sect. 8.4) where Hobson sets out his criticisms of König's assumption that there is an absolute dichotomy between those real numbers that are "finitely definable" and those that are not. Hobson's conclusion is that the "essence" of Cantor's diagonal proof is that "there exists, and can exist, at any time, no stock of words and symbols which cannot be increased for the purpose of defining new elements of the continuum" (Hobson 1921, pp. 87–88). Turing will show that this claim must be qualified in the context of his computable real numbers, where there is a (relatively) "absolute" distinction between those computable numbers that are "finitely definable" and those that are not. The reference to Hobson provides Turing with a standard source, in the Cambridge context, for thinking about the diagonal procedure.

[78]Zhao Fan (2020) has, however, begun to develop an interpretation by going through the relevant literature. As he points out, Petzold (2008), which goes through Turing's paper line by line, catches the reference to Hobson but relies on the third edition of Hobson's textbook (1927), and not the second (1921), as did Turing. This leads Petzold to miss the special application of Cantor's diagonal procedure to definability that Hobson makes on the pages Turing refers to in §63, discussed above; he assumes instead that Turing is making a straightforward reference to Cantor's diagonal argument as it is usually understood, from the extensional point of view.

The Hobson reference also helps Turing set up the specific form of argument that he will give in this section of his paper. For he shows that the *computable* real numbers cannot be diagonalized out of in the same way as the real numbers can be, à la Cantor or Hobson. In the context he has built there is no analogous expansive, creative role for "mental" or "linguistic" creativity. For this has already been accommodated in his informal, human model at its ground. Unlike the real numbers, the computable numbers *do* form an enumerable collection. One may therefore define a diagonal process for them. And yet one cannot diagonalize out of the class with this process. For any such diagonal rule is not computable in Turing's sense.

In the above-quoted opening of §8 Turing remarks that it would be natural to assume, based on the usual theory of functions of a real variable, that the limit of a sequence of computable numbers will only be computable if it is "defined by some rule": this of course is precisely what any reader of Hobson's textbook would appreciate. But this is so in the present situation, he notes, only if "the sequence of computable numbers is defined by some rule". And here an ambiguity emerges in the idea of being "defined by some rule". Turing has already argued (in §§6–7) that his Universal Turing machine "defines by some rule" all the computable real numbers. He will go on to show (in §8) that there is no hope of utilizing the diagonal method to escape this machine's characterization of the totality of computable real numbers.

This will yield his negative resolution of the *Entscheidungsproblem*. For, once the connection between the very idea of a taking a step in a "formal system of logic" and being "computable" in the sense of Turing's "computable numbers" is forged, it will be shown that there can be no general decision procedure (no "computable" procedure) defined that can decide in a finite number of purely logically, or "computable" steps, for an arbitrary sentence of a formalized theory of mathematics, whether or not that sentence does or does not follow from the axioms of the theory. This will follow from the application of the diagonal technique: no computable routine can determine, in a finite number of steps, what the digit is at a particular place of an arbitrary computable number defined on a given enumeration of all the computable numbers.

Turing's argument may be formulated as a version of the Richard Paradox.[79] Let E be the class of all "finitely definable" decimals in *Turing's* sense of "computable" ones. E has $\aleph_0$ members, for each Turing Machine may be regarded as a sequence of ordered n-tuples, rectified according to an enumeration given by the Universal Turing Machine. We now consider N, a decimal defined by a certain diagonal rule on E's enumeration. Though we may know that N differs from each decimal in E, we also know that N is not computable, i.e., is not "finitely definable" in *that* sense.

Because of the shift in context and Turing's sharpening of what a "finite definition" involves, he must negotiate with great delicacy the difficulties involved in putting together the non-extensional and the extensional points of view. The

---

[79]Compare Good (1966).

general issue raised in §8 has become, What does it *mean* to say that a "sequence of computable numbers is defined by some rule"? The answer will be that there are different sorts of "definitions" and different sorts of "rules", and we must be careful not to confuse them. Turing needs, in other words, to clarify the very idea of a decimal expression that is "calculable by finite means" (Turing 1936, p. 231).

His official definitions are as follows. A *circle-free* or "satisfactory" machine is one that, placed in a particular initial configuration, would print a potentially infinite sequence of 0's and 1's (blank spaces and other symbols are regarded by Turing as aids to memory, analogous to scratch paper; only these scratch symbols are ever erased). A *circular* or "unsatisfactory" machine fails to do this, writing down no more than a finite number of 0's and 1s. A *computable number* is a real number differing by an integer from a number computed by a circle-free machine (i.e., its decimal (binary) expansion will, in the non-integer part, coincide with an unending expansion of 0's and 1's printed by some circle-free machine). A *computable sequence* is one that can be represented (computed) by a circle-free machine.

It is important to see that Turing has framed his idea of a "computable real number" so as to avoid relying on any *particular* countable language – that is, any given "stock of words and symbols" that is assumed to be (in Hobson's word) "*complete*". All that is needed to give a "finite specification" of a computable number (with a rule for computing digits) is a finite list of quintuples each one of which has the form of a command with *some* specified alphabet of symbols, e.g. $< q_i, S_j, PS_k, L, q_m >$, where "$q_i$" refers to the initial machine configuration, "S_j" to the symbol read at an arbitrary location on the tape, "PS_k, L" to the instruction for the machine (here, print symbol k and move left one location on the tape), and "q_m" to the final machine configuration.[80]

Although in the specification of every computable number there must be a finite number of symbols, it does not matter which symbols they are, so long as they are able to be taken in meaningfully by a (human) computor.[81] Moreover, whatever symbols are used, this does not constrain the choice of symbols in specifying any other computable real number, except relative to rectification of a protocol for listing all the computable real numbers which can be carried out in a step-by-step manner. For if one is given any machine table, one may code it with finite alphabet of letters, and in this way generate a "standard description" of the computable real number; Turing abbreviates this with "S.D.".[82] The standard description may then

[80]Of course, we now use 4-tuples in our contemporary presentation, but this is an inessential point for us.

[81]See Footnote 82 below.

[82]Turing (1936), pp. 240ff. In notebook 162b from 1939 Wittgenstein presses the question whether a Hilbert-style series of strokes such as ||||||| is or is not a *Gestalt*, i.e., something that one can take in *as* a meaningful whole, and he uses the abbreviation "S.D." to allude to the strokes (162b pp. 14–15). If I am not mistaken, he may well be playfully alluding to Turing's notion of a "standard description" here: this is the only place in his corpus where this abbreviation appears, lending some plausibility to the idea that he is responding to issues raised in Turing's paper. Wittgenstein might

be uniquely coded, Gödel-style, by a natural number. Turing calls this coding the "description number", or "D.N.", of the computable real number (Turing 1936, pp. 240–242).

In §7 of his paper Turing writes down a detailed characterization of the Universal Computing Machine showing how to cobble together, in a step-by-step way, all the standard description numbers of all the machines (satisfactory or not) to produce one overarching machine, which itself has a standard description, and thus may be given a description number. This readies him to apply "the diagonal process" to the totality of machines, which includes the Universal Machine itself. Turing's point will be that the conjoining of the totality of command tables to one another in terms of "standard descriptions" is not alone enough to give one the wherewithal to *computably* define, *via* the diagonal process, a heretofore undefined computable number, i.e., one that is "satisfactory". – In this way the amalgamation of Machine Tables in an enumeration is very different from the amalgamation of decimal expansions of real numbers used in Cantor's argument.

After Turing has questioned that the assumption that the limit of a sequence of computable numbers must be computable, he moves to apply "the diagonal process" in the rest of §8. At first he writes down another intuitive way of arguing, though he puts it into quotation marks, in the voice of a hypothetical interlocutor:

> Or we might apply the diagonal process. "If the computable sequences are enumerable, let α_n be the nth computable sequence, and let β be the sequence with $1 - \phi_n(n)$ as its nth figure. Since β is computable, there exists a number K such that $1 - \phi_n(n) = \phi_K(n)$ all n. Putting $n = K$, we have $1 = 2\phi_K(K)$, i.e., 1 is even. This is impossible. The computable sequences are therefore not enumerable".

The argument Turing offers in quotation marks purports to show that the computable real numbers are not enumerable in just the same way as the real numbers are not, according to Cantor's original diagonal argument. However, Turing immediately rejects this:

> The fallacy in this argument lies in the assumption that β is computable. It would be true if we could enumerate the computable sequences by finite means [JF: i.e., by means of a circle-free machine], but the problem of enumerating computable sequences is equivalent to the problem of finding out whether a given number is the D.N. [JF: the description number] of a circle-free machine, and we have no general process for doing this in a finite number of steps. In fact, by applying the diagonal process argument correctly, we can show that there cannot be any such general process.

---

also mean, e.g., "Strich-Ding". But what he certainly has in mind is the issue of whether one *can* see (or imagine) such a series of strokes *as* the same as another, say ||||, merely at will, as a matter of experience, or as a matter of *logic* – and he emphasizes the latter. Turing assumes that a human computor can only take in a limited number of symbols "at a glance" (1936, p. 250) and says that this accords with "experience". Wittgenstein is emphasizing here that this is not merely a matter of experience, but is a logical point, and therefore connected, both with the issues of inventing vs. discovering (techniques vs. aspects, see Sect. 8.3) and with the issue of what it can mean to say that something is "indescribable" (see 162b pp 14ff.). To identify two collections of strokes that are not easy to take in, we require a procedure or technique.

Turing is emphasizing the different situation at work in the case of the computable real numbers. The "fallacy" in thinking that Cantor's diagonal argument can be directly applied to show that the computable numbers are not enumerable lies in the assumption that the diagonal sequence defined on the basis of the enumeration of the computable numbers is itself computable. But that assumption, that the diagonal sequence β_n is enumerable "by finite means", is, in *Turing's* sense, false.

Turing next states that

> The simplest and most direct proof of this is by showing that, if this general process exists, then there is a machine which computes β. This proof, although perfectly sound, has the disadvantage that it may leave the reader with a feeling that 'there must be something wrong'. The proof which I shall give has not this disadvantage, and gives a certain insight into the significance of the idea "circle-free". It depends not on constructing β directly, but on constructing β', whose n-th figure is $\phi_n(n)$.

Why does Turing move to β', rather than using "the simplest and most direct" argument he has just alluded to with β? We want to emphasize in what follows the idiosyncratic character of Turing's actual argument using β', highlighting its philosophical importance. So we shall first briefly compare it to the "Halting Argument", the diagonal argument ordinarily used today to show how Turing's machine-model resolves the *Entscheidungsproblem*.[83]

The Halting Argument is a good contemporary realization of what Turing had in mind as "the simplest and most direct proof" that there is no process for determining whether a given arbitrary number is or is not the description number of a circle-free machine, for it applies the Cantor Diagonal Method directly. But what Turing is after is something more concrete and perspicuous, and this is why he will also argue differently, in terms of β'.

In the usual Halting Argument presentation a Turing machine t has its input-output behavior described by way of a partial function $f : N \rightarrow N$. Presented with an initial configuration that codes a natural number j according to a specified protocol, t proceeds through its instructions, "moving" through the squares on its tape. In the event that t goes into a specified halt state with a configuration that codes a natural number k according to protocol, then $f(j) = k$ and f is said to *halt at* j, written "$f(j)\downarrow$". Otherwise, f is said to *fail to halt* at j, written "$f(j)\uparrow$". In general, f is partially defined because of the latter possibility.

Enumerating Turing machines as t_i, we have corresponding partial functions $f_i : N \rightarrow N$, and a partial function $g_i : N \rightarrow N$ is said to be "computable" if it is an f_i. (This differs from Turing's use of "computable sequence", which in his paper must be generated by a "satisfactory", or "circle-free" machine that does *not* stop printing out an expansion of the digits 0 and 1). Thus the set of Turing machines that compute partial functions is definable and enumerable. But because the notion

[83]The halting argument was probably first formulated by Martin Davis in a lecture of 1952; see the url: http://en.wikipedia.org/wiki/Halting_problem#History_of_the_halting_problem and Turing and Copeland (2004) p. 40 n 61.

of the partial function is basic, it is not possible to diagonalize out of the list, as it is in Cantor's argument.

Consider a binary array, conceived as indicating *via* "$\uparrow$" that Turing machine t_i fails to halt on input j. Each t_i computes a partial function $f_i : N \to N$ on the natural numbers, construed as a binary expansion:

$$\begin{array}{llllll} t_1 & \uparrow & 1 & 1 & 0 & \uparrow & \ldots \\ t_2 & 1 & 0 & 0 & 0 & 1 & \ldots \\ t_3 & 0 & 1 & \uparrow & 0 & 0 & \ldots \\ t_4 & 1 & 1 & 0 & \uparrow & 0 & \ldots \\ t_5 & 1 & 1 & 1 & 1 & 1 & \ldots \end{array}$$

Here one can clearly see "holes" in the array where machines fail to halt. This Swiss-Cheese difficulty defeats the application of Cantor's method of diagonal argument. The notion of a partial function, in being primordial in Turing's model, changes the situation.

But how does the proof go? As Turing's scare-quoted sketch of a proof indicates, Cantor's diagonal argument might have *appeared* to apply directly as follows. As Turing showed in §6, there is a universal Turing machine UT. It corresponds to a partial function $f(i, j)$ of two variables, yielding the output for t_i on input j, thereby simulating the input-output behavior of every t_i on the list. Now we may construct D, the Diagonal Machine, with corresponding one-variable function which on input i computes $UT(i, i)$. D is well-defined and corresponds to a well-defined (computable, partial) function.

We suppose now that we can define a "Contrary" Turing machine C that reverses the input-output behavior of D as follows – this corresponds to Turing's β. C, with the initial configuration coding j, first proceeds through the computation of $D(j)$ and then follows this rule, which we will call *Rule (*)*:

$$\text{If } D(j) = \downarrow, \text{ then } C(j) = \uparrow,$$

$$\text{If } D(j) = \uparrow, \text{ then } C(j) = 1.$$

Rule (*) says that if $D(j)$ halts it then it should proceed to instructions that never halt, and if $D(j)$ does not halt, then it should output the code for 1 and enter the halting state. (In Turing's sketch of the proof with β, the Diagonal Rule would subtract one from the digit in the enumeration of digits, directly mimicking the Cantor diagonal argument by "changing a digit" along the diagonal.)

If C were a Turing machine (were a computable function), it would be computed by t_k for some k. Consider t_k on input k. By Rule (*), if t_k halts on k then it fails to halt on k; but if it fails to halt on k, then it halts on k. Thus t_k halts on k if and only if it fails to halt on k: a contradiction showing that our supposition that C is a *computable* function is false.

Rule (*) assumes Halting Knowledge, i.e., that machine C can reach a conclusion about the behavior of D on any arbitrary input j and apply Rule (*). But to have such knowledge requires going through all the (possibly) infinitely many steps of the D machine. And that is not itself a procedure that we can express by a rule for a one-variable Turing machine. Halting Knowledge is not Turing computable.

All this is clear enough. Yet, as we have seen, Turing worries that this argument might give someone "an uncomfortable feeling". One might, for example, question whether the law of excluded middle really applies to every question we might pose about the behavior of a "computable real number". The Halting Argument builds negation *in* to Rule (*). And Turing's sketch of a proof with β – which imagines altering the digits of the diagonal expansion systematically, as in the classical Cantor argument – might strike one as problematic once one appreciates that the diagonal rule, though definable, is not itself computable in Turing's sense.

So let us turn to β', the more concrete and rigorous proof that Turing prefers and goes through in detail in his (Turing 1936). This argument, he says,"gives a certain insight into the significance of the idea 'circle-free'". It will work by considering how to define a machine H, using an enumeration of all Turing machines, to *directly* compute a certain sequence whose digits are drawn (without being altered or switched) from the n's along $\phi_n(n)$, the diagonal sequence issuing from the enumeration of all computable sequences.

Let us call the enumeration of all computable sequences α; α_n is the nth computable sequence in the enumeration of computable sequences, i.e., those sequences computable by a circle-free machine. $\phi_n(m)$ is the mth figure in α_n. β' is the sequence whose nth figure is the output of the nth circle-free machine on input n: it corresponds to $\phi_n(n)$, which we may think of as the *positive* diagonal sequence.

Suppose that there is a definite process for deciding whether an arbitrary number is that of a circle-free machine, i.e., that there is a machine D which, given the standard description number k of an arbitrary Turing machine M, will test to see whether k is the number of a circular machine or not. If M is circular, D outputs on input k "u" for "unsatisfactory"; and if M is circle-free, D outputs on k "s" for "satisfactory". D enumerates α_n by finite means. Combining D with the universal machine U we may construct a machine H that is designed to compute the sequence β'.

H proceeds as follows. Its motion is divided into sections. In the first $n-1$ sections the integers $1, 2, \ldots n-1$ have been tested by D. A certain number of these, say $r(n-1)$, have been marked "s" i.e., are description numbers of circle-free machines. In the nth section the machine D tests the number n. If n is the D.N. of a satisfactory machine, then $r(n) = 1 + r(n-1)$ and the first $r(n)$ figures of the sequence whose description number is n are calculated. H writes down the $r(n)$th figure of this sequence. This figure will be a figure of β', for it is the output on n of the nth circle-free Turing machine in the enumeration of α_n by finite means that D is assumed to provide. Otherwise, if n is not satisfactory, then $r(n) = r(n-1)$ and the machine goes on to the $n+1$th section of its motion. H is circle-free, by the assumption that D exists.

Now let k be the description number of H. What does k do on input k? Since k is the description number of H and H is circle-free, the verdict delivered by D cannot be "u". But the verdict also cannot be "s". For if it were, H would write down as the kth digit of β' the kth digit of the sequence computed by the kth circle-free machine in α_n, namely by H itself. But the instruction for H on input k would be "calculate the first $r(k) = r(k-1) + 1$ figures computed by the machine with description number k (that is, H) and write down the $r(k)$th". The computation of the first $r(k) - 1$ figures would be carried out without trouble. But the instructions for calculating the $r(k)$th figure would amount to "calculate the first $r(k)$ figures computed by H and write down the $r(k)$th". This digit "would never be found", as Turing says: at the kth step, it would be "circular", contrary to the verdict "s" and the original assumption that D exists.[84] The trouble is that H's instructions at the kth step amount to the "circular" order "Do what you are to do at the kth step", or "Do What I Do", and this is an empty command, or idling wheel, for H. It is not of course contradictory, but it cannot be *followed* with any particular step.

Let us call H the "Do What You Do Machine". Turing's positive diagonal argument – the one he prefers to give – reveals that the limits of logic lie, not only in contradictions, but also in circular, empty, idling constructions that we are able to put together with words. The command "Do What You Do" is neither contradictory, nor empty in all contexts. If I am teaching you to ride a bike, and you are pedaling, and I say, "Keep on doing what you are doing" then that is a useful piece of advice. But in the context of the *Entscheidungsproblem*, where we are not given any particular context of operating, it is empty.

It is not that Turing has generated a paradox or regress argument.[85] Nor has he offered any kind of verificationalist or epistemological argument. He has not depended upon any particular theory of mind or language. His argument, unlike Richard's, does not turn on any ambiguity of language or its expansive abilities. And he has accommodated Hobson's point about the expanding apparatus of our language, adapting it to that to which we can assign a systematic protocol. Turing reveals the way in which the notion of a command that can be *followed*, attached to a technique of use, is basic to logic itself. The point is deeply philosophical, not only logico-mathematical. For the fact that we can see that this command is, without further supplementation, *unusable* demonstrates that the human interface, the context of a shareable, actionable command, is fundamental to the nature of logic. "Do What You Do" is not really a command as such (imagine, say, drawing a card in a game with this written on it). It *requires* a technique of application.

A useful analogy is to be drawn with the "positive" Russell Paradox. This concerns the set of all sets that *are* members of themselves. (This is the complement of the usual Russell set of all sets that are *not* members of themselves.) In a sense, this *positive* Russell set "comes before" Russell's set, is more primordial, for there is no use of negation within its definition.

The positive Russell set is not contradictory. For we may define

$$S = \{x \mid x \in x\}.$$

[84] Turing (1936), p. 247.

[85] Although the argument might be adapted in this direction, as is nicely done by Copeland in Turing and Copeland (2004), p. 38.

And then ask

$$\text{Is } S \in S?$$

The answer is:

$$\text{If Yes, then } S \in S.$$

$$\text{If No, then } S \notin S.$$

So we have that:

$$S \in S \iff S \in S.$$

There is no consistency or paradox here. But there is a problem. For all that we can deduce here is that:

$$S \in S \iff S \in S, \text{ and also } S \notin S \iff S \notin S.$$

We are caught in a kind of circular thought of the form, "it is whatever it is". This is surely not incoherent or inconsistent. The trouble is deeper: the thought cannot be *implemented* or applied.

We have here what might be regarded as a kind of performatively by empty rule. You tell yourself to do something depending upon what the rule tells you to do, but you cannot do anything, because you get into a loop or tautological circle. This set membership question cannot be a question that can be applied, because one cannot apply the set's defining condition at every point. "Falls under the concept" and "$\in$" cannot be *used* if they are equated without a technique, or more specific orientation on, application. This, as we have seen above throughout our discussion, is a hallmark of the later Wittgenstein's philosophy.

We should recall Wittgenstein's complaints about Hardy's prose-like comments in CPM to motivate the students with illustrations. In these cases an apparently unproblematic way of thinking is applied. But two different ways of thinking are involved. In the case of the positive Russell set, there is the thinking of S as an object or element that is a member of other sets (the extensional point of view) and there is the thinking of S in terms of a concept, or defining condition (the non-extensional point of view).

An analogous line of reasoning may be applied to, e.g., "autological" in the Grelling paradox if we ask, "Is autological autological?". Without using negation, one does not get a contradiction. But one may generate a question with the concept that may be sensibly answered with either Yes or No. And in this sense it is an unanswerable question. The trouble is, one cannot get to a decision point here. One cannot *play* the game of Yes and No.[86]

Similarly, in Turing's (1936) proof, there is the unproblematic characterization of a particular machine, and then there is the difficulty that it must, at one precise point, get stuck in a loop, confronted with the command to do what it does. The lesson is that the extensional and the non-extensional viewpoints are each fine in their place. But the two ways of thinking cannot be collapsed into one another without emptiness. We must retain our hold on the differing *aspects* under which the real numbers may be seen.

Now what of the law of excluded middle? We see that its scope ends, not only in truth-value gaps or in contradictions, but also in the emptiness of certain commands – albeit commands which we may nevertheless transform into contentful ones by augmenting them with the grip of a particular context or orientation toward a task.[87] It is important to emphasize that both arguments Turing uses – i.e., both his direct use of Cantor's diagonal method in his first, sketched argument using β (or in the Halting Argument) and his "Do What You Do", circular argument with β' – are *constructive* arguments in the classical sense: neither invokes the law of the excluded middle to reason about infinite objects. But, as it turns out, it is not part of the ordinary activity of a human computer, or the general concept of a person working *within* a formal system of the kind involved, that a person must take a stance on the law of the excluded middle generally.

Turing acknowledged that classical philosophical concerns about negation and the law of excluded middle in infinite contexts – the worry about what it means to treat a completed totality of steps as just another step – might emerge in the course of his arguments. In the corrections he published to his paper, urged on by Paul Bernays, he waxed a bit more philosophical about this.[88]

His first correction remedied a flaw in a definition pointed out by Bernays, thereby narrowing a reduction class he had framed for the Decision Problem. The second made his analysis more general, showing quite explicitly that his definition of "computable number" serves independently of a choice of logic. Turing wrote to Bernays that when he wrote the original paper (Turing 1936),

> I was treating 'computable' too much as one might treat 'algebraic', with wholesale use of the principle of excluded middle. Even if this sounds harmless, it would be as well to have it otherwise.[89]

So he modified the means by which computable numbers are associated with computable sequences, citing Brouwer's notion of an overlapping choice sequence,

[86] Wittgenstein discusses the Grelling paradox at length in his MS 117pp. 221ff. (1938) making the same point, and showing that even an apparent contradiction may turn out to find a perfectly acceptable technique of application. This forms part of a response to an objection that Newman had made to Wittgenstein, asking him whether the presence of a contradiction sufficed to make something a philosophical problem. Compare the discussion in Mühlhölzer (2010), Chap. II.8.

[87] Compare Sect. 4.2 above.

[88] Turing (1937b).

[89] Turing (1937a), Turing to Bernays 22 May 1937.

as Bernays suggested he do.[90] This avoids what Turing calls a "disagreeable situation" arising in his initial arguments: although the law of the excluded middle may be invoked to show that a Turing machine *exists* that will compute a function (e.g., the Euler constant), we may not have the means to *describe* any such machine.[91]

The price of Turing's generalization is that real numbers no longer receive unique representations by means of sequences of figures: there is no computable way to work back from a description number to a unique real number. The payoff is that his definition's applicability no longer depends upon invoking the law of the excluded middle in infinite contexts. The loss, he explains, "is of little theoretical importance, since the [description numbers of Turing machines] are not unique in any case" and the "totality of computable numbers [remains] unaltered".[92] – In other words, Turing's characterization of the computable numbers is robust with respect to its representation by this or that formal system, this or that choice of logic, or any specific analysis of what a real number really *is*.

Today it is sometimes said, following Gödel, that the class of computable numbers is *absolute* with respect to its representation in this or that formal system. What is meant is that no matter how strong the axioms are of the system in which one is working, so long as one is working in a recursively axiomatizable system of the relevant, Hilbertian kind, the class of what are the "computable functions" stays the same. Differently put, Turing found a strong *parameter* by means of which to measure what counts as a "step" in a formal system in the Hilbertian context.

This was essential for his resolution of the *Entscheidungsproblem*. If there had been a decision procedure of the required kind, then all that would have been needed would have been to write that algorithm down. But since there *cannot* be such a procedure, in order to show this one must *analyze* what is meant by a "formal system" in the relevant Hilbertian sense. This could not have been done by writing down *another* formalism or system of equations: one had to do something different, something philosophical. And this is what Turing did. By relying on an intuitive comparison between a human being reckoning according to a rule the digits of a computable number and a "step" in a formal system – that is, a non-extensional perspective – he was able to reach the appropriate level of generality, but without begging the question by relying on this or that formal system.[93]

[90]Turing (1937a), Bernays to Turing 24 September 1937. The corrections using Brouwer's notion of an overlapping sequence are explained in Petzold (2008), pp. 310ff. Petzold conjectures that conversations with Church at Princeton (or with Weyl) may have stimulated Turing's interest in recasting his proof, though he suspects that "Turing's work and his conclusions are so unusual that . . . he wasn't working within *anyone's* prescribed philosophical view of mathematics" (p. 308). I agree. But in terms of possible influences on Turing, Bernays must be mentioned, and Wittgenstein too should be added to the mix. See Floyd (2017) for a longer argument.

[91]Turing (1937b), p. 546.

[92]Turing (1937b), p. 546.

[93]See Kennedy (2020, in press) for a wide-ranging discussion, not only of Turing, but of this general point about "foundations".

As Gödel wrote, with Turing's analysis of computability "one has for the first time succeeded in giving an absolute definition of an interesting epistemological notion, i.e., one not depending on the formalism chosen" (Gödel means a formal system of the relevant (recursively axiomatizeable, finitary language) kind).[94] And in a note added in 1963 to a reprinting of his famous paper (1931), Gödel called Turing's analysis "a precise and unquestionably adequate definition of the general notion of formal system", allowing a "completely general version"' of his incompleteness theorems to be proved.[95] After all, until one could really make out what a formal system *is* in general, the range of applicability of Gödel's incompleteness theorems still hung in the air.[96]

This "absoluteness", however, rides upon the back of another point. Turing stressed in his corrections that there is no unique way of correlating a real decimal expansion with its description number. This is fundamentally due to the fact that however we rectify the presentation of the Turing machines systematically, we must code their behavior with some protocol or other, some parameter which tells us what the behavior of the machine *means* in this particular context: a technique. The carrying out of that protocol-assignment should itself be computable. However, we need somewhere to *begin* that is meaningful *in a computable way*, and thus arises a certain level of inevitable circularity.[97] There is no *general* computable way of determining when a coding is or is not computable, deciding whether we have slipped in a "deviant" encoding into our assignments, or whether a demon has stepped in unbeknownst to us. We must simply acquiesce in the idea that we know how to rectify our understanding of the input and the output on a given machine's tape, step-by-step, at some point or other.[98]

Always, when we come down to the fundamentals, we must assume that the computor can "take in symbols at a glance" as meaningful, as Turing says (1936, pp. 250–51). If the strings are too complex to be so taken, or swim about ephemerally, preventing us from grasping when we have an instance of the *same* sign, we assume that we *can* break them down, step-by-step, into simpler elements that can be determined and taken in. As Turing says, the justification for *his* definition of "calculable by finite means" "lies in the fact that the human memory is necessarily limited" (1936, p. 231). But this is a logical, not a psychological point about the need for techniques: at some point mathematics comes in to establish the "sameness" of symbolic configurations. The mathematician's job it is to design and establish the worth of particular techniques, parameters and procedures allowing us to rectify and take in symbol complexes as "the same" or "different".

---

[94]Gödel (1946) and compare the detailed discussion in Kennedy (2017).

[95]Gödel (1986), p. 195; compare Gödel (1964b), p. 399 and Sieg (2006a,b).

[96]See Kennedy (2017) for an illuminating discussion of this.

[97]See Kennedy (2017) for a discussion.

[98]On deviant encodings see Shapiro (1982), Copeland and Proudfoot (2010), and Kennedy (2017), p. 68 n. 26.

As we have seen, Turing made a most illuminating adjustment to our understanding of the application of Cantor's diagonal method as a technique: he fits it into a new context. First, his application shows that it is not possible to diagonalize out of the class of computable numbers. Second, his "Do What You Do" argument turns up a remarkable feature of foundational thinking about logic: there is no way to eliminate the need for techniques – that is, invented human means of *applying* signs in logic – from the foundations of logic.

This is clear from what Turing's (1936) diagonal argument requires us to *see*: without further supplementation of a context or action *as* action-guiding, this command (or "I do what I do") is empty. "Do what you do" or "I do what I do" may be "logically true" in the manner of a tautology, difficult or impossible to imagine false (recall the truth-teller: "I am true"). But this shows us that the limits of logic lie in the failure of a command, or sentence, to find the friction it needs in a technique, embedded in ongoing forms of life, rather than in absolutely certain principles of thought.

A way to put this in Wittgensteinian terms is that is part of our concept of a *rule* that we know what in particular to *do* with it in a particular given situation, we know how to fashion a technique for embedding a rule, word or routine *in* life. We have already seen (p. 195 above) that in his 1939 Cambridge lectures on the foundations of mathematics that Turing attended [LFM], Wittgenstein stressed this point, hammering home the importance of our fashioning of techniques over and over again, not as a psychological or mathematical, but rather as a logical fact.[99]

In 1947 Wittgenstein reconstructed a Turing's diagonal argument in terms of the idea of decimal expansions and rule-following. In a moment we shall consider this extraction; it was already sketched above (on p. 145). But it is worthy of further detailed consideration as a capstone to our reading of Wittgenstein's non-extensionalist perspective on the real numbers.

First we should set Wittgenstein's version of Turing's argument into context, not only against the backdrop of Turing's "On computable numbers", but also Wittgenstein's own development. For there was a kind of mutual interplay between Wittgenstein's and Turing's philosophizing in the period 1934–1954, from the time of *The Blue and Brown Books* to Turing's final publication, in the year of his death.[100]

8.6 Rule-Following (1936–1937)

We have seen how Wittgenstein evolves after 1937, drawing into his thinking the idea of a variety of techniques for the use of a word, and the need for *Lebensformen*, forms of life, into which words must be embedded – as opposed to "cultures" which

[99]LFM pp. 73, 74, 131, 168, 174,179, 191, 200, 202, 254.

[100]See Turing (1954) and Floyd (2016, 2019b) for details of the historical reconstruction.

are in consensus on particular truths. In *The Blue and Brown Books*, dictated to his mathematics students in 1933–1934 and 1934–1935, he had reached a certain point in his thinking about the notion of logical simplicity, but he had not taken these steps. Relying on a kind of linear, step-by-step, quasi-evolutionary structure, Wittgenstein presented the idea that simplicity should be relativized to language-games, and that to imagine a language *is* to imagine a culture.

This overcame the straight conventionalism about what is simple that characterized his middle period, bringing activities of human beings with words into his construal of "logic". But as we explained above in Sects. 1.2 and 1.3, *The Blue and Brown Books* still proceed in a "false" linear way, as if language consists in a series of rule-governed conventional games, cobbled together bit by bit by the straightforward addition of routines. It lacks the critical focus on the ideas of *Regelmässigkeit* and what it is to follow a rule that are the hallmarks of Wittgenstein's post-1937 work. – That focus emerged in the fall of 1936 (see Sect. 1.3). And one reason to think Wittgenstein would have been interested in Turing's (1936) paper – aside from its making concrete mathematical sense of the cogency of his philosophy of logic – is that his attempted revision of *The Brown Book* already contains what he came to call, in the fall of 1936, a new "problem".[101] Turing's paper speaks to this problem, though Wittgenstein conceives of it in a more all-penetrating way.

What is this "problem" Wittgenstein comes to see in §47 of EPB in the fall of 1936?[102] It is a certain construal of a problem about rule-*following* – a topic that had engaged Wittgenstein from early on, arguably even in the *Tractatus*.[103] Let us explain by looking first at §41 of *The Brown Book*. Here Wittgenstein had attempted to generalize a particular kind of language-game he had framed in §§32–33 in answer to the question, "What do we call a rule"? He considers the following example (reminiscent of how Turing analyzes formal rules in terms of human computors utilizing tables in a step-by-step manner in response to commands):

> 33.). B moves about according to rules which A gives him. B is supplied with the following table:

a	→
b	←
c	↑
d	↓

[101] Wittgenstein's reference to the "problem" with §47 of EPB is on UF p. 47 (MS 152, p. 5). See Floyd (2017) for an argument that Turing's framing of the notion of a "Turing Machine" in 1935–1936 may have been influenced by his reading of BB.

[102] Joachim Schulte argues that the early remarks in MS 152 stem from the fall of 1936; it is not impossible that they were written in the spring of 1937. See Schulte's commentary on MS 152 at wittgensteinsource.org, under the Bergan Nachlass Edition.

[103] I discuss this issue of the *Tractatus* briefly in Floyd (1991, 2004), at pp. 93ff; more needs to be said about what "rule-following" means for Wittgenstein at different points in his development, and whether there is just one problem there.

A gives an order made up of the letters in the table, say: "*aacaddd*". B looks up the arrow corresponding to each letter of the order and moves accordingly; in our example thus:

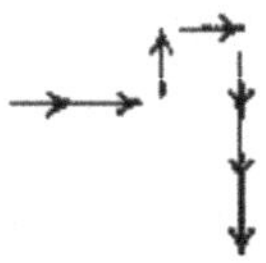

Wittgenstein explores at length the question of whether the "sentence" *aacaddd* or the diagram is an appropriate "expression of a rule", concluding that the answer will depend upon the uses made by member of the "tribe" in which the commands are given.

Later on Wittgenstein would have emphasized the more refined idea of a technique. But he does not as yet take this to be a fundamental notion of logic. Instead, he blithely moves ahead to frame an idea of "*general training*':

§41. . . . the pupil is not just trained to use a single table; but the training aims at making the pupil use *any* table correlating letters with arrows [emphasis added].

The "problem" Wittgenstein sees raised here – but only in retrospect, revisiting the remark from the perspective of his attempted revision of *The Brown Book* in the fall of 1936 – is the "*any*". What is meant by "any" such table "correlating letters with arrows"? This is not simply a matter of the range of a quantifier, but rather the whole idea of what "any" can possibly mean. We need an analysis of "table", "command", "use", and "correlation". This of course both Turing and Wittgenstein supply: Turing with an analysis suited to mathematical development, Wittgenstein with a more purely philosophical investigation of the related notions.

The issue of training a pupil, not merely to follow a *particular* table correlating letters with arrows, but *any* such table draws out a further, distinctive problem: one of novelty. Can we take the human potential to amalgamate such tables into novel games for granted as something unlimited, of indefinite extent and richness (as Hobson maintained)? Might we imagine arranging all the possibilities in a row, and then keep diagonalizing out repeatedly? If we could, would this mean something about the "inherent" nature of the objects of rule-governed activity (mathematics and logic), or something about us?

Recall Hobson's idea that the power of human symbolic expressivity is indefinitely rich. On this view, Cantor's diagonal argument shows that there can be, as Hobson puts it, no "*final and complete* stock of words and symbols, by means of which alone elements of the continuum can be defined by finite definitions".[104] Focussed as he was on characterizations of the real numbers, Hobson did not reach down to the very idea of rule-governed, symbolical routines *in general*. The ultimate

[104] Hobson (1921), pp. 87–88, quoted above on p. 226.

question for Wittgenstein, in the fall of 1936, is this more general one. What room there is for the idea of human creativity in a setting in which language-games with tables such as the ones sketched above are regarded as vehicles, or "systems" of meaning? And how might he treat this question in order to allow us to recognize what it is that falls under the concept of "any" such table?

The difficulty here is not simply that we need to be careful about what the generality of concepts consists in. Wittgenstein had already admitted that he was not offering any kind of general analysis of what it is to follow a rule (BB: §41). He highlights in *The Brown Book* (p. 88) the importance of the idea of a family resemblance among items falling under a single concept (a concept held together by affinities and analogies and not necessarily by a common property). He also stresses the complexities, traps and difficulties surrounding the idea of regarding a command in two ways: as something expressed in a sentence of some kind (a sequence of letters, a diagram) and as a causal mechanism. Here we often mistake similarities for differences, causal issues for logical ones, mental actions and images for logical steps, and so on.

No, as Wittgenstein remarked in his fall 1936 notebook, there is *another* "problem" lurking here.[105] It comes alive in his attempted revision of *The Brown Book*'s discussion of "general training", in EPB §§46–47. And it raises an even deeper challenge to the idea that following a rule is a "specific process" than any of the arguments he had raised so far. For his idea of "general training" ultimately broaches the difficulty of pinning down precisely what it is we mean in *general* by a "characteristic process" of following a rule, or taking a "step" in calculating according to a fixed rule in formal logic, according to a strictly circumscribed, symbolical, finitely expressed routine. (If, as is not unlikely, Turing had read or discussed these issues in reading or discussing the contents of *The Blue and Brown Book* in 1934–1935, then this might explain how it was that he came up with *his* novel approach to the *Entscheidungsproblem* in terms of tables and human calculators so quickly in 1935–1936.[106])

Wittgenstein's new challenge in 1936 is to face the "problem" of what it is to put together two differing tables (sentences, language-games) of the kind described

---

[105] See MS 152, p. 5; compare 152, p. 68.

[106] As Hodges (2012) explains, there is little to go on in explaining how Turing managed to come up with his machine-analogy so quickly, having only been introduced to the *Entscheidungsproblem* in a course with Newman in the spring of 1935. My suggestion is that Turing read and/or discussed *The Blue and Brown Books* (Floyd 2017, especially at pp. 112f., 124). Hodges does say that Watson and Turing were seen discussing diagonal arguments with Braithwaite at High Table in 1935. There is a 1-0 array formulation of Cantor's diagonal argument written down by Wittgenstein himself, at MS 157a, p. 17v, in the midst of a discussion of the "I" as a kind of "massless" vanishing point accompanying action. The remark was written sometime between 1934 and 1937, in a notebook that appears to form a bridge between the thoughts of *The Blue and Brown Book* and Wittgenstein's later remarks on rule following. MS 157a is also a source for Wittgenstein's 1938 lectures on aesthetics, which contain an extended criticism of the idea of a "super-rigid" mechanism capturing all logical necessity (part of the dream of the *Entscheidungsproblem*). See LA §26ff and WCL, pp. 78ff.)

in *The Brown Book*. And this is a new and different issue, requiring a different point of attack from that of his family resemblance idea. It brings us to the whole general problem of inference, the transition from one sentence (or sentence structure or game) to another. Most generally, it is the question of the limits and nature of our power to articulate symbolically, circumscribe, and clarify the domain of human, step-by-step, reckoning-according-to-a-rule behavior by means of our own sentences and words.

"General training" is reworked in EPB §46, and "the problem" emerges immediately afterwards, in EPB §47. First Wittgenstein alludes back to EPB §38:

> 38. B should draw different linear ornaments. Each ornament is the repetition of an element that A indicates. For example, if A gives the command 'c a d a', then B draws a line
>
> 
>
> and in that case, I think would call 'c a d a' the rule. ... [A] rule is repeated application. ...

And he reconstructs this in §47 this way:

> 47. The commands and their execution as in (38); but only three sentences are needed:
>
> 'a c'
>
> 'a c c'
>
> 'c a a'.
>
> We shall say that in (38) B is *led* by drawing the broken line from the composite sign of the command. – But if we ask ourselves whether the three sentences in (47) are *followed* in the execution, it seems as if we could say both 'Yes' and 'No'. ...
>
> [Our attempted] explanations [in terms of states of mind and reactions] ... do not satisfy us at all, and it is the limitation of the language-game (47) which makes them all unsatisfactory. This is expressed in the explanation which [may come] to mind: B is ... led by the combinations of the letters in the three sentences *if* he could also execute commands which are other combinations of these letters. And when we say this, this *capacity* to execute other commands also appears to us as a definite state of the one who executes the three commands in (47). But if we look at the case, as it were, from up close, no such condition can be seen.

Here the question of the generality of the process of playing such games is transposed into the problem of assigning a "definite state" of mastery or "capacity" – mental or physiological, perhaps – to the player of the game, on the basis of our assignment of a code, or series of sentences to the player's moves. This is supposed to explain the player's ability to create novel combinations of letters that are then followed systematically as a continuation of the application of the same rule. However, if we simply take the description of the game at face-value, we see that it places no conditions of any such kind upon the player who is "following" the rule he or she is given. *We* are importing talk of mental states and capacities and novelty from outside the game: as it were, from far away.

Recall that Hobson put great weight on what he called "ideas", i.e., our ability to "focus the mind" so as to "single out" mathematical objects: just that capacity which enables us to go beyond "a finite stock" of words and symbolic means of expression, according to him. Once we have so focussed we may, he maintained, always assign a new symbol "in an *ad hoc* manner" whenever we wish.

Wittgenstein parries such accounts with a different idea: the question is, *What* has been mastered? How far does the idea of an "ability to play the game indefinitely" run? And what explains our ability to go beyond a given game, expanding it? *Not* simply "being led by collections of letters" or "having the capacity" for this, or mentally singling out a specific collection of possibilities. If, for example, one is worried about whether a countable language could possibly express or capture "all" the real numbers – as Hobson was, in thinking about Cantor's diagonal argument – then one needs to say something more specific, something that entangles us in attending to the techniques that we actually use, including the diagonal technique.

In MS 152, which identifies the new problem *as* a "problem", Wittgenstein criticizes at length the idea of a mental state that could constitute the capacity to read a series of letters or words: this is an origin of the so-called "reading machine" passages in PI §§156ff. It is significant, I think, that this manuscript book begins with warmup exercises involving representations of real numbers, with Wittgenstein writing out rational numbers using the theory of continued fractions. Let us consider why he may have done this.

In the continued fraction way of representing real numbers – as opposed to our usual decimal expansion notation – rational numbers may receive unique, finite representations, while irrational numbers receive unique infinitely descending representations. (In contrast in the decimal notation this cannot be achieved: .99999 is the same as 1.0, for example.)[107] Both kinds of representations may be put into the same form, as sequences of letters. But when they are, quite different "capacities" are at work.

In general, a continued fraction is of the form:

$$a_0 + \cfrac{b_1}{a_1 + \cfrac{b_2}{a_2 + \cfrac{b_3}{a_3 + \ddots}}}$$

Where the dots terminate in a finite expression if the number is rational, but do not if it is irrational.

[107]For an elementary discussion of the geometrical and other interpretations of continued fractions, see the discussion at http://www.maths.surrey.ac.uk/hosted-sites/R.Knott/Fibonacci/cfINTRO.html, accessed November 11, 2019.

For example, as Wittgenstein himself writes out (MS 152, p. 1) $\frac{4}{13}$ may be written as its reciprocal this way:

$$0 + \frac{1}{\frac{13}{4}}$$

which is equal to

$$0 + \frac{1}{3 + \frac{1}{4}}$$

and this is of the appropriate form, normalized so that the remainder numerators descending along the right side are all equal to 1. By definition a *simple* continued fraction has all the b_i's = 1. This allows for a standard abbreviation in so-called *simplified list notation* by copying off the sequence of a_i's, ignoring the b_i's in the numerators, which are all assumed to be 1's: the above example of $\frac{4}{13}$ is represented in this way by [0; 3, 4].

Wittgenstein only deals with such rational cases in his warmup exercises in MS 152, though he writes general rule-terms for the forms of substitutions that go on in the typical step from reciprocal of a fraction to the next term. The notation of continued fractions, when fully written out, vividly displays the form of a recursive process, fractions within fractions, yielded according to patterns and rules of division – and, as Wittgenstein's sketchy examples show, it also allows for the addition and multiplication of specific terms.[108]

We may compare EPB §47 – which sets out a language-game framed by three-letter commands for the production of patterns of lines (quoted above) – with the simplified list notation of continued fractions. When we do, we see that the example of continued fractions serves as an interesting numerical analogue for Wittgenstein's simple language-game used to illustrate the "problem" of general training in EPB.

The scope of this "problem", we have said, encompasses the whole issue of decoding human meaning through the uses of symbols on which rule-governed behavior depends. In the example of continued fractions we see the problem illustrated. First, from any simplified list notation example of finite length we may easily recover the continued fraction. And from any given rational fraction we may easily, through recurrent taking of reciprocals and dividing, reach the finite expression of it in simplified list notation. This goes not only for cases of such three-element simplified list notation examples such as [0; 3, 4] but for more complicated cases such as [2; 1, 4, 3]. The latter, as we may see, is a simplified list form continued

[108]The theory of continued fractions is treated in Chap. X of Hardy and Wright (1938), whose first edition appeared in 1938. On Wittgenstein's 1932 discovery of the need for a uniqueness rule in recursive specifications of functions see Marion and Okada (2018).

fraction expression for $\frac{45}{16}$, because

$$\frac{45}{16} = 2 + \frac{13}{16} = 2 + \frac{1}{16/13} = 2 + \frac{1}{1 + \frac{3}{13}},$$

or

$$2 + \cfrac{1}{1 + \cfrac{1}{4 + \frac{1}{3}}}.$$

There is a question here about how a human being trained in the writing out of continued fractions is able to "go on" finding simplified list notation representations for rational numbers. For this one must have a capacity for "substituting in the same way" over and over again.

This issue raises itself with a vengeance when we consider the continued fraction expansions of irrational numbers, which contain the ellipsis (...). For example, π may be written as an endless continued fraction this way:

$$[3; 7, 15, 1, 292, 1, 1, 1, 2, 1, 3, 1, 14, 2, 1, 2, 2, 2, 2 \ldots].$$

But here, with the ellipsis, the list notation must be "followed" quite differently from the rational case.

To revert to an example we have discussed above in Sect. 3.5, the square root of 2 may be written in list notation as $[1; 2, 2, 2, \ldots]$. As an irrational number represented in simplified list notation, it nicely exhibits the quite vivid cycling character of fractions within fractions with the "2, 2, 2 ...".[109]

We begin with the fact that

$$(*)\ \sqrt{2} = 1 + (\sqrt{2} - 1).$$

Focussing on "$\sqrt{2} - 1$", we may easily calculate that:

$$\frac{\sqrt{2}-1}{1} \cdot \frac{\sqrt{2}+1}{\sqrt{2}+1} = \frac{2-1}{\sqrt{2}+1} = \frac{1}{1+\sqrt{2}}.$$

And, substituting this calculation's result in for "$\sqrt{2} - 1$" in our original equation (*) above, we obtain

$$\sqrt{2} = 1 + \frac{1}{1+\sqrt{2}}.$$

[109] Wittgenstein utilizes "2, 2, 2, ..." in connection with rule-following in RFM I §3, corresponding to MS 117, p. 19 (1937), TS 221, p. 149 (1938) and TS 222, p. 3 (1938). Compare LFM p. 30 and Floyd (1991) for a discussion.

This equation in turn allows us to substitute this result about $\sqrt{2}$ into itself repeatedly, to obtain a pattern exhibited, in terms of continued fractions notation, this way:

$$\begin{aligned}\sqrt{2} &= 1 + \frac{1}{1+\sqrt{2}}\\ &= 1 + \frac{1}{1+(1+\frac{1}{1+\sqrt{2}})}\\ &= 1 + \frac{1}{2+\frac{1}{1+\sqrt{2}}, \ddots}\end{aligned}$$

that is, in list notation, $\sqrt{2} = [1; 2, 2, 2 \ldots]$.

The *uses* of the simplified list notation differ markedly, depending upon whether the number represented is rational or irrational. [1; 2, 2, 2 ...] may be used to generate better and better rational approximations to $\sqrt{2}$ as we truncate the list further and further to the right, whereas in the case of a rational number, the simplified list notation does not correspond to an approximation at all. Thinking of the situation non-extensionally, the simplified list notation for irrational numbers provides us with an endless rule or procedure for generating ratios within ratios, and so a kind of expressive picture of the aspect of irrational numbers that involves them, geometrically speaking, in infinite descent. (One may compare our geometrical discussion of the proofs of the irrationality of $\sqrt{2}$ in Sect. 3.5 above).

In this way, as Wittgenstein put it in his remarks on Hardy's p. 7, "everything leans a little" when we write out an irrational number in list notation and compare this activity with writing out the list notation for a rational number (see our discussion of Wittgenstein's remark about Hardy in Sect. 3.7 above). The jump from saying that one "has the capacity" to follow *finite* lists to saying that one "has the capacity" to follow indefinitely iterated processes of division and substitution, is a real jump, whether we are dealing only with rational numbers or with irrational numbers. And this, I believe, is why Wittgenstein began MS 152 with these examples: they are concrete demonstrations of the "problem" that emerges in §47 of EPB.

8.7 Notebooks 162a–162b

Let us return to Cantor's diagonal argument, which confronts us with a different way in which we may "go out of" a game, not by running out of letters and generating new labels for new ideas in an *ad hoc* manner, as Hobson held in his quasi-extensionalist way, but instead by generating new rules through the process, procedure or rule of diagonalizing. There are connections, in Wittgenstein's mind, between the unclarity in what it means to say that the real numbers are "uncountable", the unclarity in saying that a sentence is "unprovable", and the unclarity in saying that techniques of expanding real decimals "cannot be ordered" in terms of one technique. In each case, the unclarity stems from the fact that in order to say something negative about what *cannot* be done, we need to have a clear conception of what it would have been to do it.

Of special interest, with regard to the generality of Cantor's diagonal technique, are Notebooks 162a and 162b, which Wittgenstein kept during 1939–1940, the period when he was teaching his Cambridge seminar on the foundations of mathematics – his first seminar as Professor – with Turing in the audience.[110] In them explicit references to Turing do not occur; in fact in Wittgenstein's corpus as a whole explicit references to Turing are rare.[111] However, the themes pursued in 162a–162b do appear in what was noted down by students in LFM, and reacted to by Turing. In his lectures, as we know, Wittgenstein self-consciously kept allusions to more advanced technical results to a minimum, presupposing only a nodding acquaintance with mathematics.[112] But in these pocket notebooks, Wittgenstein explored more advanced examples.

MS 162b (1939–1940) begins with a remark inaugurating Wittgenstein's investigation of the possibility of "a technique to teach one all techniques". The notebook itself is an immediate continuation of what Felix Mülhölzer rightly called the "wonderful" pocket notebook 162a (see p. 130 above).[113] Notebook 162b is less wonderful, because it is much less clear. And yet its ambition is wonderful in its own way. For here we see Wittgenstein attempting to draw lessons from his discussion of Cantor's diagonal method into the wider purview of his mature philosophy. And it is at precisely this point that the whole idea of a "technique" and its limits becomes thematically central in his work. We have already seen his complaining in 162a,

[110]See above Sect. 1.3 and also Grattan-Guiness (2013) and Floyd (2017). Notebooks 159 and 160 (both from 1938, and partly in English) are also of interest, treating of the diagonal technique in the midst of a host of other problems, though they do not focus on the real numbers as pointedly as the 1939–1940 notebooks (which also include the pocket notebook 161, kept during the spring of 1939). Portions of 159 and 160 were published in CE, but Rhees (or the editors of PO, see p. 369) unfortunately eliminated all logico-mathematical parts when he made his excerpts.

[111]But see Footnote 82 above. There is a reference to Turing in 161, p. 4r, also kept during the spring of 1939 during Wittgenstein's course of lectures on the foundations of mathematics, as well as in RPP I §1096, discussed below and also in Floyd (2012c).

[112]See the opening of LFM where Wittgenstein insists on this approach. Only two slight references to Cantor occur in LFM, for example (pp. 14,103), as well as a brief remark about "true but unprovable propositions" (p. 188), and these are quite offhand. There are extensive notes of a lecture on Gödel in SWL, apparently from early 1938; some of the general themes treated are raised in 162a and 162b.

[113]Some of the remarks about Cantor may be found in the 1940 remarks in MS 117, for example at p. 155; compare the discussion in Chap. 7 above.

p. 29 that Hardy neglected to present Cantor's argument in CPM (compare p. 127 above). His complaint here, as expressed in 162a, is that

> The general sentence may say what all special ones say, but the general technique does not teach what all the special techniques do.
>
> In teaching you to lay a brick on another, I do not teach you to build a house – though indeed I have taught you the general form of building a house.

This should be compared with his annotation to Hardy's CPM, p. 28:

> Whoever teaches someone to plane planks doesn't teach him thereby to make tables and chairs, not even in general, but rather he teaches a technique of very general application.[114]

162a and 162b have not been published, along with much of Wittgenstein's writing about diagonal argumentation.[115] For this reason it is striking that a central aim Wittgenstein had in these particular notebooks was to synthesize, extend, confront and adapt his maturing philosophical perspective to issues raised by diagonal argumentation.

It seems that these pocket notebooks served at least in part as separate brainstorming places for Wittgenstein's own thoughts or reactions to conversations related to the seminar, which is why there are touch points of connection between them and LFM. In general 162a and 162b show that in the spring of 1939 Wittgenstein was attempting to extend his early version of PI to the foundations of logic and mathematics, this time by focussing on Cantor's method as a technique.[116] There are direct points of connection with his annotations to Hardy's CPM here:

---

[114]Compare RFM V §24:

> The harmful thing about logical technique is that it makes us forget the special mathematical technique. Whereas logical technique is only an auxiliary technique in mathematics. For example, it sets up certain connections between different techniques.
>
> It is almost as if one tried to say that cabinet-making consisted in glueing.

[115]It seems the editors, perhaps attempting to protect Wittgenstein, decided against publication of such remarks. Many of Wittgenstein's remarks on Gödel after 1938 were not published in RFM$_1$; but this is true also of 162a and 162b, of which only bits and pieces of excerpts have appeared (in CE and CV), none of these having anything to do with diagonal argumentation. Compare p. 45 of Z, where a drawing is supplied for a remark about Helmholtz from 162b p. 38v about the possibility of imagining a 4-dimensional Necker cube, but without its original context.

[116]With Watson in attendance Wittgenstein lectured on Gödel briefly, in early 1938 or 1939 (see SWL Chap. 2). Some initial remarks on the topic were penned in fall of 1937, after his summer discussions with Watson and Turing, and placed into his *Frühfassung* of PI. But these do not attack diagonal argumentation in general, as do these later remarks. Wittgenstein also gave a brief series of lectures on aesthetics in the summer of 1938 (see Wittgenstein 2007, hereafter LA) and here there are condensed references to his criticizing the "charm" the Cantor diagonal argument held for one of his students, Ursell. His point seems to be that Ursell had succumbed *only* to the charm, and not to the specific proof technique (see LA, p. 28). Because this is 1938, the issue of propaganda is at stake, as well as pseudo science. Wittgenstein is pressed by his students on whether he is "making propaganda" for a certain style of thinking, and he admits that one might say he, or the class together, is doing this, in deflating the "charm" of the Cantor argument. Examples from aesthetics also occur in 162b.

Hardy's failure to frame the Cantor diagonal argument in his textbook is mentioned in 162a (p. 29; compare p. 127 above).

His attempts in 162b are partial and faltering, unpolished in the ways pocket notebook sketches always are. They would never be brought to completion. Nevertheless, though slim and sketchy, their quick and unvarnished character reveals larger motives, allowing us to reconstruct an account of at least some of his ways of reflecting on the generality of Cantor's diagonal method as a proof method, or technique.

162a ends and 162b begins in the orbit of the exploration of techniques and their importance for our understanding of sentences. Wittgenstein draws here an analogy between a mathematician and a magician who deceives himself with his own sleights of hand in a mirror (this remark was discussed above on p. 179 by Felix Mülhölzer). He warns that such a thing could only happen if the mathematician, as an expert, has "become so familiar with the *mastery* [*Griffe*] of his tricks" – or techniques – that he loses sight altogether of how he relies on them, thereby generating an experience of hocus pocus. Wittgenstein's point is that when the techniques behind the "magic" of mathematics are hidden from view, new applications are assimilated to the old ones to produce something other than sober understanding. As he also writes,

> ... from a certain place in the auditorium, you can see the magic; from the other side it looks very different and absolutely not like a piece of magic (162a, p. 87).

This idea, more generally applied to extra-mathematical sentences, forms the focus of notebooks 162a–b. As Wittgenstein puts it in PI §129:

> The aspects of things that are most important for us are hidden because of their simplicity and familiarity. (One is unable to notice something – because it is always before one's eyes.) The real foundations [*Grundlagen*] of their inquiry do not strike people at all. Unless that fact has at some time struck them. – And this means: we fail to be struck by what, once seen, is most striking and most powerful.[117]

As we have seen, misunderstandings easily arise when sentences (say, about the real numbers) are "puffed up" with a kind of absolute content or misleading sense of generality that obscures the forms of their uses, or applications.

But a new problem emerges from this. At the end of 162a Wittgenstein explores a number of different kinds of examples to critically reflect on his own claims about "techniques":

> "You cannot see a human being in the window; I will show you [nevertheless] how one looks in the window". (I then take a photograph with infrared rays.)
>
> What does it mean to say: "Before I understand the process – I do not understand the result"?
>
> Someone shows me a picture and says, "This is how you look in the dark". Do I understand it now, because I've been shown a picture? (162a pp. 93–4).

[117]The original version of this remark is in MS 110, p. 259 from 1931, in citing Frazer's *The Golden Bough* (1917); see MS 112, p. 117 (1931), BT, p. 419, FF §105(107).

The idea is that the result, or picture, requires us to keep in mind the specific techniques that help us tether our mode of expression to its significance. But of course, the result cannot be reduced to *one* technique, as if a sum of all the others. For the picture might well receive a novel application, going beyond the original context of its production. This point applies, as Wittgenstein makes clear, to the foundations of mathematics:

> Was there always the problem/problems of the foundations [*Grundlagen*] of mathematics?
>
> The problem corresponds to/grows most easily from a context in which there are strong tendencies to assimilate modes of expression with wholly different application.
>
> If these tendencies suddenly engage with a wish to take up the similarity of modes of expression to show that there is *essentially* no difference [between applications or techniques], then philosophical disturbances arise through the discrepancy among modes of expression that cannot be carried through in the old sphere [of their application] (162a, pp. 94–5).

This point about the "wish" to show that "there is essentially no difference" in uses or techniques is illustrated by further examples:

> [For example:] "He has discovered the eastern pole of Earth". – Has he perhaps discovered a reason to call something quite trivial "the discovery of the Eastern Pole?"
>
> "He has found a remedy for unemployment". "He makes proposals to eliminate unemployment." Do we understand these sentences?
>
> "He has discovered what time it is on the sun" (MS 162a, pp. 95–6).

The point is that just on their own, apart from any understanding of their techniques of application and illustration, such sentences hang in the air for us, are strange and hardly clear. In this way hocus pocus, a sense of magic or mystery, makes itself felt. Wittgenstein summarizes his general point this way:

> I do not compare this sentence with something puzzling, but with something that is not understood. I show you another way of looking at this sentence: not in the light of the enigmatic, but in the light of the misunderstood (162a, pp. 95–6).

This general point of criticism connects directly with issues we have been discussing in this book. This is made clear in MS 122, p. 30r, written in 1939:

> What is the invention of the decimal system actually? The invention of a system of abbreviations – but what is the system of abbreviations? Is it merely the system of new signs, or even a system of their applications for the sake of brevity? And if it's the second, it's a new way of seeing the old system of signs.

As in Felix Mühlhölzer's discussion above (p. 186), we may take Wittgenstein to mean here by "system" a given totality of numbers together with a rule – a method or a technique – for ordering these numbers. The important point are the multiplicities of specific techniques in relation to one another. Think of the differences between the decimal notation and the method of continued fractions we have just discussed.

Wittgenstein is offering an implicit reply to someone holding Hobson's view of our expanding linguistic capacities. On Wittgenstein's view, new techniques draw out new uses, and these in turn bring in new *aspects* of the numbers, and also bring to light new aspects of the old system. It is perfectly imaginable, as Wittgenstein imagines in RFM II §13, that someone went about writing down lists of decimal expansions of real numbers blissfully unaware that no such list could every be completed by the activity of writing expansions down. Once the diagonal technique is drawn in, the person's original activities appear differently: as something truncated, perhaps problematic, but in any case evincing a new aspect.[118]

At the end of 162a and the beginning of 162b the suppression of the role of technique is illustrated through a series of remarks bringing these ideas into play alongside Cantor's method of diagonal argumentation. Wittgenstein engages in a rather strange effort to "miniaturize" or domesticate the diagonal method, almost violently forcing his points into odd crevices of use in order to gradually lead us back to everyday situations. He does this precisely to weigh the costs of highlighting or ignoring the importance of techniques and alternatative, conflicting points of view on them. In particular, he is exploring the contrast between an extensionalist view of a sentence as a sequence of letters and the non-extensionalist conception of a sentence as a form embedded in life by us.

First he raises a problem for his own point of view, one he will develop an answer to in RFM III (composed, like notebooks 162a and 162b, in 1939–1940):

> But doesn't this amount to the fact that I can very well acknowledge [*anerkennen*] as a theorem [*Satz*] a *non*-Russellian proof [and] fail to acknowledge a Russellian one? (MS 162a, p. 100)

The question is, What is the relation between our grasp of the sentence "$2 + 2 = 4$" and these very different techniques of formalized proof? Here Wittgenstein's idea of the surveyability [*Übersichtlichkeit*] of proofs and rules raises its head: the issue of what we can run through, take up with understanding, and communicate to someone else.[119] A proof of "$2 + 2 = 4$" in *Principia Mathematica* will take hundreds of characters and lines, whereas written down in a condensed picture, with dots – say for a child – we have a different mode of proof, a different way of parameterizing our reasoning. It is the latter we use to draw a child into games with this sentence. Wittgenstein's point is that the techniques involved here are in the background of the formalized proofs in a system like that of Whitehead and Russell, though their role may be hidden from view or suppressed by an extensional conception of a formalized sentence.

Both the everyday picture proofs and the formalized proofs involve techniques: neither stands on its own without the need for an understanding or ability to embed the sentence in life.[120] But the techniques in play are quite different. And thus certain aspects of the numbers are hidden from view if we insist that these two

---

[118]See the discussion on p.162f. above.

[119]For a detailed and extensive discussion of this notion of surveyability, and Wittgenstein's answer to his question compare the commentary on RFM III, Mühlhölzer (2010). Turing takes the communicability and surveyability of a command to be basic features of his model in Turing (1936) at p. 253; see Floyd (2016), p. 28 and (2018b), p. 8 for a discussion.

[120]This is the point explored *ad nauseum* in RFM I, with numerous examples of such proofs with dots.

proofs are proofs *of the same thing*, where by this we mean a single sense expressed by the sentence "2 + 2 = 4".

We are tempted to treat the sentence's sense as if it may be understood from all sides at once, rather than admitting that we should try to regard it, to use Frege's way of putting things, many-sidedly but not all sides at once.[121] Wittgenstein would like to see his emphasis on techniques as a development, rather than a refutation, of Frege's point, but the role of a formal system of logic in individuating senses of sentences is more unclear to him than it was for Frege, given his emphasis on special techniques of using words.

So Wittgenstein calls attention to the variety of ways we might have of understanding the following questions (at LFM p. 32 a similar example is discussed with Turing):

> How many multiplications have I learned to perform? –
> How many multiplications can I perform? – (MS 162a, p. 101)

How do these questions engage with Wittgenstein's concerns? The answer is that they instantiate his general "problem" about following "*any* rule" such as that expressed in the multiplication table for natural numbers. Here the problem lies in what is to "learn to perform" *any* operation of multiplication once one has learned the natural number multiplication table. As a matter of fact, one was only taught multiplication rules through certain examples; "learning" involves being able to "go beyond" what one is strictly speaking given. But this "going beyond", this novelty, takes a variety of differing forms. Moreover, the sense of performance, the "can" or "is able to", is multi-faceted (see PI §§187ff.).

The idea of the mathematician (and at LFM pp. 32ff, it is Turing himself) is that one can straightforwardly number the number of multiplications I have learned to perform: $\aleph_0$. This seems predicated on the basis of the idea that we take ourselves always to be using a countable language, so that the sentences we "are able to" multiply, hence the results, are not more than $\aleph_0$ in number. But this argument, while perfectly acceptable in a certain way, overlooks the different surroundings, and the differing applications, in which these question may swim: the non-extensional perspective on sentences as forms that we are capable of embedding in life in novel ways. Cross-multiplying fractions is a *different* technique for establishing sameness of numbers than multiplying with small natural numbers. Multiplying with very large numbers or approximations or continued fractions is still different from this.

But, most of all, if we think of Cantor's diagonal technique, as Hobson did, as a way of showing that we cannot have a "complete and finished" stock of symbols, we see that the straightforward assignment of a number such as $\aleph_0$ to the question "How many multiplications have I learned to perform?" as a *definitive* answer is unsatisfactory, and precisely because it underestimates the variety of techniques or parametrizations of language we have learned in connection with the question. The uses of our sentences applying multiplication are diverse and many-sided. If,

[121] See, e.g., Frege (1892b), p. 27.

as Frege held, a statement of cardinal number is an assignment of a number to a concept, then it must be emphasized, from Wittgenstein's mature point of view, that such assignments must not be regarded one-sidedly.

Wittgenstein proceeds to draw out the connection with Cantor. This is appropriate in the context of preparing for his seminar, for any mathematician like Turing will have learned to "multiply" with transfinite numbers. (In fact, while he was listening to Wittgenstein's lectures Turing was working on his dissertation, in which a series of novel ordinal notations are generated to "escape" Gödelian incompleteness.[122]) First Wittgenstein considers the following everyday analogy:

> "You cannot put all boxes of the world in a box." Why? Because they are too many?
> – I will prove to you that there are an infinite number of boxes; because no box, no matter how big you make it, can house all boxes (MS 162a, p. 101).

This is the "miniaturized" Wittgenstein version of diagonal thinking illustrated in a down-to-earth way. Note that the remarks are all contained in single-quotes, a sure sign that they represent possible interlocutory interventions, rather than principles endorsed.

Begin from ordinary life, without prejudices. There are circumstances in which we would *not* wish to say of a box that it contains all boxes (if we are sitting in a large shoe store, with thousands of boxes in it arranged higgly piggly in disarray). Of course, if the shoe store itself is cube-shaped, we might say this all the same. But there are nevertheless also circumstances in which we would be inclined naturally to say of a box that it contains all boxes (if we are looking at a nested collection of boxes, held within one larger box sitting on the table before us). We might imagine playing a language-game with a child, in which both responses might be solicited naturally, through different (imagined) circumstances.[123] What we would *not* say is that there are "too many" boxes to box, unless this meant something about the capacity of a *particular* box, i.e., its size in relation to others. – The important point here is that we are able to adjust what we say to particular circumstances, using our words as techniques.

Mathematically educated readers are of course likely to object to these homespun remarks with contempt, supposing that they reduce the difficulties to childish or merely empirical ones about linguistic usage. What could such a remark have to do with Cantor? But let us imagine instead that Wittgenstein has in mind here an illustration, and not an application of a given sentence. As he writes elsewhere,

> In a certain surrounding these problems become overwhelmingly difficult, in other surroundings they lose their difficulty. What I am trying to do is to teach you to make transitions from the one to the other.[124]

---

[122]Turing (1938).

[123]Compare Chap. 9 of Diamond (1991).

[124]MS 159, 14r, from 1938. The surrounding context of this remark is the Gödel sentence "I am not provable", on which Wittgenstein was very tentatively lecturing in early 1938. See SWL, Chap. 2, which is however different in important ways from the passages we are looking at from 1939–1940.

The contrasting images of the illustration would then be these: there might be a diffuse aggregate of boxes (analogously, techniques of operating with real numbers, what we have called "expansions"), reaching out in all directions without much order (the disorganized shoe-store) or even the idea of an end. Or the image might be of a well-organized shoe store, with all the boxes neatly placed into rows. Finally, we might picture a tight, ordered series of boxes, beginning with the smallest and heading up to the largest (the nested collection which, however (as we know from Cantor's argument) may always be further nested).

Taking his remark to serve as a metaphor for the collection of real numbers and our efforts to "box" them into whole extensions – rather than as a flat-footed remark on linguistic usage – Wittgenstein's images are rather powerful. As we have seen in considering Hobson and Turing's arguments, there are variants of Cantor's diagonal argument that apply to rules or techniques of "boxing" numbers, rather than being proofs utilizing or applying to actually infinite totalities. We can, must and should distinguish Cantor's diagonal technique from his proof.

8.8 Wittgenstein's Diagonal Argument (1947)

Let us now turn to Wittgenstein's diagonal argument, which he wrote down in 1947 as a language-game rendition of Turing's diagonal argument in "On computable numbers". The argument is interesting both for its connection with Turing's paper and with Wittgenstein's conception of Cantor's method.[125] One of Wittgenstein's main points is to stress the human-inspired foundation of Turing's analysis of computable ("finitely definable") real numbers.[126] This allows him to shed light on the rule-following problem through a rendition of diagonalization in terms of rules and techniques for expanding real numbers.

Wittgenstein begins by remarking that

> Turing's 'Machines'. These machines are *humans* who calculate. And one might express what he says also in the form of *games*. And the interesting games would be such as brought one *via* certain rules to nonsensical instructions! ... One has received the command "Go on in the same way" when this makes no sense, say because one has got into a circle. For that command makes sense only in certain positions. (Watson) (RPP §1096, compare MS 135, p. 118).

The idea here is clearly to recapitulate Turing's positive, circular diagonal argument (the "Do What You Do" argument, as I called it in the last Section) in terms of language-games, using it to draw a moral, namely, that "an order makes sense only in certain positions [*Stellen*]". (The allusion to Watson is presumably to his 1938, p. 449 where Turing's "Do What You Do" argument is written down.) Wittgenstein makes explicit the human source of Turing's model: its non-

[125] The argument is analyzed in detail in Floyd (2012c).

[126] On the human inspiration of Turing's model see Floyd (2017).

extensional, procedural aspect. The argument Wittgenstein gives is brief, and, as has been stated above (at p. 145), it highlights the centrality of our formation of the diagonal, rather than the switching of items along the diagonal. Let us recapitulate it, this time using the later version from RPP I §§1096–7[127]:

> 1097. A variant of [C]antor's diagonal proof: Let $N = F(K, n)$ be the form of the laws for the expansion of decimal fractions. N is the nth decimal place of the Kth development. The diagonal law then is $N = F(n, n) = Def\ F'(n)$.
>
> To prove that $F'n$ cannot be one of the rules $F(k, n)$. Assume it is the 100th. Then the formation rule of
>
> $F'(1)$ runs $F(1, 1)$
>
> of $F'(2)$ runs $F(2, 2)$ etc.
>
> But the rule for the formation of the 100th place of $F'(n)$ is/becomes $F(100, 100)$; that is, it tells us only that the 100th place is supposed to be equal to itself, and so for $n = 100$ is *not* a rule.

As we see, it is Turing's Diagonal Argument which Wittgenstein is translating into the vocabulary of language-games in 1947.[128] Even supposing that there is, extensionally speaking, something which *is* the digit value of $F(100, 100)$ in itself, if we ask *which* digit it is, we end up with the answer, "$F(100, 100)$", which doesn't say one way or the other what it is, because that will depend upon the assumption that this sequence is the value of $F'(100)$ at 100. The diagonal rule, in other words, cannot be *applied* at this step: we cannot *follow* it.

This draws out the more general moral that, as Wittgenstein puts it, a command must be issued from one or another stance, or position. There is no trouble at all in *defining* the diagonal rule. The trouble only enters when we try to say, with a number, *which* rule it is. As soon as we attempt to parametrize our diagonal reasoning, we are defeated. For when we suppose that it is, say, the 100th, it becomes clear that at the 100th step our characterization will not allow the command to issue into any genuine directive. "Calculate $F(100, 100)$ by calculating $F(100, 100)$" is empty. It is not a rule that we can follow as we can the others on the list. And in this sense, as Wittgenstein writes, it is *not* a rule. – And this argument works for *any* natural number we might try to pick, not just 100. And this in turn can be seen without relying on a proof by mathematical induction.

But *what*, precisely, does this argument prove? It shows that there is a new rule that is not like the other rules on the list: that the technique of enumerating techniques will always generate something that is *not* just another technique. Differently put, it shows that our techniques are not exhausted by any technique of enumerating digits. And it shows this by drawing out something that is fundamental to our very idea of logic – more fundamental, in fact, than the idea of any particular

[127] See the discussion above on p. 145 for further details of the manuscripts.

[128] Moreover, the argument had a legacy. Wittgenstein was later credited by Kreisel with "a very neat way of putting the point" of Gödel's use of the diagonal argument to prove the incompleteness of arithmetic, in terms of the empty command, "Write what you write" (Kreisel (1950), p. 28n). Compare Floyd (2019).

logical law holding or not holding. We must *see* how to apply the rules and commands we work with. This is not a matter of community-wide agreement on philosophical theses or logical laws. Nor is it a matter of stipulated conventions. It is not a question of consensus at all, but rather of the need for forms of life: the embeddings of words in techniques that are somehow recognizable, communicable, and applicable.

Herein lies the moral of Wittgenstein's annotations to Hardy's CPM and the surrounding manuscripts and notebooks on which he worked in the period 1937–1942. His aim is clearly *not* to refute the extensional point of view, Dedekind's notion of a cut, or Cantor's diagonal argument. Rather, he seeks a sober understanding of the terrain of the real numbers: the techniques by means of which we reveal them, the aspects that they are capable of showing us, the *reality* of the real numbers. His non-extensionalism refutes nothing. But it draws out what the real numbers, from a non-extensional point of view, *are*.[129]

[129]I am grateful to Zhao Fan, Mathieu Marion, and especially Felix Mühlhölzer and the students in the annual Ludwig Wittgenstein Gesellschaft Summer School at Kirchberg am Wechsel, Austria, 2018, for many stimulating discussions of the material contained in this chapter.

Part IV
Wittgenstein's Annotations to Hardy

Chapter 9
Images, Transcripts and Translations of the Annotations

Abstract The images are from Wittgenstein's personal copy of the 8th edition of G. H. Hardy's *A Course of Pure Mathematics* (1941). The whereabouts of this are presently unknown.

In this chapter we provide images of the pages annotated by Wittgenstein, with thanks to the Wren Library, Trinity College Cambridge and its Librarian, Jonathan Smith, for permission to reproduce them.

Our convention is this. In *italics* we transcribe the words we take Wittgenstein to have written down. In regular font, we translate these, when necessary, from German to English.

J. Floyd, F. Mühlhölzer, *Wittgenstein's Annotations to Hardy's Course of Pure Mathematics*, Nordic Wittgenstein Studies 7,
https://doi.org/10.1007/978-3-030-48481-1_9

2. REAL VARIABLES [I

which has already occurred in a simpler form, but then the problem of determining the precise position of p/q becomes more complicated.]

2. The representation of rational numbers by points on a line. It is convenient, in many branches of mathematical analysis, to make a good deal of use of geometrical illustrations.

The use of geometrical illustrations in this way does not, of course, imply that analysis has any sort of dependence upon geometry: they are illustrations and nothing more, and are employed merely for the sake of clearness of exposition. This being so, it is not necessary that we should attempt any logical analysis of the ordinary notions of elementary geometry; we may be content to suppose, however far it may be from the truth, that we know what they mean.

Assuming, then, that we know what is meant by a *straight line*, a *segment* of a line, and the *length* of a segment, let us take a straight line Λ, produced indefinitely in both directions, and a segment A_0A_1 of any length. We call A_0 the *origin*, or *the point* 0, and A_1 *the point* 1, and we regard these points as representing the numbers 0 and 1.

In order to obtain a point which shall represent a positive rational number $r = p/q$, we choose the point A_r such that

$$A_0A_r/A_0A_1 = r,$$

A_0A_r being a stretch of the line extending in the same direction along the line as A_0A_1, a direction which we shall suppose to be from left to right when, as in Fig. 1, the line is drawn horizontally across the paper. In order to obtain a point to represent a

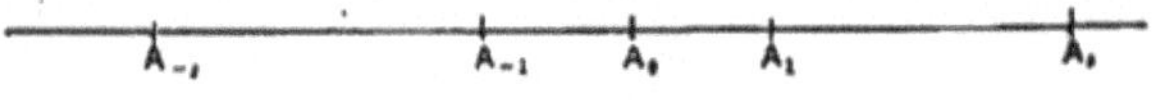

Fig. 1

negative rational number $r = -s$, it is natural to regard length as a magnitude capable of sign, positive if the length is measured in one direction (that of A_0A_1), and negative if measured in the other, so that $AB = -BA$; and to take as the point representing r the point A_{-s} such that

$$A_0A_{-s} = -A_{-s}A_0 = -A_0A_s.$$

Illustration ist allerdings unwesentlich, nicht aber Applikation.

Illustration ist allerdings unwesentlich, nicht aber Application.
Illustration is indeed inessential, not, however, application.

however large, we can find more than n positive integers'. This is plainly true whatever n may be, e.g. for $n = 100,000$ or $100,000,000$. The assertion means exactly the same as 'we can find *as many positive integers as we please*'.

The reader will easily convince himself of the truth of the following assertion, which is substantially equivalent to what was proved in the second paragraph of this section: given any rational number r, and any positive integer n, we can find another rational number lying on either side of r and differing from r by less than $1/n$. It is merely to express this differently to say that we can find a rational number lying on either side of r and differing from r by *as little as we please*. Again, given any two rational numbers r and s, we can interpolate between them a chain of rational numbers in which any two consecutive terms differ by as little as we please, that is to say by less than $1/n$, where n is any positive integer assigned beforehand.

From these considerations the reader might be tempted to infer that an adequate view of the nature of the line could be obtained by imagining it to be formed simply by the rational points which lie on it. And it is certainly the case that if we imagine the line to be made up solely of the rational points, and all other points (if there are any such) to be eliminated, the figure which remained would possess most of the properties which common sense attributes to the straight line, and would, to put the matter roughly, look and behave very much like a line.

A little further consideration, however, shows that this view would involve us in serious difficulties.

Let us look at the matter for a moment with the eye of common sense, and consider some of the properties which we may reasonably expect a straight line to possess if it is to satisfy the idea which we have formed of it in elementary geometry.

The straight line must be composed of points, and any segment of it by all the points which lie between its end points. With any such segment must be associated a certain entity called its *length*, which must be a *quantity* capable of *numerical measurement* in terms of any standard or unit length, and these lengths must be capable of combination with one another, according to the ordinary rules of algebra, by means of addition or multiplication.

Unsinn.
Nonsense.

"Suppose, if possible," heisst, glaube ich, gar nichts. Es scheint mir wie eine gezierte Ausflucht. – Es könnte etwa [sollte] heissen wie: try to suppose, + statt dessen könnte man dann einfach sagen "try". Etwa: "Prüfe $(p/q)^2 = 2$ ......"

6 REAL VARIABLES [I

ABC such that $AB = AC = 1$. Then if $BC = x$, $x^2 = 2$. Or we may determine the length x by means of Euclid's construction (Euc. VI. 13) for a mean proportional to 1 and 2, as indicated in the figure. Our requirements therefore involve the existence of a length measured by a number x, and a point P on Λ, such that

$$A_0P = x, \quad x^2 = 2.$$

But it is easy to see that *there is no rational number such that its square is* 2. In fact we may go further and say that there is no rational number whose square is m/n, where m/n is any positive fraction in its lowest terms, unless m and n are both perfect squares.

For suppose, if possible, that

$$p^2/q^2 = m/n,$$

p having no factor in common with q, and m no factor in common with n. Then $np^2 = mq^2$. Every factor of q^2 must divide np^2 and, as p and q have no common factor, every factor of q^2 must divide n. Hence $n = \lambda q^2$, where λ is an integer. But this involves $m = \lambda p^2$; and, as m and n have no common factor, λ must be unity. Thus $m = p^2$, $n = q^2$, as was to be proved. In particular it follows, by taking $n = 1$, that an integer cannot be the square of a rational number, unless that rational number is itself integral.

It appears then that our requirements involve the existence of a number x and a point P, not one of the rational points already constructed, such that $A_0P = x$, $x^2 = 2$; and (as the reader will remember from elementary algebra) we write $x = \sqrt{2}$.

The following alternative proof that no rational number can have its square equal to 2 is interesting. (But in what way is it interesting?)

Suppose, if possible, that p/q is a positive fraction, in its lowest terms, such that $(p/q)^2 = 2$ or $p^2 = 2q^2$. It is easy to see that this involves $(2q-p)^2 = 2(p-q)^2$; and so $(2q-p)/(p-q)$ is another fraction having the same property. But clearly $q < p < 2q$, and so $p - q < q$. Hence there is another fraction equal to p/q and having a smaller denominator, which contradicts the assumption that p/q is in its lowest terms.

Examples II. 1. Show that no rational number can have its cube equal to 2.

Wie, wenn man hier Ziffern statt der Buchstaben gebraucht? Also: daß $5^2 = 2 \times 3^2$ folgt $(2\cdot 3 - 5)^2 = 2\times(5-3)^2$. Es soll aber $3 < 5 < 2\cdot 3$ sein, + also $5 - 3 < 3$, — also $\frac{5^2}{3^2} = \frac{(<5)^2}{(<3)^2}$

[Center]: *But in what way is it interesting?*

[Left and Top Margin]: *"Suppose, if possible," heißt, glaube ich, gar nichts. Es scheint mir wie eine feige Ausflucht. – Es könnte etwa soviel heißen wie: try to suppose, und statt dessen könnte man dann einfach sagen "Try". Etwa: "Prüfe* $(p/q)^2 = 2\ldots$"
"Suppose, if possible," means, I believe, nothing at all. It appears to me like a cowardly evasion. – It might mean, say, something like: try to suppose, and instead of this one might simply say "Try". For example: "Check $(p/q)^2 = 2\ldots$".

[Bottom Margin]: *Wie, wenn man hier Ziffern statt der Buchstaben gebraucht? Also: Aus* $5^2 = 2 \times 3^2$ *folgt* $(2 \times 3 - 5)^2 = 2 \times (5 - 3)^2$. Es soll aber $3 < 5 < 2 \times 3$ *sein und also* $5 - 3 < 3$, *also* $5^2/3^2 = (< 5)^2/(< 3)^2$.
How [would it be] if one used here numerals instead of the letters? That is: From $5^2 = 2 \times 3^2$ follows $(2 \times 3 - 5)^2 = 2 \times (5 - 3)^2$. But there should be $3 < 5 < 2 \times 3$ and hence $5 - 3 < 3$, hence $5^2/3^2 = (< 5)^2/(< 3)^{2*}$.

*We have corrected Wittgenstein's formula "$5 - 3 < 3$".

Hier steht alles ein bißchen schief. Gerade genug um zu zeigen, daß das Ganze kein Organismus ist.

2. Prove generally that a rational fraction p/q in its lowest terms cannot be the cube of a rational number unless p and q are both perfect cubes.

3. A more general proposition, which is due to Gauss and includes those which precede as particular cases, is the following: *an algebraical equation*

$$x^n + p_1 x^{n-1} + p_2 x^{n-2} + \dots + p_n = 0,$$

with integral coefficients, cannot have a rational but non-integral root.

[For suppose that the equation has a root a/b, where a and b are integers without a common factor, and b is positive. Writing a/b for x, and multiplying by b^{n-1}, we obtain

$$-\frac{a^n}{b} = p_1 a^{n-1} + p_2 a^{n-2} b + \dots + p_n b^{n-1},$$

a fraction in its lowest terms equal to an integer, which is absurd. Thus $b = 1$, and the root is a. It is evident that a must be a divisor of p_n.]

4. Show that if $p_n = 1$ and neither of

$$1 + p_1 + p_2 + p_3 + \dots, \quad 1 - p_1 + p_2 - p_3 + \dots$$

is zero, then the equation cannot have a rational root.

5. Find the rational roots (if any) of

$$x^4 - 4x^3 - 8x^2 + 13x + 10 = 0.$$

[The roots can only be integral, and so ± 1, ± 2, ± 5, ± 10 are the only possibilities: whether these are roots can be determined by trial. It is clear that we can in this way determine the rational roots of any such equation.]

4. Irrational numbers (*continued*). The result of our geometrical representation of the rational numbers is therefore to suggest the desirability of enlarging our conception of 'number' by the introduction of further numbers of a new kind.

The same conclusion might have been reached without the use of geometrical language. One of the central problems of algebra is that of the solution of equations, such as

$$x^2 = 1, \quad x^2 = 2.$$

The first equation has the two rational roots 1 and -1. But, if our conception of number is to be limited to the rational numbers, we can only say that the second equation has no roots; and the same is the case with such equations as $x^3 = 2$, $x^4 = 7$. These facts are plainly sufficient to make some generalisation of our idea of number desirable, if it should prove to be possible.

Hier steht alles ein bißchen schief. Gerade genug um zu zeigen, daß das Ganze kein Organismus ist.
Here everything leans a little. Just enough to show that the whole thing is no organism.

and in which any two consecutive numbers differ by as little as we please. Let us then take a member x of L and a member y of R, and interpolate between them a chain of rational numbers of which x is the first and y the last, and in which any two consecutive numbers differ by less than δ, δ being any positive rational number as small as we please, such as ·01 or ·0001 or ·000001. In this chain there must be a last which belongs to L and a first which belongs to R, and these two numbers differ by less than δ.

We can now prove that *an x can be found in L and a y in R such that $2-x^2$ and y^2-2 are as small as we please*, say less than δ. Substituting $\frac{1}{4}\delta$ for δ in the argument which precedes, we see that we can choose x and y so that $y-x<\frac{1}{4}\delta$; and we may plainly suppose that both x and y are less than 2. Thus

$$y+x<4, \quad y^2-x^2=(y-x)(y+x)<4(y-x)<\delta;$$

and since $x^2<2$ and $y^2>2$ it follows *a fortiori* that $2-x^2$ and y^2-2 are each less than δ.

It follows also that *there can be no largest member of L or smallest member of R*. For if x is any member of L, then $x^2<2$. Suppose that $x^2=2-\delta$. Then we can find a member x_1 of L such that x_1^2 differs from 2 by less than δ, and so $x_1^2>x^2$ or $x_1>x$. Thus there are larger members of L than x; and, since x is *any* member of L, it follows that no member of L can be larger than all the rest. Hence L has no largest member, and similarly R has no smallest.

5. Irrational numbers (*continued*). We have thus divided the positive rational numbers into two classes, L and R, such that (i) every member of R is greater than every member of L, (ii) we can find a member of L and a member of R whose difference is as small as we please, (iii) L has no greatest and R no least member. Our common-sense notion of the attributes of a straight line, the requirements of our elementary geometry and our elementary algebra, alike demand *the existence of a number x greater than all the members of L and less than all the members of R, and of a corresponding point P on Λ such that P divides the points which correspond to members of L from those which correspond to members of R.*

Let us suppose for a moment that there is such a number x, and that it may be operated upon in accordance with the laws of algebra, so that, for example, x^2 has a definite meaning. Then x^2

[illegible] sich nähernden [illegible], möchte man sagen, bestimmen einen Ort – aber keinen rationalen. Und ich müsse von diesem Ort sagen, er liege links von allen rechten, rechts von allen linken Gliedern. Das aber ist eine irreführende Ausdrucksweise. Denn die Ortsbestimmung [illegible]

Diese beiden sich nähernden Reihen, möchte man sagen, bestimmen einen Ort – aber keinen rationalen. Ich müsse von diesem Ort sagen, er liege links von allen rechten und rechts von allen linken Gliedern. Das aber ist eine irreführende Ausdrucksweise, denn die Ortsbestimmung kommt zu keinem Ende.
These two approaching series, one wants to say, determine a place – but not a rational one. I should have to say of this place that it is located left of all the right and right of all the left members. But this is a misleading form of expression, for the determination of the place does not come to an end.

Es hat keinen Sinn zu sagen "die Länge dieses Stabes ist $\sqrt{2}$ cm – es sei denn, man wolle ihn damit sagen er sei die Hypotenuse in einem Dreieck etc.

Es wäre wie wenn man von einem (irgend) Stab sagen würde, das ca $3\frac{1}{7}$ Fuß lang ist, zu sagen: "es ist rund π Fuß lang".

cannot be either less than or greater than 2. For suppose, for example, that x^2 is less than 2. Then it follows from what precedes that we can find a positive rational number ξ such that ξ^2 lies between x^2 and 2. That is to say, we can find a member of L greater than x; and this contradicts the supposition that x divides the members of L from those of R. Thus x^2 cannot be less than 2, and similarly it cannot be greater than 2. We are therefore driven to the conclusion that $x^2 = 2$, and that x is the number which in algebra we denote by $\sqrt{2}$. And this number $\sqrt{2}$ is not rational, for no rational number has its square equal to 2. It is the simplest example of what is called an *irrational* number.

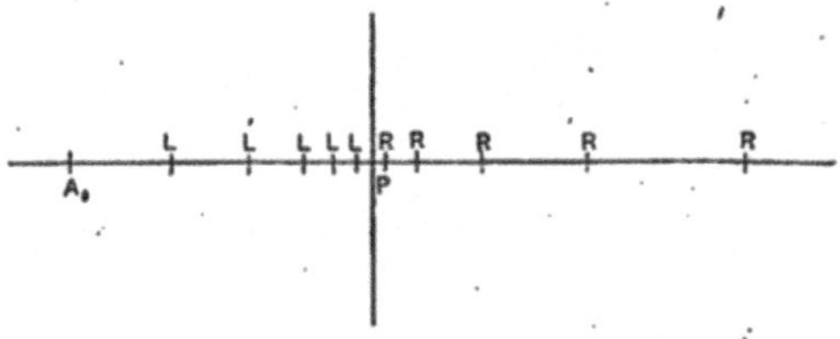

Fig. 3

But the preceding argument may be applied to equations other than $x^2 = 2$, almost word for word; for example to $x^2 = N$, where N is any integer which is not a perfect square, or to

$$x^3 = 3, \quad x^3 = 7, \quad x^4 = 23,$$

or, as we shall see later on, to $x^3 = 3x + 8$. We are thus led to believe in the existence of irrational numbers x and points P on Λ such that x satisfies equations such as these, even when these lengths cannot (as $\sqrt{2}$ can) be constructed by means of elementary geometrical methods.

The reader will no doubt remember that in treatises on elementary algebra the root of such an equation as $x^q = n$ is denoted by $\sqrt[q]{n}$ or $n^{1/q}$, and that a meaning is attached to such symbols as

$$n^{p/q}, \quad n^{-p/q}$$

by means of the equations

$$n^{p/q} = (n^{1/q})^p, \quad n^{p/q} n^{-p/q} = 1.$$

Bedenke daß eine 'bestimmte Länge' nicht durch direkte Messung definiert sein muß!

[Left Margin]: *Es wäre seltsam von irgendeinem Stück Holz, das ca.* $3\frac{1}{7}$ *Fuß lang ist, zu sagen: "es ist rund* π *Fuß lang".*
It would be strange to say of any piece of wood which is ca. $3\frac{1}{7}$ feet long: "it is about π feet long".

[Top Margin]: *Es hat keinen Sinn zu sagen "die Länge dieses Stabes ist* $\sqrt{2}$ *cm" – es sei denn, man wollte etwa damit sagen es sei die Hypothenuse in einem Dreieck etc.*
It makes no sense to say "the length of this rod is $\sqrt{2}$ cm" – except you might want to say with this that it is the hypotenuse of a triangle etc.

[Bottom Margin]: *Bedenke daß eine "bestimmte Länge" nicht durch direkte Messung definiert sein muß.*
Bear in mind that a "determinate length" need not be defined by a direct measurement.

And he will remember how, in virtue of these definitions, the 'laws of indices' such as

$$n^r \times n^s = n^{r+s}, \quad (n^r)^s = n^{rs}$$

are extended so as to cover the case in which r and s are any rational numbers.

The reader may now follow one or other of two alternative courses. He may, if he pleases, be content to assume that 'irrational numbers' such as $\sqrt{2}$, $\sqrt[3]{3}$, ... exist and are amenable to the algebraical laws with which he is familiar*. If he does this he will be able to avoid the more abstract discussions of the next few sections, and may pass on at once to §§ 13 *et seq.*.

If, on the other hand, he is not disposed to adopt so naïve an attitude, he will be well advised to pay careful attention to the sections which follow, in which these questions receive fuller consideration.

Examples III. 1. Find the difference between 2 and the squares of the decimals given in § 4 as approximations to $\sqrt{2}$.

2. Find the differences between 2 and the squares of

$$\tfrac{1}{1},\ \tfrac{3}{2},\ \tfrac{7}{5},\ \tfrac{17}{12},\ \tfrac{41}{29},\ \tfrac{99}{70}.$$

3. Show that if m/n is [Any] a good approximation to $\sqrt{2}$, then $(m+2n)/(m+n)$ is a better one, and that the errors in the two cases are in opposite directions. Apply this result to continue the series of approximations in the preceding example.

4. If x and y are approximations to $\sqrt{2}$, by defect and by excess respectively, and $2-x^2<\delta$, $y^2-2<\delta$, then $y-x<\delta$.

5. The equation $x^2=4$ is satisfied by $x=2$. Examine how far the argument of the preceding sections applies to this equation (writing 4 for 2 throughout). [If we define the classes L, R as before, they do not include *all* rational numbers. The rational number 2 is an exception, since 2^2 is neither less than nor greater than 4.]

6. Irrational numbers (*continued*). In § 4 we discussed a special mode of division of the positive rational numbers x into two classes, such that $x^2<2$ for the members of one class and $x^2>2$ for those of the other. Such a mode of division is called a *section* of the numbers in question. It is plain that we could

* This is the point of view which was adopted in the first edition of this book.

7, 8] REAL VARIABLES 15

occurs is sufficient to fix its interpretation. When we say (see § 9) that $\frac{1}{2} < \sqrt{(\frac{1}{2})}$, we *must* mean by '$\frac{1}{2}$' the real number $\frac{1}{2}$.

The reader should observe, moreover, that no particular logical importance is to be attached to the precise form of definition of a 'real number' that we have adopted. We defined a 'real number' as being a section, i.e. a pair of classes. We might equally well have defined it as being the lower, or the upper, class; indeed it would be easy to define an infinity of classes of entities each of which would possess the properties of the class of real numbers. What is essential in mathematics is that its symbols should be capable of *some* interpretation; generally they are capable of *many*, and then, so far as mathematics is concerned, it does not matter which we adopt. Bertrand Russell has said that 'mathematics is the science in which we do not know what we are talking about, and do not care whether what we say about it is true', a remark which is expressed in the form of a paradox but which in reality embodies a number of important truths. It would take too long to analyse the meaning of Russell's epigram in detail, but one at any rate of its implications is this, that the symbols of mathematics are capable of varying interpretations, and that we are in general at liberty to adopt whichever we prefer.

There are now three cases to distinguish. It may happen that all negative rational numbers belong to the lower class, and zero and all positive rational numbers to the upper. We describe this section as the *real number zero*. Or again it may happen that the lower class includes some positive numbers. Such a section we describe as a *positive real number*. Finally it may happen that some negative numbers belong to the upper class. Such a section we describe as a *negative real number**.

The difference between our present definition of a positive real number x and that of § 7 amounts to the addition to the lower class of zero and all the negative rational numbers. An example of a negative real number is given by taking the property P of § 6 to be $x + 1 < 0$ and Q to be $x + 1 \geqslant 0$.

* There are also sections in which every number belongs to the lower or to the upper class. The reader may be tempted to ask why we do not regard these sections also as defining numbers, which we might call the *real numbers positive and negative infinity*.

There is no logical objection to such a procedure, but it proves to be inconvenient in practice. The most natural definitions of addition and multiplication do not work in a satisfactory way. Moreover, for a beginner, the chief difficulty in the elements of analysis is that of learning to attach precise senses to phrases containing the word 'infinity'; and experience seems to show that he is likely to be confused by any addition to their number.

Es ist hier unklar, was "interpretation" heißt.
Here it is unclear what "interpretation" means.

24 REAL VARIABLES [I

Deduce that any polynomial in x with rational coefficients (i.e. any expression of the form

$$a_0x^n + a_1x^{n-1} + \ldots + a_n,$$

where $a_0, \ldots, a_n$ are rational numbers) can be expressed in the form $P + Q\sqrt{q}$.

7. If $a + \sqrt{b}$, where b is not a perfect square, is the root of an algebraical equation with rational coefficients, then $a - \sqrt{b}$ is another root of the same equation.

8. Express $1/(p + \sqrt{q})$ in the form prescribed in Ex. 6. [Multiply numerator and denominator by $p - \sqrt{q}$.]

9. Deduce from Exs. 6 and 8 that any expression of the form $G(x)/H(x)$, where $G(x)$ and $H(x)$ are polynomials in x with rational coefficients, can be expressed in the form $P + Q\sqrt{q}$, where P and Q are rational.

10. If p, q, and $p^2 - q$ are positive, we can express $\sqrt{(p + \sqrt{q})}$ in the form $\sqrt{x} + \sqrt{y}$, where

$$x = \tfrac{1}{2}\{p + \sqrt{(p^2 - q)}\}, \quad y = \tfrac{1}{2}\{p - \sqrt{(p^2 - q)}\}.$$

11. Determine the conditions that it may be possible to express $\sqrt{(p + \sqrt{q})}$, where p and q are rational, in the form $\sqrt{x} + \sqrt{y}$, where x and y are rational.

12. If $a^2 - b$ is positive, then a necessary and sufficient condition that

$$\sqrt{(a + \sqrt{b})} + \sqrt{(a - \sqrt{b})}$$

should be rational is that $a^2 - b$ and $\tfrac{1}{2}\{a + \sqrt{(a^2 - b)}\}$ should both be squares of rational numbers.

15. The continuum. The aggregate of all real numbers, rational and irrational, is called the *arithmetical continuum*.

It is convenient to suppose that the straight line A of § 2 is composed of points corresponding to all the numbers of the arithmetical continuum, and of no others*. The points of the line, the aggregate of which may be said to constitute the *linear continuum*, then supply us with a convenient image of the arithmetical continuum.

We have considered in some detail the chief properties of a few classes of real numbers, such, for example, as rational

* This supposition is merely a hypothesis adopted (i) because it suffices for the purposes of our geometry and (ii) because it provides us with convenient geometrical illustrations of analytical processes. As we use geometrical language only for purposes of illustration, it is not part of our business to study the foundations of geometry.

Das eigentlich irreleitende an diesem Bild ist der Unterschied zwischen dem Gebrauch von "ein Punkt existiert" + "eine Zahl existiert".

Das eigentlich irreleitende an diesem Bild ist der Unterschied zwischen dem Gebrauch von "ein Punkt existiert" und "eine Zahl existiert".

What is actually misleading about this picture is the difference between the use of "a point exists" and "a number exists".

irrational numbers which we have so far considered. This number π is not an isolated or exceptional case. Only special classes of irrational numbers are roots of such algebraical equations; and only still more special classes are expressible by means of surds.

16. The continuous real variable. The 'real numbers' may be regarded from two points of view. We may think of them *as an aggregate*, the 'arithmetical continuum' defined in the preceding section, or *individually*. And when we think of them individually, we may think either of a particular *specified* number (such as 1, $-\frac{1}{2}$, $\sqrt{2}$, or π) or we may think of *any* number, *an unspecified* number, *the number* x. This last is our point of view when we make such assertions as 'x is a number', 'x is the measure of a length', 'x may be rational or irrational'. The x which occurs in propositions such as these is called *the continuous real variable*: and the individual numbers are called the *values* of the variable.

A 'variable', however, need not necessarily be continuous. Instead of considering the aggregate of *all* real numbers, we might consider some partial aggregate contained in the former aggregate, such as the aggregate of rational numbers, or the aggregate of positive integers. Let us take the last case. Then in statements about *any* positive integer, or *an unspecified* positive integer, such as 'n is either odd or even', n is called the variable, a *positive integral variable*, and the individual positive integers are its values.

Naturally 'x' and 'n' are only examples of variables, the variable whose 'field of variation' is formed by all the real numbers, and that whose field is formed by the positive integers. These are the most important examples, but we have often to consider other cases. In the theory of decimals, for instance, we may denote by x any figure in the expression of any number as a decimal. Then x is a variable, but a variable which has only ten different values, viz. 0, 1, 2, 3, 4, 5, 6, 7, 8, 9. The reader should think of other examples of variables with different fields of varia-

Wer Schreinerhobeln lehrt, lehrt ihn damit nicht Tische & Stühle machen, auch nicht allgemein, aber er lehrt eine Technik von sehr allgemeiner Anwendung.

tion. He will find interesting examples in ordinary life: policeman x, the driver of cab x, the year x, the xth day of the week. The values of these variables are naturally not numbers.

17. Sections of the real numbers. In §§ 4–7 we considered 'sections' of the rational numbers, i.e. modes of division of the rational numbers (or of the positive rational numbers only) into two classes L and R possessing the following characteristic properties:

(i) that every number of the type considered belongs to one and only one of the two classes;

(ii) that both classes exist;

(iii) that any member of L is less than any member of R.

It is plainly possible to apply the same idea to the aggregate of all real numbers, and the process is, as the reader will find in later chapters, of very great importance.

Let us then suppose* that P and Q are two properties which are mutually exclusive, and one of which is possessed by every real number. Further let us suppose that any number which possesses P is less than any number which possesses Q. We call the numbers which possess P the *lower* or *left-hand class* L, and those which possess Q the *upper* or *right-hand class* R.

Thus P might be $x \leqq \sqrt{2}$ and Q be $x > \sqrt{2}$. It is important to observe that a pair of properties which suffice to define a section of the rational numbers may not suffice to define one of the real numbers. This is so, for example, with the pair '$x < \sqrt{2}$' and '$x > \sqrt{2}$' or (if we confine ourselves to positive numbers) with '$x^2 < 2$' and '$x^2 > 2$'. Every rational number possesses one or other of the properties, but not every real number, since in either case $\sqrt{2}$ escapes classification.

There are now two possibilities†. Either L has a greatest member l, or R has a least member r. *Both* of these events cannot

* The discussion which follows is in many ways similar to that of § 6. We have not attempted to avoid a certain amount of repetition. The idea of a 'section', first brought into prominence in Dedekind's famous pamphlet *Stetigkeit und irrationale Zahlen*, is one which must be grasped by every reader of this book, even if he be one of those who prefer to omit the discussion of the notion of an irrational number contained in §§ 6–12.

† There were three in § 6.

Man kann die rat. Zahlen so in größere & kleinere einteilen, daß die Einteilung aussieht wie ein Schnitt, man kann aber nicht die reellen Zahlen ebenso einteilen. Man kann sie schneiden, aber der Schnitt bewirkt keine reale Einteilung.

A rule, in form, is examined which will later be used.

Man könnte sagen: Überlegungen nach diesem Schema werden später oft von Nutzen sein.

[Left Margin:] *A mode, or form, of reasoning is expounded which will later be used.*
Man könnte sagen: Überlegungen nach diesem Schema werden später oft von Nutzen sein.
One might say: considerations according to this schema will often be useful later on.

[Top Margin:] *Wer einen Bretter hobeln lehrt, lehrt ihn damit nicht Tische und Stühle machen, auch nicht allgemein, aber er lehrt eine Technik von sehr allgemeiner Anwendung.*
Whoever teaches someone to plane planks doesn't teach him thereby to make tables and chairs, not even in general, but rather he teaches a technique of very general application.

[Bottom of Page:] *Man kann die rat. Zahlen so in größer und kleiner einteilen, daß die Einteilung wirkt wie ein Schnitt, man kann aber nicht die reellen Zahlen ebenso einteilen. Man kann sie schneiden, aber der Schnitt bewirkt keine solche Einteilung.*
One can divide the rat. numbers in[to] greater and smaller in such a way that the division functions like a cut, but one cannot divide the real numbers in the same way. One can cut them, but the cut doesn't bring about such a division.

Der Beweis scheint zu zeigen, daß wir jetzt am [illegible] sind & es auf der Linie keine anderen Punkte mehr gibt, denn wenn wir sie teilen nennen wir dafür einen reellen Punkt [illegible] hin.

16, 17] REAL VARIABLES 29

occur. For if L had a greatest member l, and R a least member r, the number $\frac{1}{2}(l+r)$ would be greater than all members of L and less than all members of R, and so could not belong to either class. On the other hand *one* event must occur*.

For let L_1 and R_1 denote the classes formed from L and R by taking only the rational members of L and R. Then the classes L_1 and R_1 form a section of the rational numbers. There are now two cases to distinguish.

It may happen that L_1 has a greatest member α. In this case α must be also the greatest member of L. For if not we can find a greater, say β. There are rational numbers lying between α and β, and these, being less than β, belong to L, and therefore to L_1; and this is plainly a contradiction. Hence α is the greatest member of L.

On the other hand it may happen that L_1 has no greatest member. In this case the section of the rational numbers formed by L_1 and R_1 is a real number α. This number α must belong to L or to R. If it belongs to L we can show, precisely as before, that it is the greatest member of L; and similarly that, if it belongs to R, it is the least member of R.

Thus in any case either L has a greatest member or R a least. Any section of the real numbers therefore 'corresponds' to a real number in the sense in which a section of the rational numbers sometimes, but not always, corresponds to a rational number. This conclusion is of very great importance; for it shows that the consideration of sections of all the real numbers does not lead to any further generalisation of our idea of number. Starting from the rational numbers, we found that the idea of a section of the rational numbers led us to a new conception of a number, that of a real number, more general than that of a rational number; and it might have been expected that the idea of a section of the real numbers would have led us to a conception more general still. The discussion which precedes shows that this is not the case, and that the aggregate of real numbers, or the continuum

Das würde an einen Beweisfälschung!

* This was not the case in § 6.

Hier ist angenommen, daß der Schnitt nicht eine reelle Zahl & zugleich eine [Zahl ein] ganz andere ist ~~[illegible]~~ treffen kann.

[Top Margin]: *Der Beweis scheint zu zeigen, daß wir jetzt am Ende sind und es auf der Linie keine anderen Punkte mehr gibt; denn wenn wir sie teilen müssen wir durch einen reellen Punkt schneiden.*
The proof seems to show that we now are finished and that on the line there are no more points; for if we divide them we must cut through a real point.

[Right Margin]: *Das Prüde an dieser Beweisführung!*
The prudishness in this line of proof!

[Bottom Margin]: *Hier ist angenommen, daß der Schnitt nicht eine reelle Zahl und zugleich eine "Zahl" einer ganz anderen Art treffen kann.*
Here it is supposed that the section cannot hit a real number and at the same time a "number" of quite another kind.

has a kind of completeness which the aggregate of the rational numbers lacked, a completeness which is expressed in technical language by saying that the continuum is closed.

The result which we have just proved may be stated as follows:

Dedekind's theorem. *If the real numbers are divided into two classes L and R in such a way that*

(i) *every number belongs to one or other of the two classes,*

(ii) *each class contains at least one number,*

(iii) *any member of L is less than any member of R,*

then there is a number α, which has the property that all the numbers less than it belong to L and all the numbers greater than it to R. The number α itself may belong to either class.

In applications we have often to consider sections not of *all* numbers but of all those contained in an *interval* (β, γ), that is to say of all numbers x such that $\beta \leqq x \leqq \gamma$. A 'section' of such numbers is of course a division of them into two classes possessing the properties (i), (ii), and (iii). Such a section may be converted into a section of *all* numbers by adding to L all numbers less than β and to R all numbers greater than γ. It is clear that the conclusion stated in Dedekind's theorem still holds if we substitute 'the real numbers of the interval (β, γ)' for 'the real numbers', and that the number α in this case satisfies the inequalities $\beta \leqq \alpha \leqq \gamma$.

18. Points of accumulation. A system of real numbers, or of the points on a straight line corresponding to them, defined in any way whatever, is called an *aggregate* or *set* of numbers or points. The set might consist, for example, of all the positive integers, or of all the rational points.

It is most convenient here to use the language of geometry*. Suppose then that we are given a set of points, which we will denote by S. Take any point ξ, which may or may not belong to S. Then there are two possibilities. Either (i) it is possible to choose a positive number δ so that the interval $(\xi - \delta, \xi + \delta)$ does not contain any point of S, other than ξ itself†, or (ii) this is not possible.

* The reader will hardly require to be reminded that this course is adopted solely for reasons of linguistic convenience.

† This clause is of course unnecessary if ξ does not itself belong to S.

Dies scheint ein ungeheuerliches Resultat. Denn es scheint als hätten wir etwas Entscheidendes über die letzte Struktur des Continuums herausgefunden, wobei wir uns das Continuum etwa als den Raum vorstellen.
This seems to be an enormous result. For it seems as if we had found out something crucial about the ultimate structure of the continuum, where we imagine the continuum as space, say.

H.'s Prozedur ist hier ganz anders als bei den Zahlen. Dort fing er mit den rationalen Z. an & führte nach & nach neue Zahlenarten ein. Hier führt er zuerst einen vagen allgemeinen Begriff der Funktion ein, + erst später Arten von Funktionen!

CHAPTER II

FUNCTIONS OF REAL VARIABLES

20. The idea of a function. Suppose that x and y are two continuous real variables, which we may suppose to be represented geometrically by distances $A_0P = x$, $B_0Q = y$ measured from fixed points A_0, B_0 along two straight lines A, M. And let us suppose that the positions of the points P and Q are not independent, but are connected by a relation which we can imagine expressed as a relation between x and y: so that, when P and x are known, Q and y are also known. We might, for example, suppose that $y = x$, or $2x$, or $\frac{1}{2}x$, or $x^2 + 1$. In all of these cases the value of x determines that of y. Or again we might suppose that the relation between x and y is given, not by means of an explicit formula for y in terms of x, but by means of a geometrical construction which enables us to determine Q when P is known.

In these circumstances y is said to be a *function* of x. This notion of functional dependence of one variable upon another is perhaps the most important in the whole range of higher mathematics. In order to enable the reader to be certain that he understands it clearly, we shall, in this chapter, illustrate it by means of a large number of examples.

But before we proceed to do this, we must point out that the simple examples of functions mentioned above possess three characteristics which are by no means involved in the general idea of a function, viz.:

(1) y is determined *for every value of* x;

(2) to each value of x for which y is given corresponds *one and only one value of* y;

(3) the relation between x and y is expressed by means of *an analytical formula*, from which the value of y corresponding to a

Das Wichtige ist, daß es hier eine mathematische & eine nicht-mathematische Zuordnung gibt!

[Left Margin and Top Margin]: *H.'s Prozedur ist hier ganz anders als bei den Zahlen. Dort fing er mit den rationalen Zahlen an und führte nach und nach neue Zahlenarten ein. Hier führt er zuerst einen vagen allgemeinen Begriff der Funktion ein und erst später Arten von Funktionen!*
H.'s procedure is here quite different from the case of the numbers. There he began with the rational numbers and then by and by introduced new kinds of numbers. Here, he introduces first a vague general notion of a function and only later kinds of functions!

[Bottom Margin]: *Das Wichtige ist, daß es hier eine mathematische und eine nicht-mathematische Zuordnung gibt!*
The important thing here is that there is a mathematical and a non-mathematical correspondence!

Durch diese Erklärungen wird die Idee der Funktion nicht definiert, sondern bloß illustriert.

given value of x can be calculated by direct substitution of the latter.

It is indeed the case that these particular characteristics are possessed by many of the most important functions. But the consideration of the following examples will make it clear that they are by no means essential to a function. All that is essential is that there should be some relation between x and y such that to some values of x at any rate correspond values of y.

Examples X. 1. Let $y = x$ or $2x$ or $\frac{1}{2}x$ or x^2+1. Nothing further need be said at present about cases such as these.

2. Let $y = 0$ whatever be the value of x. Then y is a function of x, for we can give x any value, and the corresponding value of y (viz. 0) is known. In this case the functional relation makes the same value of y correspond to all values of x. The same would be true were y equal to 1 or $-\frac{1}{2}$ or $\sqrt{2}$ instead of 0. Such a function of x is called *a constant*.

3. Let $y^2 = x$. Then if x is positive this equation defines *two* values of y corresponding to each value of x, viz. $\pm\sqrt{x}$. If $x = 0$, $y = 0$. Hence to the particular value 0 of x corresponds *one* and only one value of y. But if x is negative there is *no* value of y which satisfies the equation. That is to say, the function y is not defined for negative values of x. This function therefore possesses the characteristic (3), but neither (1) nor (2).

4. Consider a volume of gas maintained at a constant temperature and contained in a cylinder closed by a sliding piston*.

Let A be the area of the cross-section of the piston and W its weight. The gas, held in a state of compression by the piston, exerts a certain pressure p_0 per unit of area on the piston, which balances the weight W, so that

$$W = Ap_0.$$

Let v_0 be the volume of the gas when the system is thus in equilibrium. If additional weight is placed upon the piston the latter is forced downwards. The volume (v) of the gas diminishes; the pressure (p) which it exerts upon unit area of the piston increases. Boyle's experimental law asserts that the product of p and v is very nearly constant, a correspondence which, if exact, would be represented by an equation of the type

$$pv = a \qquad \text{(i)},$$

where a is a number which can be determined approximately by experiment.

* I borrow this instructive example from Prof. H. S. Carslaw's *Introduction to the calculus*.

Durch diese Erklärungen wird die Idee der Funktion nicht definiert, sondern bloß illustriert.
By these explanations the idea of a function is not defined, but merely illustrated.

54–56] POSITIVE INTEGRAL VARIABLE 117

the meaning of which we have explained above. Later on we shall learn how to attach a meaning to other phrases involving the symbol ∞, but the reader will always have to bear in mind

(1) that ∞ *by itself* means nothing, although *phrases containing it* sometimes mean something,

(2) that in every case in which a phrase containing the symbol ∞ means something it will do so simply because we have previously attached a meaning to this particular phrase by means of a special definition.

Now it is clear that if $\phi(n)$ has the property P for large values of n, and if n 'tends to ∞', in the sense which we have just explained, then n will ultimately assume values large enough to ensure that $\phi(n)$ has the property P. And so another way of putting the question 'what properties has $\phi(n)$ for sufficiently large values of n?' is 'how does $\phi(n)$ behave as n tends to ∞?'

56. The behaviour of a function of n as n tends to infinity. We shall now proceed, in the light of the remarks made in the preceding sections, to consider the meaning of some kinds of statements which recur continually in higher mathematics. Let us consider, for example, the two following statements: (*a*) $1/n$ *is small for large values of* n, (*b*) $1-(1/n)$ *is nearly equal to* 1 *for large values of* n. Obvious as they may seem, there is a good deal in them which will repay the reader's attention. Let us take (*a*) first, as being slightly the simpler.

We have already considered the statement '$1/n$ *is less than* ·01 *for large values of* n'. This, we saw, means that the inequality $1/n < \cdot 01$ is true for all values of n greater than some definite value, in fact greater than 100. Similarly it is true that '$1/n$ *is less than* ·0001 *for large values of* n': in fact $1/n < \cdot 0001$ if $n > 10000$. And instead of ·01 or ·0001 we might take ·000001 or ·00000001, or any positive number we like.

It is obviously convenient to have some way of expressing the fact that *any* such statement as '$1/n$ *is less than* ·01 *for large values of* n' is true, when we substitute for ·01 any smaller

Muß denn eine Funktion ein Verhalten für $x \to \infty$ haben? Die extensionale Vorstellung läßt es so erscheinen.

Muß denn eine Funktion ein Verhalten für $x \to \infty$ haben?
Die extensionale Vorstellung läßt es so erscheinen.
Must a function have a behavior for $x \to \infty$?
The extensional view makes it appear so.

a function of δ. We shall sometimes emphasise this fact by writing n_0 in the form $n_0(\delta)$.

The reader should imagine himself confronted by an opponent who questions the truth of the statement. He would name a series of numbers growing smaller and smaller. He might begin with ·001. The reader would reply that $1/n < \cdot 001$ as soon as $n > 1000$. The opponent would be bound to admit this, but would try again with some smaller number, such as ·0000001. The reader would reply that $1/n < \cdot 0000001$ as soon as $n > 10000000$: and so on. In this simple case it is evident that the reader would always have the better of the argument.

We shall now introduce yet another way of expressing this property of the function $1/n$. We shall say that '*the* **limit** *of* $1/n$ *as (or when)* n *tends to* ∞ *is* 0', a statement which we may express symbolically in the form

$$\lim_{n\to\infty} \frac{1}{n} = 0,$$

or simply $\lim(1/n) = 0$. We shall also sometimes write

$$\text{'}1/n \to 0 \text{ as } n \to \infty\text{'}$$

which may be read '$1/n$ tends to 0 as n tends to ∞'; or simply '$1/n \to 0$'. In the same way we shall write

$$\lim_{n\to\infty}\left(1-\frac{1}{n}\right) = 1, \quad \lim\left(1-\frac{1}{n}\right) = 1,$$

or $1-(1/n) \to 1$.

57. Now let us consider a different example: let $\phi(n) = n^2$. Then '*n^2 is large when n is large*'. This statement is equivalent to the more formal statements.

'if Δ is any positive number, however large, then $n^2 > \Delta$ for sufficiently large values of n',

'we can find a number $n_0(\Delta)$ such that $n^2 > \Delta$ for all values of n greater than or equal to $n_0(\Delta)$'.

And it is natural in this case to say that 'n^2 tends to ∞ as n tends to ∞', or 'n^2 tends to ∞ with n', and to write

$$n^2 \to \infty.$$

Finally consider the function $\phi(n) = -n^2$. In this case $\phi(n)$ is large, but negative, when n is large, and we naturally say that

Ich kann jedes Δ durch ein n^2 übersteigen; ~~für jedes n gibt es noch ein Δ~~ + jedes n muss ich für irgend ein Δ noch übersteigen.

und der Leser würde dadurch nichts beweisen, es sei denn, dass er dem Gegner die Technik lehrt, eine immer kleinere Zahl zu erzeugen.

[Right Margin]: *Und der Leser würde dadurch nichts beweisen, es sei denn, daß er seinen Gegner die Technik lehrt eine immer kleinere Zahl zu erzeugen.*
And the reader would prove nothing by this, unless he taught his opponent the technique of producing a number that is always smaller.

[Bottom Margin]: *Ich kann jedes Δ durch ein n^2 übersteigen, and jedes n muß ich für irgendeine Δ noch übersteigen.*
I can exceed any Δ by an n^2 and I must still exceed any n by some Δ.

120 LIMITS OF FUNCTIONS OF A [IV

'$-n^2$ tends to $-\infty$ as n tends to ∞' and write

$$-n^2 \to -\infty.$$

And the use of the symbol $-\infty$ in this sense suggests that it will sometimes be convenient to write $n^2 \to +\infty$ for $n^2 \to \infty$, and generally to use $+\infty$ instead of ∞, in order to secure greater uniformity of notation.

But we must once more repeat that in all these statements the symbols ∞, $+\infty$, $-\infty$ mean nothing whatever by themselves, and only acquire a meaning when they occur in certain special connections in virtue of the explanations which we have just given.

58. Definition of a limit. After the discussion which precedes the reader should be in a position to appreciate the general notion of a *limit*. Roughly we may say that $\phi(n)$ ***tends to a limit*** l ***as*** n ***tends to*** ∞ ***if*** $\phi(n)$ ***is nearly equal to*** l ***when*** n ***is large.*** But although the meaning of this statement should be clear enough after the preceding explanations, it is not, as it stands, precise enough to serve as a strict mathematical definition. It is, in fact, equivalent to a whole class of statements of the type '*for sufficiently large values of* n, $\phi(n)$ *differs from* l *by less than* δ'. This statement has to be true for $\delta = \cdot 01$ or $\cdot 0001$ or *any* positive number; and for any such value of δ it has to be true for *any* value of n after a certain definite value $n_0(\delta)$, though the smaller δ is the larger, as a rule, will be this value $n_0(\delta)$.

We accordingly frame the following formal definition:

DEFINITION I. *The function $\phi(n)$ is said to tend to the limit l as n tends to ∞, if,* for any *~~however small be the~~ positive number δ, $\phi(n)$ differs from l by less than δ for sufficiently large values of n; that is to say if,* any *~~however small be the~~ positive number δ, we can determine a number $n_0(\delta)$ corresponding to δ, such that $\phi(n)$ differs from l by less than δ for all values of n greater than or equal to $n_0(\delta)$.*

It is usual to denote the difference between $\phi(n)$ and l, taken positively, by $|\phi(n) - l|$. It is equal to $\phi(n) - l$ or to $l - \phi(n)$, whichever is positive, and agrees with the definition of the *modulus* of $\phi(n) - l$, as given in Ch. III, though at present we are only considering real values, positive or negative.

. . . if, ~~however small be the~~ for some positive number δ . . .
. . . if for any positive number δ. . .

57–59] POSITIVE INTEGRAL VARIABLE 121

With this notation the definition may be stated more shortly as follows: '*if, given any positive number, δ, ~~however small~~, we can find $n_0(\delta)$ so that $|\phi(n)-l|<\delta$ when $n \geqq n_0(\delta)$, then we say that $\phi(n)$ tends to the limit l as n tends to ∞, and write*

$$\lim_{n\to\infty} \phi(n) = l\text{'}.$$

Sometimes we may omit the '$n\to\infty$'; and sometimes it is convenient, for brevity, to write $\phi(n)\to l$.

The reader will find it instructive to work out, in a few simple cases, the explicit expression of n_0 as a function of δ. Thus if $\phi(x) = 1/n$ then $l = 0$, and the condition reduces to $1/n<\delta$ for $n \geqq n_0$, which is satisfied if $n_0 = 1+[1/\delta]$*. There is one and only one case in which *the same* n_0 will do for *all* values of δ. If, from a certain value N of n onwards, $\phi(n)$ is constant, say equal to C, then it is evident that $\phi(n)-C = 0$ for $n \geqq N$, so that the inequality $|\phi(n)-C|<\delta$ is satisfied for $n \geqq N$ and all positive values of δ. And if $|\phi(n)-l|<\delta$ for $n \geqq N$ and all positive values of δ, then it is evident that $\phi(n) = l$ when $n \geqq N$, so that $\phi(n)$ is constant for all such values of n.

59. The definition of a limit may be illustrated geometrically as follows. The graph of $\phi(n)$ consists of a number of points corresponding to the values $n = 1, 2, 3, \ldots$.

Draw the line $y = l$, and the parallel lines $y = l-\delta$, $y = l+\delta$ at distance δ from it. Then

$$\lim_{n\to\infty} \phi(n) = l$$

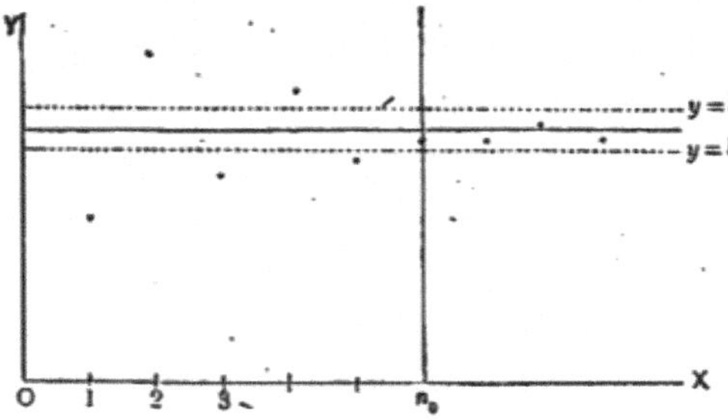

Fig. 25

* Here and henceforward we shall use $[x]$ in the sense of Ch. II, i.e. as the greatest integer not greater than x.

~~however small~~

References

Ambrose, Alice (1935a). Finitism in Mathematics (I). *Mind*, 44 (174): 186–203.

Ambrose, Alice (1935b). Finitism in Mathematics (II). *Mind*, 44 (175): 317–40.

Ambrose, Alice (1937). Finitism and "The Limits of Empiricism". *Mind*, 46 (183): 379–385.

Anscombe, G.E.M. (1957). Report on *Analysis* Problem, no. 10, "It is Impossible to be told Anyone's Name". *Analysis*, 17: 49–51.

Beaney, Michael, editor (2013). *The Oxford Handbook of the History of Early Analytic Philosophy*. Oxford University Press, New York/London.

Beaney, Michael, Harrington, Brendan, and Shaw, Dominic, editors (2018). *Aspect Perception After Wittgenstein: Seeing-As and Novelty*. Routledge, New York.

Becker, Oskar (1975). *Grundlagen der Mathematik in geschichtlicher Entwicklung*. Suhrkamp, Frankfurt am Main.

Bernays, Paul (1957). Bemerkungen zum Paradoxon von Thoralf Skolem. *Avhandlinger utgitt av Det Norske Videnskaps-Akademi i Oslo, I. Mat.-Naturv. Klasse*, 2nd series: 3–9. English translation by Dirk Schllimm and Steve Awodey, "Considerations regarding the paradox of Thoralf Skolem", Text No. 22 of the Bernays Project, at http://www.phil.cmu.edu/projects/bernays/.

Bernays, Paul (1959). Betrachtungen zu Ludwig Wittgensteins *Bemerkungen über die Grundlagen der Mathematik*/Comments on Ludwig Wittgenstein's *Remarks on the Foundations of Mathematics*. *Ratio*, 2 (1): 1–22.

Biesenbach, H. (2014). *Anspielungen und Zitate im Werk Ludwig Wittgensteins*, Erweiterte Neuausgabe. Sofia: St. Kliment Ohridski University Press.

Biggs, Michael and Pichler, Alois (1993). Wittgenstein: Two Source Catalogues and a Bibliography, Catalogues of the Published Texts and of the Published Diagrams, each Related to its Sources. Report available at http://wab.uib.no/wp-no7.pdf. Original publication "Two Source Catalogues and a Bibliography" in *Working Papers from the Wittgenstein Archives at the University of Bergen*, No. 7, 1993.

Black, Max (1964). *A Companion to Wittgenstein's Tractatus*. Cornell University Press, Ithaca, N.Y.

Blackwell, A. and MacKay, D., editors (2006). *Power*. Cambridge, Cambridge University Press.

Bourbaki, Nicolas (1942). *Élements de mathématique, fascicule iii, livre II. Topologie Génerale, chapitre 2. Groupes topologiques, et chapitre 3. Nombres réels*. Hermann, Paris. English translation by N. Bourbaki and P. Spain, *General Topology: Chapters 1–4*, 1995, Springer, Berlin.

J. Floyd, F. Mühlhölzer, *Wittgenstein's Annotations to Hardy's Course of Pure Mathematics*, Nordic Wittgenstein Studies 7,
https://doi.org/10.1007/978-3-030-48481-1

Bourbaki, Nicolas (1949). *Élements de mathématique, fascicule ix, livre IV, Fonctions d'une Variable Réelle.* Hermann, Paris. English translation by N. Bourbaki and P. Spain, *Functions of a Real Variable: Elementary Theory*, 2003, Springer, Berlin.

Brouwer, L. E. J. (1929). Mathematik, Wissenschaft und Sprache. *Monatshefte für Mathematik und Physik*, 36 (1): 153–164.

Burgess, John P. (2012). *Rigor and Structure*. Oxford University Press, New York.

Cantini, Andrea and Bruni, Riccardo (2017). Paradoxes and contemporary logic. In *The Stanford Encyclopedia of Philosophy*, ed. Edward N. Zalta, https://plato.stanford.edu/archives/fall2017/entries/paradoxes-contemporary-logic/.

Cantor, Georg (1872). Über die Ausdehnung eines Satzes aus der Theorie der trigonometrischen Reihen. *Mathematische Annalen*, 5: 123–132.

Cantor, Georg (1874). Über eine Eigenschaft des Inbegriffes aller reellen algebraischen Zahlen. *Journal für die reine und angewandte Mathematik*, 77: 258–262.

Cantor, Georg (1883). *Grundlagen einer allgemeinen Mannigfaltigkeitslehre. Ein mathematisch-philosophischer Versuch in der Lehre des Unendlichen.* B.G. Teubner, Leipzig. English translation in Ewald (1996), Vol. 2, pp. 878–920.

Cantor, Georg (1891). Über eine elementare Frage der Mannigfaltigkeitslehre. *Jahresbericht der Deutschen Mathematiker-Vereinigung*, I: 75–78. English translation in Ewald (1996), vol. 2, pp. 920–922.

Cantor, Georg (1915). *Contributions to the Founding of the Theory of Transfinite Numbers.* Open Court. English translation, edted by E.B. Jourdain. Reprint 1955, Dover Publications, New York.

Cantor, Georg and Dedekind, Richard (1937). *Briefwechsel Cantor-Dedekind*, volume 516 of *Actualités scientifiques et industrielles.* Hermann, Paris. Edited by Emmy Noether and Jean Cavaillès.

Cohen, Paul (1963). The independence of the continuum hypothesis I. *Proceedings of the National Academy of Sciences of the United States of America*, 50: 1143–1148.

Cohen, Paul (1964). The independence of the continuum hypothesis II. *Proceedings of the National Academy of Sciences of the United States of America*, 51: 105–110.

Conway, John Horton (2006). The Power of Mathematics. In Blackwell and MacKay (2006), pp. 36–50.

Copeland, B. Jack, Posy, Carl J., and Shagrir, Oron (2013). *Computability: Turing, Gödel, Church, and Beyond.* MIT Press, Cambridge, MA.

Copeland, B. Jack and Proudfoot, Diane (2010). Deviant encodings and Turing's analysis of computability. *Studies in History and Philosophy of Science*, 41 (3): 247–252.

Copeland, B. Jack and Shagrir, Oron (2007). Physical Computation: How General are Gandy's Principles for Mechanisms? *Minds & Machines*, 17: 217–231.

Corry, Leo (2009). Writing the Ultimate Mathematical Textbook: Nicolas Bourbaki's *Éléments de Mathématique*. In Robson and Stedall (2009), pp. 565–588.

Crary, Alice, editor (2007). *Wittgenstein and the Moral Life: Essays in Honor of Cora Diamond.* MIT Press, Cambridge, MA.

Crocco, Gabriella and Engelen, Eva-Maria, editors (2016). *Kurt Gödel: Philosopher-Scientist*, volume 1. Presses Universitaires de Provence, Aix-en-Provence.

Davis, Martin (2004). *The Undecidable: Basic Papers on Undecidable Propositions, Unsolvable Problems and Computable Functions.* Dover Publications, New York. Reprint of original book published in 1965 by Raven Press, Hewlett, New York.

Davis, Martin and Sieg, Wilfried (2015). Conceptual Confluence in 1936: Post and Turing. In Sommaruga and Strahm (2015), pp. 3–28.

De Pellegrin, E., editor (2011). *Interactive Wittgenstein.* Synthese Library. Springer Verlag, New York.

Dedekind, Richard (1872). *Stetigkeit und irrationale Zahlen.* Friedrich Vieweg und Sohn, Wiesbaden. English translation *Continuity and Irrational Numbers* in Ewald (1996), Vol. 2, pp. 766–779.

Dedekind, Richard (1888). *Was sind und was sollen die Zahlen?* Vieweg, Braunschweig. English translation in Ewald (1996), Vol. 2, pp. 790–833.

Dedekind, Richard (1930–1932). *Gesammelte Mathematische Werke*. F. Vieweg und Sohn, Braunschweig. 3 Volumes. Edited by Robert Fricke, Emmy Noether, and Ore Øystein.

Dedekind, Richard (1932). Aus Briefen an R. Lipschitz. In Dedekind (1930 to 1932), Vol. 3, pp. 464–482. Republished in the appendices to Dugac (1976), pp. 215–20.

Dedekind, R. (1985). *Vorlesung über Differential- und Integralrechnung 1861/62*. Braunschweig: F. Vieweg und Sohn.

Deiser, Oliver (2008). *Reelle Zahlen*. Springer, Berlin, 2nd edition.

Detlefsen, Michael and Arana, Andrew (2011). Purity of Methods. *Philosophers' Imprint*, 11 (2): 1–20.

Diamond, Cora (1991). *The Realistic Spirit: Wittgenstein, Philosophy, and the Mind*. MIT Press, Cambridge, MA.

Diamond, Cora (1999). How Old Are These Bones? Putnam, Wittgenstein and Verification. *Aristotelian Society Supplementary Volume*, 73: 99–134.

Diamond, Cora (2018). Wittgenstein, Mathematics, and Ethics: Resisting the Attractions of Realism. In Sluga and Stern (2018), pp. 209–244.

Dieudonné, Jean (1960). *Foundations of Modern Analysis*. Pure and Applied Mathematics. Academic Press, Cambridge, MA.

Dixon, Alfred Cardew (1906). On 'well-ordered' aggregates. *Proceedings of the London Mathematical Society*, Second Series: 18–20.

Dreben, Burton and Floyd, Juliet (1991). Tautology: How Not to Use A Word. *Synthese*, 87 (1): 23–50.

Dugac, Pierre (1976). *Richard Dedekind et les Fondaments des Mathématiques*. Vrin, Paris.

Dummett, Michael (1991a). *Frege: Philosophy of Mathematics*. Harvard University Press, Cambridge, MA.

Dummett, Michael (1991b). *The Logical Basis of Metaphysics*. Harvard University Press, Cambridge, MA.

Dummett, Michael (1993). *The Seas of Language*. Oxford University Press, Oxford.

Dybjer, P., Lindström, S., Palmgren, E., and Sundholm, G., editors (2012). *Epistemology versus Ontology, Logic, Epistemology: Essays in Honor of Per Martin-Löf*. Logic, Epistemology, and the Unity of Science. Springer Science+Business Media, Dordrecht.

Ebbinghaus, Heinz-Dieter (1996). *Numbers*. Springer, New York. Co-authored with Hans Hermes, Friedrich Hirzebruch, Max Koecher, Klaus Lamotke, Klaus Mainzer, Jürgen Neukirch, Alexander Prestel and Reinhold Remmert.

Eilenberg, Samuel and MacLane, Saunders (1942). Natural Isomorphisms in Group Theory. *Proceedings of the National Academy of Sciences USA*, 28: 537–543.

Eilenberg, Samuel and MacLane, Saunders (1945). General Theory of Natural Equivalences. *Transactions of the American Mathematical Society*, 58: 231–294.

Eliot, T.S., Dickey, F., Formichelli, J., and Schuhard, R., editors (2015). *The Complete Prose of T.S. Eliot: The Critical Edition: Literature Politics, Belief 1927–1929*. Johns Hopkins University Press, Baltimore.

Eliot, T. S. (1928). *The Sacred Wood*. Methuen and Co. Ltd., London. Preface to 2nd ed. reprinted in Eliot et al. (2015), pp. 413–415.

Ellis, Jonathan and Guevara, Daniel, editors (2012). *Wittgenstein and the Philosophy of Mind*. Oxford University Press, New York/Oxford U.K.

Engelmann, Mauro Luiz (2013). *Wittgenstein's Philosophical Development: Phenomenology, Grammar, Method, and the Anthropological View*. Palgrave History of Analytic Philosophy. Palgrave Macmillan, Basingstoke, UK.

Epple, Moritz (1999). Das Ende der Größenlehre: Grundlagen der Analysis 1860–1910. in Jahnke (1999), pp. 371–410.

Ewald, William, editor (1996). *From Kant to Hilbert: A Source Book in the Foundations of Mathematics*. Oxford University Press, New York. 2 Volumes.

Fan, Zhao (2020). Hobson's Conception of Definable Numbers. In *History and Philosophy of Logic*, DOI 10.1080/01445340.2020.1731784.

Feferman, Solomon (1988). Weyl Vindicated: Das Kontinuum Seventy Years Later. *Temi e prospettive dalla logica e della scienza contemporanee*, I: 59–93. Reprinted in Feferman (1998), pp. 249–283.

Feferman, Solomon (1998). *In the Light of Logic*. Oxford University Press, Oxford.

Floyd, Juliet (1991). Wittgenstein on 2,2,2...: On the Opening of *Remarks on the Foundations of Mathematics*. *Synthese*, 87 (1): 143–180. Reprinted in Hintikka (1991), pp. 23–50.

Floyd, Juliet (1995). Wittgenstein, Gödel and the Trisection of the Angle. In Hintikka (1995), pp. 373–426.

Floyd, Juliet (2001a). Number and Ascriptions of Number in the *Tractatus*. In Floyd and Shieh (2001), pp. 145–192.

Floyd, Juliet (2001b). Prose versus Proof: Wittgenstein on Gödel, Tarski and Truth. *Philosophia Mathematica*, 3 (9): 280–307.

Floyd, Juliet (2004). Wittgenstein on Philosophy of Logic and Mathematics. *The Graduate Faculty Philosophy Journal, The New School of Social Research*, 25 (2): 227–287. Reprinted in Shapiro (2005), pp. 75–128.

Floyd, Juliet (2010). On Being Surprised: Wittgenstein on Aspect Perception, Logic and Mathematics. In Krebs and Day (2010), pp. 314–337.

Floyd, Juliet (2012a). Das Überraschende: Wittgenstein on the Surprising in Mathematics. In Ellis and Guevara (2012), pp. 224–258.

Floyd, Juliet (2012b). Wittgenstein, Carnap, and Turing: Contrasting Notions of Analysis. In Wagner (2012), pp. 34–46.

Floyd, Juliet (2012c). Wittgenstein's Diagonal Argument: A Variation on Cantor and Turing. In Dybjer et al. (2012), pp. 25–44.

Floyd, Juliet (2013a). The Varieties of Rigorous Experience. In Beaney (2013), pp. 1003–1042.

Floyd, Juliet (2013b). Turing, Wittgenstein and Types: Philosophical Aspects of Turing's 'The Reform of Mathematical Notation' (1944–5). In Turing (2013), pp. 250–253.

Floyd, Juliet (2015). Depth and Clarity: Critical Study, Felix Mühlhölzer, *Braucht die Mathematik eine Grundlegung? Eine Kommentar des Teils III von Wittgensteins* BEMERKUNGEN ÜBER DIE GRUNDLAGEN DER MATHEMATIK (Vittorio Klostermann, Frankfurt am Main 2010). *Philosophia Mathematica*, 23, Special Issue on Mathematical Depth (2): 255–277.

Floyd, Juliet (2016). Chains of Life: Turing, *Lebensform*, and the Emergence of Wittgenstein's Later Style. *Nordic Wittgenstein Review*, 5 (2): 7–89.

Floyd, Juliet (2017). Turing on "Common Sense": Cambridge Resonances. In Floyd and Bokulich (2017), pp. 103–152.

Floyd, Juliet (2018a). Aspects of Aspects. In Sluga and Stern (2018), pp. 361–388.

Floyd, Juliet (2018b). *Lebensformen*: Living logic. In Martin (2018), pp. 59–92.

Floyd, Juliet (2019a). "Ultimate" Facts? Zalabardo on the Metaphysics of Truth. *Australasian Philosophical Review*, 2: 299–314.

Floyd, Juliet (2019b). Wittgenstein and Turing. In Mras et al. (2019), pp. 263–296.

Floyd, J. (2020, in press). Aspects of the Real Numbers: Putnam, Wittgenstein and Non-Extensionalism. The Monist.

Floyd, Juliet and Bokulich, Alisa, editors (2017). *Philosophical Explorations of the Legacy of Alan Turing — Turing 100*. Boston Studies in the Philosophy of Science. Springer, Dordrecht.

Floyd, Juliet and Kanamori, Akihiro (2016). Gödel vis-à-vis Russell: Logic and Set Theory to Philosophy. In Crocco and Engelen (2016), pp. 243–326.

Floyd, Juliet and Putnam, Hilary (2000). A Note on Wittgenstein's 'Notorious Paragraph' about the Gödel Theorem. *Journal of Philosophy*, 45 (11): 624–632.

Floyd, Juliet and Putnam, Hilary (2012). Wittgenstein's 'Notorious' Paragraph about the Gödel Theorem: Recent Discussions. In Putnam (2012), pp. 458–481.

Floyd, Juliet and Shieh, Sanford, editors (2001). *Future Pasts: Perspectives on the Analytic Tradition in Twentieth Century Philosophy*. Oxford University Press, New York.

Fraenkel, Abraham (1935). Zum Diagonalverfahren Cantors. *Fundamenta Mathematicae*, 25: 45–50.

Franchella, Miriam (2016). In the Footsteps of Julius König's Paradox. *Historia Mathematica*, 43: 65–86.
Frazer, James George (1917). *The Golden Bough*. Macmillan, London. 3rd edition, revised and enlarged; 1st ed. 1894, 2nd ed. revised and enlarged 1900; compare the 1922 abridged edition, Macmillan, New York.
Frege, Gottlob (1879). *Begriffsschrift, eine der arithmetischen nachgebildete Formelsprache des reinen Denkens*. Louis Nebert, Halle. Reprinted 1964 by G. Olms, Hildesheim. English translation by Stefan Bauer-Mengelberg in van Heijenoort (1967), pp. 5–82 and Preface and Part I by Michael Beaney in Frege (1997), pp. 47–78.
Frege, Gottlob (1891). *Funktion und Begriff*. Hermann Pohle, Jena. English translation by Peter Geach, in Frege (1997), pp.130–148.
Frege, Gottlob (1892a). Über Begriff und Gegenstand. *Vierteljahrsschrift für Wissenschaftliche Philosophie*, 16: 192–205. English translation by Peter Geach, "On Concept and Object", in Frege (1997), pp. 182–194. Quotations are from this translation.
Frege, Gottlob (1892b). Über Sinn und Bedeutung. *Zeitschrift für Philosophie und philosophische Kritik*, 100: 25–50. English Translation by Max Black, "On *Sinn* and *Bedeutung*", in Frege (1997), pp. 151–171.
Frege, Gottlob (1893/1903). *Grundgesetze der Arithmetik; Begriffsschriftlich Abgeleitet, Volumes I and II*. Verlag von Hermann Pohle, Jena. English translation, and edited by, Philip A. Ebert, Marcus Rossberg and Crispin Wright, Oxford University Press, Oxford, 2013.
Frege, Gottlob (1897). Logik. In Frege (1969), pp. 137–163. English translation by Peter Long and Roger White, "Logic (1897)", in Frege (1997), pp. 227–250.
Frege, Gottlob (1969). *Frege: Nachgelassene Schriften, Band I*. Felix Meiner Verlag, Hamburg. Edited by Hans Hermes, Friedrich Kambartel and Friedrich Kaulbach.
Frege, Gottlob (1997). *The Frege Reader*. Blackwell Publishers, Malden, MA. Edited by Michael Beaney.
Frege, Gottlob (2011). Letters to Ludwig Wittgenstein. In De Pellegrin (2011), pp. 15–73. English translation by Burton Dreben and Juliet Floyd, edited with an introduction by Juliet Floyd. Original German publication in McGuinness and Haller (1989), pp. 5–33.
Friederich, Simon (2011). Motivating Wittgenstein's Perspective on Mathematical Sentences as Norms. *Philosophia Mathematica*, III (19): 1–19.
Friederich, Simon (2014). Warum die Mathematik keine ontologische Grundlegung braucht: Wittgenstein und die axiomatische Methode. *Wittgenstein-Studien*, 5: 163–178.
Gabbay, Dov M., Kanamori, Akihiro, and Woods, J.H., editors (2012). *Sets and Extensions in the Twentieth Century*, volume 6 of *Handbook for the History of Logic*. Cambridge University Press, Cambridge.
Gandy, R.O. (1988). The Confluence of Ideas in 1936. In Herken (1988), pp. 55–112.
Gaskin, Richard and Hill, Daniel J. (2013). Reach's Puzzle and Mention. *Dialectica*, 67 (2): 201–222.
George, Alexander and Velleman, Daniel J. (2002). *Philosophies of Mathematics*. Blackwell, Oxford.
Gibson, Arthur (2010). Francis Skinner's Original Wittgenstein *Brown Book* Manuscript. In Munz et al. (2010), pp. 351–366.
Gödel, Kurt (1931). Über formal unentscheidbare Sätze der Principia Mathematica und verwandter System I. *Monatshefte für Mathematik und Physik*, 38: 173–198. Reprinted with English translation in Gödel (1986), p. 144–195.
Gödel, Kurt (1938). The Consistency of the Axiom of Choice and of the Generalized Continuum-Hypothesis. *Proceedings of the National Academy of Sciences of the United States of America*, 24: 556–557. Reprinted in Gödel (1990), p. 234–237.
Gödel, Kurt (1946). Remarks Before the Princeton Bicentennial Conference on Problems in Mathematics. Reprinted with English translation in Gödel (1990), pp. 150–153.
Gödel, Kurt (1958). Über eine bisher noch nicht benützte Erweiterung des finiten Standpunktes. *Dialectica*, 12 (7): 280–287. Reprinted with English translation as "On a hitherto unutilized extension of the finitary standpoint" in Gödel (1990), pp. 240–251.

Gödel, Kurt (1964a). On Undecideable Propositions of Formal Mathematical Systems. Lectures given at the Institute for Advanced Study, February-May 1934. Transcribed by S.C. Kleene and J.B. Rosser, later published in Davis (2004, pp. 39–74) and republished in Gödel (1986, pp. 346–369), with a Postscriptum, namely Gödel (1964b).

Gödel, Kurt (1964b). Postscriptum to the 1934 Princeton Lectures, 1964. Postscriptum added 3 June 1964 to the reprinting of Gödel (1964a) in Davis (2004, pp. 71–74), published as a Postscript to Gödel(1964a) in Gödel (1986, pp. 369–371).

Gödel, Kurt (1986). *Collected Works, Volume I: Publications 1929–1936*. Clarendon Press, Oxford. Edited by Solomon Feferman *et al.*

Gödel, Kurt (1990). *Collected Works, Volume II: Publications 1938–1974*. Clarendon Press, Oxford. Edited by Solomon Feferman *et al.*

Goldfarb, Warren (1992). Wittgenstein on Understanding. *Midwest Studies In Philosophy*, 17 (1): 109–122.

Goldstein, Catherine, Schappacher, Norbert, and Schwemmer, Joachim, editors (2007). *The Shaping of Arithmetic after C.F. Gauss's Disquisitiones Arithmeticae*. Springer, Berlin, Heidelberg.

Good, I. J. (1966). A Note on Richard's Paradox. *Mind*, 75 (299): 431–431.

Gowers, Timothy (2017a). A Dialogue Concerning the Need for the Real Number System. At https://www.dpmms.cam.ac.uk/~wtg10/reals.html, accessed January 30, 2017.

Gowers, Timothy (2017b). What is so wrong with thinking of real numbers as infinite decimals? At https://www.dpmms.cam.ac.uk/~wtg10/decimals.html, accessed January 30, 2017.

Grattan-Guiness, Ivor (2001). The Interest of G. H. Hardy, F.R.S., in the Philosophy and the History of Mathematics. *Notes of the Royal Society of London*, 55: 411–424.

Grattan-Guiness, Ivor (2013). The Mentor of Alan Turing: Max Newman (1897–1984) as a Logician. *The Mathematical Intelligencer*, 35: 55–63.

Grauert, Hans and Lieb, Ingo (1967). *Differential- und Integralrechnung I*. Springer, Berlin.

Gray, Robert (1994). Georg Cantor and Transcendental Numbers. *American Mathematical Monthly*, 101: 819–832.

Grève, Sebastian and Mühlhölzer, Felix (2014). Interview: Wittgenstein's Philosophy of Mathematics, Felix Mühlhölzer in Conversation with Sebastian Grève. *Nordic Wittgenstein Review*, 3 (2): 151–180.

Grève, Sebastian Sunday and Mácha, Jakub, editors (2016). *Wittgenstein and the Creativity of Language*. Palgrave Macmillan, Basingstoke/New York.

Hairer, E. and Wanner, G. (1996). *Analysis by Its History*. Springer, New York.

Haller, Rudolf and Puhl, Klaus, editors (2002). *Wittgenstein and the Future of Philosophy. A Reassessment after 50 years*. Proceedings of the 24th International Wittgensein Symposium, 2001. öbv&hpt, Vienna.

Hallett, Michael (2010). Frege and Hilbert. In Potter and Ricketts (2010), pp. 413–464.

Halmos, Paul R (1960). *Naive Set Theory*. D. Van Nostrand, New York.

Hammel-Haider, Gabriele and Swennen, Leonardus, editors (1989). *Wittgenstein: Biographie-Philosophie-Praxis, Wiener Secession*. Wiener Secession, Vienna. Catalogue of an Exhibition held at the Wiener Secession 13 September-29 October 1989.

Hardy, G. H. (1929). Mathematical Proof. *Mind*, 38 (149): 1–25.

Hardy, G. H. (1934). Obituary of Professor E.W. Hobson, F.R.S. *Nature*, 133: 938–939.

Hardy, G. H. (1940). *A Mathematician's Apology*. The University Press, Cambridge Eng.

Hardy, G. H. (1941). *A Course of Pure Mathematics*. Cambridge University Press, Cambridge, 8th edition. 10th edition (1952, with index) republished as Hardy (2008).

Hardy, G. H. (2008). *A Course of Pure Mathematics*. Cambridge University Press, Cambridge, 10th (1952, with index) edition.

Hardy, G. H. and Wright, E.M. (1938). *An Introduction to the Theory of Numbers*. Oxford University Press, Oxford. 2nd ed. 1945, 3rd ed. 1954, 4th ed. 1960.

Havil, Julian (2012). *The Irrationals*. Princeton University Press, Princeton, NJ.

Hellman, G. and Shapiro, D., editors (2020, in press). *The History of the Continuum: Philosophical and Mathematical Perspectives*. New York: Oxford University Press.

Herken, Rolf, editor (1988). *The Universal Turing Machine: A Half-Century Survey*. Oxford Science Publications. Oxford University Press, New York.

Hesseling, Dennis E. (2003). *Gnomes in the Fog: The Reception of Brouwer's Intuitionism in the 1920s*. Birkhäuser, Basel and Boston.

Heuser, Harro (2009). *Lehrbuch der Analysis Teil 1*. Vieweg + Teubner, Wiesbaden, 17th edition.

Hewitt, Edwin and Stromberg, Karl (1965). *Real and Abstract Analysis: A Modern Treatment of the Theory of Functions of a Real Variable*. Springer, Berlin. With corrections in the second printing of 1969.

Hilbert, David (1926). Über das Unendliche. *Mathematische Annalen*, 95: 161–190. French translation by André Weil in *Acta Mathematica* 48:91–122, 1926; English translation in van Heijenoort (1967), pp. 367–392.

Hintikka, Jaakko, editor (1991). *Wittgenstein in Florida*. Springer Netherlands, Dordrecht.

Hintikka, Jaakko, editor (1995). *From Dedekind to Gödel: Essays on the Development of the Foundations of Mathematics*. Synthese Library. Kluwer Academic, Dordrecht.

Hintikka, Jaakko and Puhl, Klaus, editors (1995). *The British Tradition in 20th Century Philosophy: Proceedings of the 17th International Wittgenstein Symposium, Kirchberg am Wechsel*. Hölder-Pichler-Tempsky, Vienna.

Hobson, E.W. (1905). On the general theory of transfinite numbers and order types. *Proceedings of the London Mathematical Society*, 3: 170–188.

Hobson, E.W. (1906). On the arithmetic continuum. *Proceedings of the London Mathematical Society, 2nd Series*, 4: 21–28. Delivered in 1905, submitted in 1906, first published in the March 1907 volume of this journal.

Hobson, E. W. (1907). *The Theory of Functions of a Real Variable and the Theory of Fourier's Series*, volume I. Cambridge University Press, Cambridge, U.K. 1st edition.

Hobson, Ernest William (1913). *"Squaring the Circle"; a History of the Problem*. Cambridge University Press, Cambridge.

Hobson, Ernest William (1921). *The Theory of Functions of a Real Variable and the Theory of Fourier's Series*, volume I. University Press, Cambridge. Revised and enlarged 2nd edition.

Hobson, Ernest William (1926). *The Theory of Functions of a Real Variable and the Theory of Fourier's Series*, volume II. University Press, Cambridge. Revised and enlarged 2nd edition.

Hobson, E. W. (1927). *The Theory of Functions of a Real Variable and the Theory of Fourier's Series*, volume I. Cambridge University Press, Cambridge, U.K. Revised and enlarged 3rd edition.

Hodges, Andrew (2012). *Alan Turing: The Enigma: the Centenary Edition*. Princeton University Press, Princeton, NJ. First edition 1983.

Hodges, Wilfrid (1998). An Editor Recalls Some Hopeless Papers. *Bulletin of Symbolic Logic*, 4 (1): 1–16.

Hrbacek, Karel and Jech, Thomas J. (1999). *Introduction to Set Theory*. Monographs and Textbooks in Pure and Applied Mathematics 220. CRC / Taylor and Francis, Boca Raton, 3rd revised and expanded edition.

Irvine, A.D., editor (2009). *Philosophy of Mathematics*. Handbook of the Philosophy of Science. Elsevier, Amsterdam/Oxford.

Jahnke, Hans Niels, editor (1999). *Geschichte der Analysis*. Spektrum Akademischer Verlag, Wiesbaden.

Jordan, Marie Ennemond Camille (1893). *Cours d'analyse de l'École Polytechnique, Deuxiéme édition, Entièrement Refondue*. Gauthier-Villars et Fils, Paris.

Kac, Mark and Ulam, Stanislaw M. (1968). *Mathematics and Logic*. Praeger, New York.

Kahane, Guy, Kanterian, Edward, and Kuusela, Oskari, editors (2007). *Wittgenstein and his Interpreters: Essays in Memory of Gordon Baker*. Blackwell Publishing, Malden, MA/Oxford.

Kanamori, Akihiro (2004). Zermelo and set theory. *The Bulletin of Symbolic Logic*, 10: 487–553.

Kanamori, Akihiro (2012a). Set Theory from Cantor to Cohen. In Gabbay et al. (2012), pp. 1–71. Revised version of "The mathematical development of set theory from Cantor to Cohen", *The Bulletin of Symbolic Logic* Vol. 2, No. 1 (March 1996): 1–71 and "Set Theory from Cantor to Cohen" in Irvine (2009), pp. 395–459.

Kanamori, Akihiro (2012b). The Mathematical Infinite as a Matter of Method. *Annals of the Japan Association for Philosophy of Science*, 20: 1–13.

Kanamori, Akihiro (2018a). Aspect-perception and the history of mathematics. In Beaney et al. (2018), pp. 109–132.

Kanamori, Akihiro (2018b). Kreisel and Wittgenstein. To appear in a conference proceedings edited by Paul Weingartner, College Publications, 2019.

Kanamori, Akihiro (2020, in press). Cantor and Continuity. In Shapiro (2020).

Kant, Immanuel (1998). *Critique of Pure Reason*. The Cambridge Edition of the Works of Immanuel Kant. Cambridge University Press, Cambridge, U.K./New York. Edited and translated into English by Allen W. Wood and Paul Guyer. 1st German edition 1781, 2nd edition 1787, Friedrich Hartknoch, Riga.

Kennedy, Juliette (2017). Turing, Gödel and the "Bright Abyss". In Floyd and Bokulich (2017), pp. 63–92.

Kennedy, Juliette (2020, in press). *Gödel, Tarski and the Lure of Natural Language: Logical Entanglement, Formalism Freeness*. Cambridge University Press, Cambridge, U.K. Manuscript, expected publication 2020.

Kennedy, Juliette and Kossak, Roman (2011). *Set Theory, Arithmetic and Foundations of Mathematics: Theorems, Philosophies*. Lecture Notes in Logic: Association for Symbolic Logic. Cambridge University Press, Cambridge, U.K./New York.

Khan, Ilyas (2014). Web publication. Manuscript circulated on the web, accessed on October 29, 2018 at https://www.jbs.cam.ac.uk/fileadmin/user_upload/fellow_associate/downloads/2014-khan-wittgensteininconsistencymathematics.pdf; cf. related lecture at the University of Navarra, a YouTube video at https://www.youtube.com/watch?v=N1EI5Vt9UM0.

Kienzler, Wolfgang (1997). *Wittgensteins Wende zu seiner Spätphilosophie 1930–1932: Eine historische und systematische Darstellung*. Suhrkamp, Frankfurt am Main.

Kienzler, Wolfgang (2001). About the Dividing Line between Early and Late Wittgenstein. In Oliveri (2001), pp. 125–130.

Kienzler, Wolfgang and Grève, Sebastian Sunday (2016). Wittgenstein on Gödel: Reading *Remarks on the Foundations of Mathematics*, Part I, Appendix III, Carefully. In Grève and Mácha(2016), pp. 76–116.

Klagge, James Carl and Nordmann, Alfred (1993). *Ludwig Wittgenstein: Philosophical Occasions, 1912–1951*. Hackett Publishing Company, Indianapolis.

Klagge, James C. and Nordmann, Alfred (2003). *Ludwig Wittgenstein: Public and Private Occasions*. Rowman and Littlefield Publishers, Lanham, MD.

Kleene, Stephen C. (1952). *Introduction to Metamathematics*. van Nostrand, New York.

Kölbel, Max and Weiss, Bernhard, editors (2004). *Wittgenstein's Lasting Significance*. Routledge, London.

König, Julius (1905). Über die Grundlagen der Mengenlehre und das Kontinuumproblem. *Mathematische Annalen*, 61: 156–160.

Körner, T. W. (2004). *A Companion to Analysis: A Second First and First Second Course in Analysis*. American Mathematical Society.

Krebs, V. and Day, W., editors (2010). *Seeing Wittgenstein Anew: New Essays on Aspect Seeing*. Cambridge University Press, New York.

Kreisel, Georg (1950). Note on arithmetic models for consistent formulae of the predicate calculus. *Fundamenta Mathematicae*, 37: 265–285.

Kreisel, Georg (1989). Zu einigen Gesprächen mit Wittgenstein: Erinnerungen und Gedanken. In Hammel-Haider and Swennen(1989), pp. 131–143.

Kross, Matthias (2008). *«Ein Netz von Normen»: Wittgenstein und die Mathematik*. Wittgensteinsteiniana. Parerga Verlag, Berlin.

Kuusela, Oskari and McGinn, Marie, editors (2012). *The Oxford Handbook of Wittgenstein*. Oxford University Press.

Landau, Edmund (1930). *Grundlagen der Analysis*. Akademische Verlagsgesellschaft, Leipzig. English translation by F. Steinhardt, *Foundations of Analysis*, American Mathematical Society, Chelsea Publishing, 2001.

Lemco, Ian (2007). Wittgenstein's Aeronautical Investigation. *Notes and Records of the Royal Society of London*, 61: 39–51.

Lewy, Casimir (1967). A Note on the Text of the *Tractatus*. *Mind*, 76 (303): 416–423.

Link, Godehard, editor (2014). *Formalism and Beyond. On the Nature of Mathematical Discourse*. Walter de Gruyter, Berlin.

Littlewood, J.E. (1926). *The Elements of the Theory of Real Functions, Being Notes of Lectures delivered in the University of Cambridge, 1925*. W. Heffer and Sons, Cambridge, 2nd. edition.

Luna, Laureano (2017). Rescuing Poincaré from Richard's Paradox. *History and Philosophy of Logic*, 38: 57–71.

Majer, Ulrich (1988). Zu einer bemerkenswerten Differenz zwischen Brouwer und Weyl. In , pp. 543–552.

Mancosu, Paolo, editor (1998). *From Brouwer to Hilbert: the Debate on the Foundations of Mathematics in the 1920s*. Oxford University Press, New York.

Marion, Mathieu (1998). *Wittgenstein, Finitism and the Foundations of Mathematics*. Oxford, Clarendon Press, New York.

Marion, Mathieu (2003). Wittgenstein and Brouwer. *Synthese*, 137: 103–127.

Marion, Mathieu and Okada, Mitsuhiro (2018). Wittgenstein, Goodstein and the Origin of the Uniqueness Rule. In Stern (2018), pp. 253–271.

Martin, Christian, editor (2018). *Language, Form(s) of Life, and Logic: Investigations after Wittgenstein*. de Gruyter, Berlin.

McGuinness, Brian (1988). *Wittgenstein: A Life: Young Ludwig 1889–1921*. University of California Press, Berkeley, CA.

McGuinness, Brian (2008). *Wittgenstein in Cambridge: Letters and Documents, 1911–1951*. Blackwell, Malden, MA, Oxford, 4th edition.

McGuinness, Brian and Haller, Rudolf, editors (1989). *Wittgenstein in Focus - Im Brennpunkt Wittgenstein*, volume 33–34 of *Grazer Philosophische Studien*. Rodopi, Amsterdam.

Meltzer, B. and Michie, D., editors (1969). *Machine Intelligence 5*. Edinburgh University Press, Edinburgh.

Menger, Karl (1994). *Reminiscences of the Vienna Circle and the Mathematical Colloquium*, volume 20 of *Vienna Circle Collection*. Kluwer Academic/Springer Science+Business Media, Dordrecht.

Miller, Steven J. and Montague, David (2009). Irrationality from the book. Electronic publication at <http://arxiv.org/abs/0909.4913v1>.

Misak, Cheryl (2016). *Cambridge Pragmatism: From Peirce and James to Ramsey and Wittgenstein*. Oxford University Press, New York.

Monk, Ray (1990). *Ludwig Wittgenstein: The Duty of Genius*. Free Press/Jonathan Cape, New York/London.

Monk, Ray (2007). Bourgeois, Bolshevist or Anarchist? The Reception of Wittgenstein's Philosophy of Mathematics. In Kahane et al.(2007), pp. 269–294.

Moore, G.E. (2016). *Wittgenstein: Lectures, Cambridge 1930–1933, From the Notes of G.E. Moore*. Cambridge University Press, Cambridge. Edited by David Stern, Brian Rogers and Gabriel Citron.

Moore, Gregory H. (1982). *Zermelo's Axiom of Choice: Its Origins, Development, and Influence*. Springer-Verlag, New York.

Mras, Gabriele M., Ritter, Bernhard, and Weingartner, Paul, editors (2019). *Philosophy of Logic and Mathematics Proceedings of the 41st International Ludwig Wittgenstein Symposium*. DeGruyter, Berlin and Boston.

Mundici, Daniel and Sieg, Wilfried (2017). Turing, the Mathematician. In Floyd and Bokulich (2017), pp. 39–62.

Munz, Volker, Puhl, Klaus, and Wang, J., editors (2010). *Language and World, Part One, Essays on the Philosophy of Wittgenstein*, volume 14 of *Publications of the Austrian Ludwig Wittgenstein Society New Series*. Ontos Verlag, Vienna.

Mühlhölzer, Felix (2001). Wittgenstein and the Regular Heptagon. *Grazer Philosophische Studien*, 62: 215–247.

Mühlhölzer, Felix (2002). Wittgenstein and Surprises in Mathematics. In Haller and Puhl(2002), pp. 306–315.
Mühlhölzer, Felix (2006). 'A mathematical proof must be surveyable': What Wittgenstein Meant by This and What it Implies. *Grazer Philosophische Studien*, 71: 57–86.
Mühlhölzer, Felix (2010). *Braucht die Mathematik eine Grundlegung? Ein Kommentar des Teils III von Wittgensteins BEMERKUNGENG ÜBER DIE GRUNDLAGEN DER MATHEMATIK*. Vittorio Klostermann, Frankfurt am Main.
Mühlhölzer, Felix (2012). Wittgenstein and Metamathematics. In Stekeler-Weithofer (2012), pp. 103–128.
Mühlhölzer, Felix (2014a). How Arithmetic is about Numbers: A Wittgensteinian Perspective. *Grazer Philosophische Studien*, 89: 39–59.
Mühlhölzer, Felix (2014b). On Live and Dead Signs in Mathematics. In Link (2014), pp. 183–208.
Mühlhölzer, Felix (2014c). 'Seltsame Frage. –' – Eine Interpretation von PU §263. *Wittgenstein-Studien*, 05: 145–162.
Nedo, Michael (2008). Anmerkungen zu Wittgensteins *Bemerkungen über die Grundlagen der Mathematik*. In Kross (2008), pp. 80–103.
Nedo, Michael and Ranchetti, Michele (1983). *Ludwig Wittgenstein: Sein Leben in Bildern und Texten*. Wittgenstein. Surhkamp, Frankfurt am Main, 1st. edition.
Oliveri, Gianluigi, editor (2001). *From the Tractatus to the Tractatus and Other Essays*. Peter Lang, Sonderdruck.
Olszewski, A., Wolenski, J., and Janusz, R., editors (2006). *Church's Thesis After 70 Years*. Ontos Verlag, Frankfurt/Paris/Ebikon/Lancaster/New Brunswick.
Peano, Giupseppe (1906). Super theorema de cantor-bernstein. *Revista de mathematica*, 8: 136–143. English translation in Peano (1973), Chapter XVIII, pp. 206–218.
Peano, Giuseppe (1973). *Selected Works of Giuseppe Peano*. University of Toronto Press, Toronto. English translation, edited by Hubert C. Kennedy.
Penrose, Roger (1994). *Shadows of the Mind*. Oxford University Press, New York/Oxford.
Petri, Birgit and Schappacher, Norbert (2007). On Arithmetization. In Goldstein et al. (2007), pp. 343–374.
Pettigrew, Richard (2008). Platonism and Aristotelianism in Mathematics. *Philosophia Mathematica*, 16: 310–322.
Petzold, Charles (2008). *The Annotated Turing: A Guided Tour through Alan Turing's Historic Paper on Computability and the Turing Machine*. Wiley Publishing, Inc., Indianapolis, IN.
Pichler, Alois (2004). *Wittgensteins Philosophische Untersuchungen: Vom Buch zum Album*. Studien zur Österreichischen Philosophie 36. Rodopi, Amsterdam/New York.
Pichler, Alois, Biggs, Michael A.R., and Szeltner, Sarah Anna (2011). Bibliographie der deutsch- und englischsprachigen Wittgenstein-Ausgaben. Technical report, University of Bergen, University of Hertfordshire.
Poincaré, Henri (1905). Les mathématiques et la logique I. *Revue de Métaphysique et de Morale*, 6 (13): 815–835. English translation of selections in Ewald (1996), Vol. 2, pp. 1021–1038.
Poincaré, Henri (1906a). Les mathématiques et la logique II. *Revue de Métaphysique et de Morale*, 3: 17–34. English translation of selections in Ewald (1996), Vol. 2, pp. 1038–1052.
Poincaré, Henri (1906b). Les mathématiques et la logique III. *Revue de Métaphysique et de Morale*, 3: 294–317. English translation of selections in Ewald (1996), Vol. 2, pp. 1052–1071.
Poincaré, Henri (1909). La logique de l'infini. In Poincaré (1917), pp. 101–139. A 1909 lecture given in Göttingen.
Poincaré, Henri (1917). *La Logique de l'Infini*. Flammarion, Paris.
Potter, Michael and Ricketts, Tom, editors (2010). *The Cambridge Companion to Frege*. Cambridge University Press, Cambridge.
Priest, Graham (2004). Wittgenstein's Remarks on Gödel's Theorem. In Kölbel and Weiss (2004), pp. 207–227.
Putnam, Hilary (2007). Wittgenstein and the Real Numbers. In Crary (2007), pp. 235–250.
Putnam, Hilary (2012). *Philosophy in An Age of Science: Physics, Mathematics, and Skepticism*. Harvard University Press, Cambridge, MA. Edited by Mario De Caro and David Macarthur.

Ramsey, Frank Plumpton (1923). Review of *Tractatus Logico-Philosophicus* by Ludwig Wittgenstein. *Mind*, 32 (128): 465–478.
Ramsey, Frank Plumpton (1926). The Foundations of Mathematics. *Proceedings of the London Mathematical Society, Series 2*, 25 (1): 338–384. Reprinted in Ramsey (1931), pp. 1–61.
Ramsey, Frank Plumpton (1929). Last Papers. In Ramsey (1931), pp. 212–269.
Ramsey, Frank Plumpton (1931). *The Foundations of Mathematics and other Logical Essays*. Routledge and Kegan Paul, London. Edited by R.B. Braithwaite.
Ramsey, Frank Plumpton (1991a). *Notes on Philosophy, Probability, and Mathematics*. Bibliopolis, Napoli. Edited by Maria Galivotti.
Ramsey, Frank Plumpton (1991b). *On Truth: Original Manuscript Materials (1927–1929) from the Ramsey Collection at the University of Pittsburgh*, volume 16 of *Episteme: a Series in the Foundational, Methodological, Philosophical, Psychological, Sociological, and Political Aspects of the Sciences, Pure and Applied*. Kluwer Academic Publishers, Dordrecht and Boston.
Rayo, A., editor (2006). *Absolute Generality*. Oxford University Press, Oxford.
Reach, K. (1938). The name relation and the logical antinomies. *Journal of Symbolic Logic*, 3 (3): 97–111.
Read, Sephen (1997). Quotation and Reach's Puzzle. *Acta Analytica*, 19: 9–20.
Richard, Jules (1905). Les principes des mathématiques et le problème des ensembles. *Revue génerale des sciences pures et appliquées*, 16: 541. English translation by Jean van Heijenoort in van Heijenoort (1967), pp. 142–144. References are to this translation.
Richards, I.A. (1928). *Principles of Literary Criticism*. Routledge & Kegan Paul, London.
Ricketts, Thomas (2014). Analysis, Independence, Simplicity and the General Sentence-Form. *Philosophical Topics*, 42 (2): 263–288.
Robson, Eleanor and Stedall, Jacqueline A., editors (2009). *The Oxford Handbook of the History of Mathematics*. Oxford University Press, Oxford.
Rogers, Brian and Wehmeier, Kai (2012). Tractarian First-Order Logic: Identity and the N-Operator. *Review of Symbolic Logic*, 5 (4): 538–573.
Rogers, Hartley (1967). *Theory of Recursive Functions and Effective Computability*. McGraw-Hill, New York.
Russell, Bertrand (1903). *The Principles of Mathematics*. The Norton Library. Cambridge University Press/Norton, Cambridge/New York. 1st edition 1903, 2nd edition 1938.
Russell, Bertrand (1905/6). On some difficulties in the theory of transfinite numbers and order types. *Proceedings of the London Mathematical Society*, 4: 29–53. Reprinted in Russell (2014), pp. 62–89.
Russell, Bertrand (1920). *Introduction to Mathematical Philosophy*. Allen and Unwin/Macmillan, London/New York, 2nd edition. 1st edition 1919.
Russell, Bertrand (1936). The Limits of Empiricism. *Proceedings of the Aristotelian Society*, 36: 131–150.
Russell, B. (2014). *Toward "Principia Mathematica", 1905–08*, volume 5 of *The Collected Papers of Bertrand Russell*. Routledge, London. Edited by Gregory H. Moore.
Sahlin, Nils-Eric (1995). On the Philosophical Relations between Ramsey and Wittgenstein. In Hintikka and Puhl (1995), pp. 150–163.
Schappacher, Norbert (2008). Bemerkungen zu Wittgensteins Philosophie der Mathematik in ihrer Beziehung zur Mathematikgeschichte. In Kross (2008), pp. 199–234.
Schopenhauer, Arthur (1988). *Sämtliche Werke, nach er ersten, von Julius Frauenstädt besorgten Gesamtausgabe neu bearbeitet und herausgegeben von Arthur Hübscher, vierte Auflage, durchgesehen von Angelika Hübscher, 7 vols.* Brockhaus, Mannheim.
Schopenhauer, Arthur (2012a). *On the The Fourfold Root of the Principle of Sufficient Reason and Other Writings*. Cambridge University Press, New York/Cambridge. English translation by David E. Cartwright of Schopenhauer (1988), vol. 7, *Über die vierfache Wurzel des Satzes vom zureichenden Grunde* (1st edition 1816, 2nd ed. 1847).
Schopenhauer, Arthur (2012b). *The World as Will and Representation, Volume I*. The Cambridge Edition of the Works of Schopenhauer. Cambridge University Press, New York/Cambridge.

English translation by Judith Norman, Alistair Welchman and Christopher Janaway of *Die Welt als Wille und Vorstellung*, Schopenhauer (1988), Vol. 3, 3rd edition (1st edition 1818, 2nd edition 1844, 3rd ed. 1859).

Schroeder, Severin (2006). *The Way Out of the Fly-Bottle*. Polity Press, Oxford/Boston.

Schröter, A. and Bäumler, M., editors (1927). *Handbuch der Philosophie*, volume II. R. Oldenbourg, Berlin.

Shapiro, Stewart (1982). Acceptable notation. *Notre Dame Journal of Formal Logic*, 23 (1): 14–20.

Shapiro, Stewart, editor (2005). *The Oxford Handbook to the Philosophy of Logic and Mathematics*. Oxford University Press, New York.

Shapiro, Stewart (2008). Identity, Indiscernibiity, and *ante rem* Structuralism: the Tale of i and $-i$. *Philosophia Mathematica*, 16: 285–309.

Shapiro, Stewart (2012). An "i" for an "i": Singular Terms, Uniqueness and Reference. *The Review of Symbolic Logic*, 5: 380–415.

Shapiro, Stewart and Wright, Crispin (2006). All Things Indefinitely Extensible. In Rayo (2006), pp. 255–304.

Shieh, Sanford (2014). In What Way Does Logic Involve Necessity? *Philosophical Topics*, 42 (2): 298–337.

Sieg, Wilfried (2006a). Gödel on Computability. *Philosophia Mathematica*, 14 (2): 189–207.

Sieg, Wilfried (2006b). Step by Recursive Step: Church's Analysis of Effective Calculability. In Olszewski et al. (2006), pp. 456–485.

Sieg, Wilfried (2009). On Computability. In Irvine (2009), pp. 535–630.

Sieg, Wilfried and Schlimm, Dirk (2005). Dedekind's Analysis of Number: Systems and Axioms. *Synthese*, 147: 121–170.

Sluga, Hans and Stern, David, editors (2018). *The Cambridge Companion to Wittgenstein*. Cambridge University Press, New York, 2nd revised. edition.

Smithies, Frank (1934). Notes of Max Newman's Foundations of Mathematics Lectures, Cambridge, Easter Term 1934. St. John's College Library, University of Cambridge, Archive. Papers of Frank Smithies, GB 275, Notebook H57.

Sommaruga, Giovanni and Strahm, Thomas, editors (2015). *Turing's Revolution: the Impact of his Ideas About Computability*. Birkhäuser/Springer, Basel.

Spelt, P.D.M. and McGuinness, Brian (2001). Marginalia in Wittgenstein's Copy of Lamb's *Hydrodynamics*. In Oliveri (2001), pp. 131–148.

Spivak, Michael (1994). *Calculus*. Cambridge University Press, Cambridge, 3rd edition.

Stekeler-Weithofer, Pirmin, editor (2012). *Wittgenstein: Zu Philosophie und Wissenschaft*. Verlag Felix Meiner, Hamburg.

Stern, David (2011). Private Language. In Kuusela and McGinn (2012), pp. 333–350.

Stern, David G. (2018). *Wittgenstein in the 1930s: Between the Tractatus and the Investigations*. Cambridge University Press, Cambridge.

Sullivan, Peter (1995). Wittgenstein on The Foundations of Mathematics, June 1927. *Theoria*, 61: 105–142.

Säätelä, Simo (2011). From Logical Method to "Messing About": Wittgenstein on "Open Problems" in Mathematics. In Kuusela and McGinn (2012), pp. 162–182.

Tao, Terence (2006). *Analysis I*. Hindustan Book Agency, Gurugram, Haryana, India.

Turing, A.M. (1936). On computable numbers, with an application to the *Entscheidungsproblem*. *Proceedings of the London Mathematical Society*, Series 2, 1936/7 (42): 230–265. Corrections in Turing (1937b).

Turing, A.M. (1937a). Correspondence with Paul Bernays. Zürich: Eidgenössische Technische Hochschule Bibliothek/Swiss Federal Institute of Technology Library.

Turing, A.M. (1937b). On computable numbers, with an application to the *Entscheidungsproblem*: a correction. *Proceedings of the London Mathematical Society*, Series 2, 1937 (43): 544–546.

Turing, A.M. (1938). *Alan Turing's Systems of Logic: The Princeton Thesis*. Princeton University Press, Princeton, N.J. Published in 2012, edited by Andrew W. Appel. Fascimile reprinting of Turing's 1938 Dissertation at Princeton, which was published as Turing (1939).

Turing, A.M. (1939). Systems of logic based on ordinals. *Proceedings of the London Mathematical Society*, 2: 161–228. Reprinted in Turing (2001a), pp. 81–148 and (with excerpts from Turing's 1938 correspondence) in Turing and Copeland (2004), pp. 125–204 as well as (with commentary) in Turing (2013), pp. 151–197.

Turing, A.M. (1948). Intelligent Machinery. Unpublished report for the National Physical Laboratory. In Meltzer and Michie (1969), pp. 3–23, Turing (1992), pp. 106–113, Turing (2013), pp. 501–516 and Turing and Copeland (2004), pp. 410–423.

Turing, A.M. (1954). Solvable and unsolvable problems. *Science News*, 31: 7–23.

Turing, A.M. (1992). *Collected Works of A.M. Turing: Mechanical Intelligence*. North-Holland, Amsterdam. Edited by D.C. Ince.

Turing, A.M. (2001a). *Collected Works of A.M. Turing: Mathematical Logic*. North Holland, Amsterdam. Edited by R.O. Gandy and C.E.M. Yates.

Turing, A.M. (2001b). The Reform of Mathematical Notation and Phraseology [1944–45]. Unpublished manuscript. In Turing (2001a), pp. 211–222; reprinted with commentary in Turing (2013), pp. 245–249; original online at http://www.turingarchive.org/search/, King's College Online Archive.

Turing, A.M. (2013). *Alan Turing — His Work and Impact*. Elsevier, Amsterdam/Burlington, MA. Edited by S. Barry Cooper and Jan van Leeuwen; collected papers of Turing with commentary by experts.

Turing, A.M. and Copeland, B. Jack (2004). *The Essential Turing: Seminal Writings in Computing, Logic, Philosophy, Artificial Intelligence, and Artificial Life*. Oxford Clarendon Press, Oxford.

Ulam, S.M. (1991). *Adventures of a Mathematician*. University of California Press, Berkeley, CA.

Väänänen, Jouko (2011). Tennenbaum's Proof of the Irrationality of $\sqrt{2}$. In Kennedy and Kossak (2011), pp. 226–227.

van Dalen, Dirk (2013). *L.E.J. Brouwer – Topologist, Intuitionist, Philosopher: How Mathematics is Rooted in Life*. Springer, New York.

van Heijenoort, Jean, editor (1967). *From Frege to Gödel: A Source Book in Mathematical Logic, 1879–1931*. Harvard University Press, Cambridge, MA. Reprinted 2002.

von Wright, G. H. (1969). The Wittgenstein Papers. *Philosophical Review*, 78 (4): 483–503.

Wagner, Pierre, editor (2012). *Carnap's Ideal of Explication and Naturalism*. History of Analytic Philosophy. Palgrave Macmillan, Basingstoke, U.K.

Watson, A. G. D. (1938). Mathematics and Its Foundations. *Mind*, 47 (188): 440–451.

Weihrauch, Klaus (2000). *Computable Analysis*. Springer, Berlin.

Weil, André (1992). *The Apprenticeship of a Mathematician*. Birkhäuser, Basel.

Weiss, Max (2013). *Logic in the Tractatus*. Thesis, University of British Columbia, Vancouver.

Weiss, Max (2017). Logic in the *Tractatus*. *Review of Symbolic Logic*, 10 (1): 1–50.

Weyl, Hermann (1918). *Das Kontinuum: Kritische Untersuchungen über die Grundlagen der Analysis*. Verlag von Veit und Comp, Leipzig. English translation by S. Pollard and T. Bole, *The Continuum*, Dover Publications, New York: 1994. References are to this translation.

Weyl, Hermann (1919). Der *circulus vitiosus* in der heutigen Begründung der Analysis. *Jahresbericht der Deutschen Mathematikervereinigung*, 28: 85–92.

Weyl, Hermann (1925). Die heutige Erkenntnislage in der Mathematik. *Symposion*, 1: 1–32. English translation by B. Müller, "The Current Epistemological Situation in Mathematics", in Mancosu (1998), pp. 123–142. References are to this translation.

Weyl, Hermann (1927). *Philosophie der Mathematik und Naturwissenschaft*. In Schröter and Bäumler (1927).

Weyl, Hermann (1948). Wissenschaft als Symbolische Konstruktion des Menschen. *Eranos-Jahrbuch*, 1948: 374–431. Reprinted in Weyl (1968), pp. 631–649.

Weyl, Hermann (1968). *Weyls Gesammelte Abhandlungen*. Springer, Berlin. Edited by Komaravolu Chandrasekharan.

Whitehead, Alfred North and Russell, Bertrand (1910). *Principia Mathematica*. University Press, Cambridge. 3 vols., 2nd edition with a new introduction 1927.

Whittaker, E. T. (1902). *A Course of Modern Analysis*. University Press, Cambridge.

Wittgenstein, Ludwig (1929). Some Remarks on Logical Form. *Aristotelian Society Supplementary Volume*, 9: 162–171.

Wittgenstein, Ludwig (1931-). Remarks on Frazer's *Golden Bough*. In English in Klagge and Nordmann (2003), pp. 115–115; drawn and edited from MS 110, TS 221 (the second part), and pencilled notes.

Wittgenstein, Ludwig (1956). *Remarks on the Foundations of Mathematics/Begründung über die Grundlagen der Mathematik*. Oxford University Press, Oxford, 1st edition. Edited by G. H. von Wright, R. Rhees, G. E. M. Anscombe. German text with facing English translation by G.E.M. Anscombe. 2nd edition 1967; 3rd revised editions Wittgenstein (1974a) and Wittgenstein (1978).

Wittgenstein, Ludwig (1965a). A Lecture on Ethics. *The Philosophical Review*, 74: 3–12. Reprinted in Klagge and Nordmann (1993), pp. 36–44.

Wittgenstein, Ludwig (1965b). *Preliminary Studies for the 'Philosophical Investigations': Generally Known as The Blue and Brown Books*. Harper and Row, New York.

Wittgenstein, Ludwig (1969a). *Das Blaue Buch, Eine Philosophische Betrachtung (Das Braune Buch)*. In Wittgenstein (1974a), volume 5. Edited by Rush Rhees.

Wittgenstein, Ludwig (1969b). *Eine Philosophische Betrachtung*. In Wittgenstein (1969a), pp. 117–282. Wittgenstein's enlargement of *The Brown Book* (MS 115, TS 310). Edited by Rush Rhees, German translation by von Petra von Morstein. Based on Wittgenstein's manuscript revision of TS 310 contained in MS 115; the revision is not complete and the published EPB concludes with a German translation of the remainder of TS 310. Not translated into English.

Wittgenstein, Ludwig (1974a). *Bemerkungen über die Grundlagen der Mathematik*. Volume 6 of Wittgenstein (1974a); German text of Wittgenstein (1978). Edited by G. H. von Wright, R. Rhees, G. E. M. Anscombe. First published in 1974 in the 1st edition of Wittgenstein (1974a).

Wittgenstein, Ludwig (1974b). *On Certainty/Über Gewissheit*. Blackwell Publishing, Oxford/Malden, MA, 2nd edition. Edited by G.E.M. Anscombe and G. H. von Wright, English translation by Denis Paul and G.E.M. Anscombe.

Wittgenstein, Ludwig (1974c). *Philosophical Grammar*. Basil Blackwell, Oxford. Edited by Rush Rhees, English translation by Anthony Kenny.

Wittgenstein, Ludwig (1976). *Wittgenstein's Lectures on the Foundations of Mathematics: Cambridge, 1939*. The Harvester Press, Ltd./University of Chicago Press, Hassocks, Sussex/Chicago, IL. Edited by Cora Diamond.

Wittgenstein, Ludwig (1978). *Remarks on the Foundations of Mathematics*. M. I. T. Press, Cambridge, Mass, 3rd edition. Edited by G. H. von Wright, R. Rhees, G. E. M. Anscombe; English translation by G. E. M. Anscombe of German edition of Wittgenstein (1974a) (i.e., Wittgenstein (1974a), Vol. 6).

Wittgenstein, Ludwig (1979). *Wittgenstein's Lectures, Cambridge, 1932–1935: From the Notes of Alice Ambrose and Margaret Macdonald*. University of Chicago Press.

Wittgenstein, Ludwig (1980a). *Philosophical Remarks*. University of Chicago Press, Chicago. Edited by Rush Rhees, English translation by Raymond Hargreaves and Roger White.

Wittgenstein, Ludwig (1980b). *Remarks on the Philosophy of Psychology. Vol 1*. Blackwell, Oxford. Edited by G. H. von Wright and Heikki Nyman, English translation by G.E.M. Anscombe.

Wittgenstein, Ludwig (1980c). *Remarks on the Philosophy of Psychology Vol. 2*. University of Chicago Press/Basil Blackwell, Chicago/Oxford. Edited by G. H. von Wright and Heikki Nyman, English translation by C.G. Luckhardt and M.A.E. Aue.

Wittgenstein, Ludwig (1980d). *Wittgenstein's Lectures, Cambridge 1930–32, From the Notes of John King and Desmond Lee*. Blackwell, Oxford.

Wittgenstein, Ludwig (1981a). *Tractatus Logico-Philosophicus*. Kegan Paul, London. English translation by C.K. Ogden. 1st German edition in *Annalen der Naturphilosophie*, Wilhelm Ostwald (ed.), 14, 1921, pp. 12–262, Leipzig. Several on-line free versions of the text are available with hypertext and translation comparison capability. See Kevin Klement's "Side-by-Side" edition, available at http://people.uMAedu/klement/tlp/, the University of Iowa *Tractatus*

Map http://tractatus.lib.uiowa.edu/, and Luciano Bazzocchi's multiple language site http://www.bazzocchi.com/wittgenstein/.

Wittgenstein, Ludwig (1981b). *Zettel*. Blackwell Publishers Ltd., Oxford, 2nd edition. Edited by G.E.M. Anscombe and G.H. von Wright, English translation by G.E.M. Anscombe.

Wittgenstein, Ludwig (1988). *Wittgenstein's Lectures on Philosophical Psychology, 1946–47*. Harvester, Brighton.

Wittgenstein, Ludwig (1989). *Werkausgabe in 8 Bänden*. Suhrkamp Verlag, Frankfurt am Main, 2nd edition. 1st edition was 1974.

Wittgenstein, Ludwig (2001). *Philosophische Untersuchungen: Kritisch-Genetische Edition*. Suhrkamp Verlag, Frankfurt am Main. Edited by J. Schulte, H. Nyman, E. von Savigny, G. H. von Wright.

Wittgenstein, Ludwig (2005a). *The Big Typescript, TS 213*. Blackwell Publishing, Malden, MA.

Wittgenstein, Ludwig (2005b). *Gesamtbriefwechsel/Complete Correspondence*. Electronic Publication/CD Rom, Intelex Corporation, Charlottesville, VA. Edited under the auspices of the Brenner-Archiv's Research Institute (University of Innsbruck) by Monika Seekircher, Brian McGuinness and Anton Unterkircher, Allan Janik and Walter Methlagl. Electronic edition 2005.

Wittgenstein, Ludwig (2007). *Lectures and Conversations on Aesthetics, Psychology, and Religious Belief*. University of California Press, Berkeley.

Wittgenstein, Ludwig (2009). *Philosophische Untersuchungen = Philosophical investigations*. Wiley-Blackwell, Chichester, U.K./Malden, MA, revised 4th edition. English translation by G.E.M. Anscombe, P.M.S. Hacker and Joachim Schulte.

Wittgenstein, Ludwig (2015). *Wittgenstein's Nachlass. The Bergen Electronic Edition*. Wittgenstein Archive, University of Bergen, Bergen, Norway. Edited by the Wittgenstein-Archiv, University of Bergen under the direction of Claus Huitfeldt. Incorporates *Wittgenstein-Source Bergen Facsimile Edition*, ed. Alois Pichler, with the assistance of H. W. Kruger, D. C. P. South, T. M. Bruvik, A. Lindebjerg, V. Olstad, Wittgenstein-Archiv University of Bergen (www.wittgensteinsource.org). Originally published by Intelex Corporation and Oxford University Press as *The Published Works of Ludwig Wittgenstein* (1993), *The Collected Works of Ludwig Wittgenstein* (1998) *Wittgenstein Letters and Lectures* (2001), and *Wittgenstein: Tagebücher und Briefe* (2002).

Wittgenstein, Ludwig (2017). *Wittgenstein's Whewell's Court Lectures, Cambridge, 1938–1941: from the Notes by Yorick Smythies*. Wiley-Blackwell, Oxford. Edited by Volker A. Munz, Bernhard Ritter and Yorick Smythies.

Wittgenstein, Ludwig and Waismann, Friedrich (1979). *Wittgenstein and the Vienna Circle: Conversations recorded by Friedrich Waismann*. Basil Blackwell/Harper and Row, Oxford/New York. Edited by Brian McGuinness. German edition is Wittgenstein (1974a), vol. 3.

Wittgenstein, Ludwig and Waismann, Friedrich (2003). *The Voices of Wittgenstein, the Vienna Circle: Original German Texts and English Translations*. Routledge, London. Edited by Gordon P. Baker.

Woodin, Hugh (2017). In Search of Ultimate-L: The 19th Midrasha Mathematicae Lectures. *The Bulletin of Symbolic Logic*, 23 (1): 1–109.

Wright, Crispin (1980). *Wittgenstein on the Foundations of Mathematics*. Harvard University Press.

Zermelo, Ernst (1904). Beweis, dass jede Menge wohlgeordnet werden kann. *Mathematische Annalen*, 59: 514–516. Reprinted in Zermelo (2010), with English translation, p. 114; also translated into English in van Heijenoort (1967), pp. 139–141.

Zermelo, Ernst (2010). *Ernst Zermelo Collected Works/Gesammelte Werke Vol. I: Set Theory, Miscellanea/Mengenlehre, Varia*. Schriften der Mathematisch-naturwissenschaftlichen Klasse der Heidelberger Akademie der Wissenschaften Nr. 21. Springer, Heidelberg. Edited by Heinz-Dieter Ebbinghaus, Craig G. Fraser, and Akihiro Kanamori.

Index

J. Floyd, F. Mühlhölzer, *Wittgenstein's Annotations to Hardy's Course of Pure Mathematics*, Nordic Wittgenstein Studies 7,
https://doi.org/10.1007/978-3-030-48481-1

D

Z

The manufacturer's authorised representative in the EU is Springer Nature Customer Service Centre GmbH, Europaplatz 3, 69115 Heidelberg, Germany. If you have any concerns regarding our products, please contact ProductSafety@springernature.com

Printed and bound by CPI Group (UK) Ltd, Croydon, CR0 4YY
30/09/2026
02225696-0016